How Biology Shapes Philosophy

How Biology Shapes Philosophy is a seminal contribution to the emerging field of biophilosophy. It brings together work by philosophers who draw on biology to address traditional and not-so-traditional philosophical questions and concerns. Thirteen essays by leading figures in the field explore the biological dimensions of ethics, metaphysics, epistemology, gender, semantics, rationality, representation, and consciousness, as well as the misappropriation of biology by philosophers, allowing readers to interrogate critically the relevance of biology for philosophy. Both rigorous and accessible, the essays illuminate philosophy and help us to acquire a deeper understanding of the human condition. This volume will be of interest to philosophers, biologists, social scientists, and other readers with an interest in bringing science and the humanities together.

DAVID LIVINGSTONE SMITH is Professor of Philosophy at the University of New England. His most recent book is *Less Than Human: Why We Demean, Enslave and Exterminate Others* (2011).

How Biology Shapes Philosophy

New Foundations for Naturalism

Edited by

DAVID LIVINGSTONE SMITH
University of New England, Biddeford, Maine

CAMBRIDGE
UNIVERSITY PRESS

University Printing House, Cambridge CB2 8BS, United Kingdom

One Liberty Plaza, 20th Floor, New York, NY 10006, USA

477 Williamstown Road, Port Melbourne, VIC 3207, Australia

314-321, 3rd Floor, Plot 3, Splendor Forum, Jasola District Centre, New Delhi - 110025, India

79 Anson Road, #06-04/06, Singapore 079906

Cambridge University Press is part of the University of Cambridge.

It furthers the University's mission by disseminating knowledge in the pursuit of
education, learning and research at the highest international levels of excellence.

www.cambridge.org
Information on this title: www.cambridge.org/9781107628205

© Cambridge University Press 2017

First published 2017
First paperback edition 2018

A catalogue record for this publication is available from the British Library

Library of Congress Cataloging in Publication data
Smith, David Livingstone, 1953– editor.
How biology shapes philosophy : new foundations for
naturalism / [edited by] David Livingstone Smith.
New York : Cambridge University Press, 2016. | Includes index.
LCCN 2016026630 | ISBN 9781107055834
LCSH: Naturalism. | Philosophy and science. | Biology.
LCC B828.2 .H69 2016 | DDC 113/.8–dc23
LC record available at https://lccn.loc.gov/2016026630

ISBN 978-1-107-05583-4 Hardback
ISBN 978-1-107-62820-5 Paperback

For Maxwell MacFarland Smith
Welcome to the biosphere!

He who understands baboon would do more towards metaphysics than Locke.

– *Charles Darwin*

Contents

Figures

Contributors

RICHARD N. BOYD, Department of Philosophy, Cornell University, Professor Boyd's recent publications include "Semantic Externalism and Knowing Our Own Minds: Ignoring Twin-Earth and Doing Naturalistic Philosophy" (2013); "Realism, Natural Kinds, and Philosophical Methods" (2010); and "Homeostasis, Higher Taxa, and Monophyly" (2010).

PATRICIA CHURCHLAND, University of California, San Diego, Professor Churchland's recent publications include *Brain-Wise* (2002), *Touching a Nerve* (2014), and *Braintrust* (2011).

DANIEL DENNETT, Center for Cognitive Studies, Tufts University, Professor Dennett's recent publications include "Turing's 'Strange Inversion of Reasoning'" (2013), "The Evolution of Reasons" (2014), and "Our Transparent Future: No Secret Is Safe in the Digital Age. The Implications for Our Institutions Are Downright Darwinian" (2015).

RONALD DE SOUSA, Department of Philosophy, University of Toronto, Professor de Sousa's recent publications include *Why Think? Evolution and the Rational Mind* (2007), *Emotional Truth* (2011), and *Love: A Very Short Introduction* (2015).

JOHN DUPRÉ, Department of Philosophy, University of Exeter, Professor Dupré's recent publications include *Processes of Life: Essays in the Philosophy of Biology* (2012), *Humans and Other Animals* (2002), and *Human Nature and the Limits of Science* (2001).

LUC FAUCHER, Département de philosophie, Université du Québec à Montréal, Professor Faucher's recent publications include "RDoC: Thinking Outside the DSM Box Without Falling into the Reductionist Trap" (2015), "Revisionism and Moral Responsibility" (2016), and "Mother Culture, Meet Mother Nature" (2017).

PETER GODFREY-SMITH, Department of Philosophy, City University of New York Graduate Center, Professor Godfrey-Smith's recent publications include *Philosophy of Biology* (2014), "Reproduction, Symbiosis, and the Eukaryotic Cell" (2015), and "Dewey and the Question of Realism" (2016).

PHILIP KITCHER, Department of Philosophy, Columbia University, Professor Kircher's recent publications include *The Ethical Project* (2011), *Preludes to Pragmatism* (2012), and "Experimental Animals" (2015).

EDOUARD MACHERY, Department of History and Philosophy of Science, University of Pittsburgh, Professor Machery's recent publications include *Doing Without Concepts* (2009), *Arguing About Human Nature* (2013), and *Current Controversies in Experimental Philosophy* (2014).

KAREN NEANDER, Department of Philosophy, Duke University, Professor Neander's recent publications include "Functional Analysis and the Species Design" (2015), "Biological Functions" (2013), and "Content for Cognitive Science" (2006).

SAMIR OKASHA, Department of Philosophy, University of Bristol, Professor Okasha's recent publications include "The Relation Between Kin and Multi-Level Selection: An Approach Using Causal Graphs" (2015), "Hamilton's Rule, Inclusive Fitness Maximization and the Goal of Individual Behaviour in Symmetric Two-Player Games" (2016), and "The Evolution of Bayesian Updating" (2013).

DAVID PAPINEAU, Department of Philosophy, King's College London, and Philosophy Program, Graduate Center, City University of New York, Professor Papineau's recent publications include *The Roots of Reason* (2003), "What Exactly Is the Explanatory Gap?" (2011), and *Philosophical Devices* (2011).

ALEXANDER ROSENBERG, Department of Philosophy, Duke University, Professor Rosenberg's recent publications include *The Atheist's Guide to Reality* (2011), "The Biological Character of Social Science" (2015), and "Functionalism" (2016).

DAVID LIVINGSTONE SMITH, Department of History and Philosophy, University of New England, Professor Smith's recent publications include *Less Than Human: Why We Demean, Enslave, and Exterminate Others* (2011), "Aping the Human Essence: Simianization as Dehumanization" (2016), and "Paradoxes of Dehumanization" (2016).

Acknowledgments

This volume has been a long time coming. As a child growing up in southwest Florida, I was passionately interested in the local fauna and wanted nothing more than to become a herpetologist until life diverted me along a different pathway. Many years further down the road, I stumbled into philosophy after· a previous career as a psychotherapist. Jim Hopkins, my *Doktorvater* at Kings College London, rekindled my biological interests by introducing me to the work of Ruth Garratt Millikan. I owe Jim a huge debt of gratitude both for introducing me to philosophy and (most pertinently here) for pointing me in such a richly rewarding philosophical direction. Ruth initially signed on to contribute a chapter but sadly had to withdraw for reasons of ill health. Nonetheless, her intellectual fingerprints are all over this book. In fact, the book's subtitle, *New Foundations for Naturalism*, is intended as a homage to her groundbreaking 1984 volume, *Language, Thought, and Other Biological Categories: New Foundations for Realism*. Thank you, Ruth. Your work is a gift that keeps on giving.

Profound thanks are due to the stellar band of philosophers who took time off from their busy professional lives to contribute original chapters on an astonishingly broad range of topics. Dan, Alex, Peter, Pat, David, Karen, Ronnie, Samir, Philip, Edouard, John, Luc, and Dick – I hope that you are all as delighted by the result of our collective efforts as I am. Thank you all for coming through with your chapters with only minimal badgering from me.

How Biology Shapes Philosophy would never have seen the light of day were it not for Hilary Gaskin, senior commissioning editor for philosophy at Cambridge University Press, who saw the need for a first-rate collection of papers on this topic and who made valuable suggestions regarding contributors. Thank you, Hilary – and thanks to the rest of the team at Cambridge University Press.

Finally, my deepest gratitude goes to Subrena Smith, with whom I have had literally hundreds of conversations about the relation between biology and philosophy. Our ongoing marriage of minds sparked the conception of this book and was crucial for bringing it to fruition. Thank you, Subrena, for your clarity, patience, commitment, and intellectual tenacity. I am unbelievably fortunate to have you in my life.

Introduction: Biophilosophy

DAVID LIVINGSTONE SMITH

This book is a collection of papers on what I call "biophilosophy." Because this term will be unfamiliar to most philosophers, and given that it has been used occasionally in the past in a variety of senses, it is appropriate to begin this book with a discussion of what I take it to mean and to justify its use. This discussion will prepare the ground for considering how, as this book's title suggests, biology *shapes* philosophy and the sense in which biophilosophy provides foundations for naturalism.[1]

Biophilosophy is easily confused with the philosophy of biology. Although biophilosophers and philosophers of biology are both concerned with the interface between philosophy and biology, their orientations toward that interface, as I stipulatively define them, are different. Philosophers of biology do not, as such, do biology. Instead, they reflect on biological concepts, biologists' patterns of inference, and the conceptual relations that obtain between biological concepts and those belonging to other scientific disciplines, among other things. One can think of philosophy of biology as higher-order biological theorizing: just as biologists use the theoretical concepts enshrined in their discipline to map the empirical landscape of the biosphere, philosophers of biology use philosophical resources to draw and redraw the conceptual topography of the biological sciences. Whereas a biologist might inquire into the question of whether a certain phenotype contributes to the fitness of the organisms that possess it, relative to a certain environment, the philosopher of biology might inquire into the question of how the notions of "phenotype," "fitness," and "environment" ought to be understood and what entailments each of these understandings has for theoretical biology.

---

[1] For example, Bunge (1979), Mahner and Bunge (1979), Allen and Bekoff (1995), Gilson (2009), Koutrofinis (2014).

In contrast, biophilosophers invert the relation between philosophy and biology. Instead of using philosophy as a resource for biology, as philosophers of biology do, they use biology as a resource for philosophy.[2] In this respect, biophilosophy is a mirror image of philosophy of biology even though, as I will explain later, the former is ultimately subordinate to the latter.

Some philosophers explicitly address the difference between biology as explanandum and biology as explanans. For example, Paul Griffiths partitions philosophy of biology into three kinds. One kind applies general considerations from philosophy of science to the special case of biology (e.g. in discussions of the question of whether there are biological laws and what implications this has for the nature of biological explanation). Another is concerned with conceptual issues (or, as Griffiths puts it, "puzzles") that are specific to biology (e.g. the question of whether species are kinds or individuals or whether they exist at all). Griffiths' third kind of philosophy of biology, which appeals to biology for help in addressing what he calls "traditional" (by which he means something like "paradigmatic") philosophical concerns, corresponds to what I call "biophilosophy."

Griffiths' terminology is not ideal because it places two very different sorts of philosophical projects under the single taxonomic umbrella of "philosophy of biology." Also, conventionally, expressions of the form "philosophy of x" use x to stand for whatever it is that's being philosophized *about*.[3] "Philosophy of biology" suggests that it is biology that is being philosophized about, even though this is not at all what Griffiths means to convey. In contrast, "biophilosophy" isn't a "philosophy of" designation. Instead (like "neurophilosophy"), it suggests a biologically informed *approach* to doing philosophy.

Peter Godfrey-Smith makes a similar distinction between philosophy of science and what he calls "philosophy of nature," writing that

In a broad sense, all of philosophy of biology is part of "the philosophy of science." But ... we can also distinguish *philosophy of science*, in a narrower sense, from *philosophy of nature*. Philosophy of science in this narrower sense is an attempt to understand the activity and the

---

[2] For a somewhat different interpretation, see Luc Faucher's contribution to this volume (Chapter 12).

[3] Curiously, this does not always apply the other way around. "Political philosophy" is the philosophy of politics. It does not refer to a politically informed approach to doing philosophy.

products of science itself. When doing philosophy of nature, we are trying to understand the universe and our place in it. The science of biology becomes an instrument – a lens – through which we look at the natural world. Science is then a resource for philosophy rather than a subject-matter. (2014, p. 4)

Godfrey-Smith's broad notion of "philosophy of science" applied to the biological sphere covers the same territory as Griffiths' broad notion of "philosophy of biology," and his more restricted sense of "philosophy of science" corresponds to Griffiths' first and second kinds of philosophy of biology. His "philosophy of nature" (again, applied to biology) includes Griffiths' third kind of philosophy of biology, as well as my "biophilosophy." However, Godfrey-Smith's category is considerably broader that what I mean to designate as "biophiloso-phy." Philosophy of nature uses *science* – by which presumably is meant the methodological and theoretical apparatuses of science plus the body of facts discovered by the application of those methods – as a resource of philosophy. As such, it is not specifically biological. The philosopher of nature might equally make use of physics, or chemistry, or psychology as a resource. So, in terms of Godfrey-Smith's vocabulary, biophilosophy turns out to be *special case* of the philosophy of nature. Of course, in common with Griffiths' terminol-ogy, "philosophy of nature" also has "philosophy of *x*" form. It also risks confusion with nineteenth-century German *Naturphilosophie* as well as a less unfortunate but nonetheless misleading associations with *philosophia naturalis*.

These sorts of considerations lead me to nominate "biophilosophy" as a name for the kind of philosophical work that these writers have in mind.

Having conceptually distinguished biophilosophy from philosophy of biology, it is important to recognize the crucial connection between them. As Godfrey-Smith points out in a discussion of the relation between philosophy of nature and (narrow-scope) philosophy of science, "These two kinds of philosophical work interact. What you think science is *telling* us about the world will depend upon how you think that part of science *works*" (2014, p. 4). To do biophilosophy well, it is necessary to get the science right. Doing that requires literacy in the relevant sectors of biological science as well as an understanding of the ways in which philosophers of science interrogate those biologi-cal claims.

Now for a cautionary note. In her book *Freud's Dream: A Complete Interdisciplinary Science of Mind* (Kitcher 1992a), Patricia Kitcher shows that appropriating scientific claims for interdisciplinary purposes can be a risky business. If the science moves on and the interdisciplinary scholar does not keep abreast of it, she finds herself left in the lurch, having grounded her work in assumptions that are no longer empirically credible (see also Sullaway 1992). Kitcher argues that this unfortunate fate overtook Freud's efforts to develop a complete interdisciplinary science of mind. Freud's "metapsychology" – his account of the unintrospectible neurological systems and processes underpinning human behavior – was grounded in what were, in the late nineteenth century, cutting-edge scientific ideas. As the new century progressed, though, most of these were shown to be false, and psychoanalytic theory was left mired in theoretical anachronisms. Kitcher plausibly argues that cognitive science may be in danger of succumbing to the same problem. "It appears to be quite easy," she observes, "to have more faith in a related discipline than its practitioners, particularly when one's theory relies on its basic concepts or needs to be supplemented by its potential results" (1992a, p. 183). There is an obvious lesson here for biophilosophy. To do biophilosophy well, one needs not only to be familiar with relevant work in philosophy of biology, as Godfrey-Smith emphasizes, but also to keep up with the changing face of the biological sciences.

The contributions to this volume demonstrate that biophilosophical work can be immensely varied. However, there are some broad metaphilosophical constraints that must be honored in order for biophilosophy to be done well – constraints that fall out from the very nature of philosophy. First, biological premises do not (all on their own) *entail* philosophical conclusions. It is a truism that data do not entail theories – so any collection of empirical evidence is consistent with any number of theoretical explanations (although, of course, not all of these will be projectable). It follows that *philosophical* theories are underdetermined by data, and if we think of philosophical theories as metatheoretical structures, then scientific theories underdetermine philosophical ones. If this is right, then there is no *straight* path from biology to philosophy. The path leading from biology to philosophy is more circuitous and, for that reason, more hazardous to negotiate.

I'll approach the question of the role of biology in the philosophical enterprise by considering a very general problem confronting anyone doing philosophy of any kind. To do philosophy is to blaze a trail through an immensely complex conceptual decision space. As Michael Rosen (2012) so brilliantly describes it, "Philosophy is a holistic discipline. All of its theories and problems relate, in the end, to all the rest."

So to address one problem we must have – if not *resolved* all the others, at least be prepared to "put them on hold" for the time being ... For a rough analogy, compare the philosopher with a chess player.[4] If her argument were to be conclusive, the philosopher would have to be able, when she makes a move (that is, puts forward an argument or advocates a position), to meet all the counter-moves that might be made, and all the counter-moves to her own counter-moves – in fact, to address the whole exponentially expanding tree-structure of possibilities that lie beneath that single move ... So ... the philosopher faces a repeated series of uncomfortable choices about what to take for granted and what to put on the table for debate at any stage. (pp. xi–xii)

Deciding which questions to beg and which ones to pursue, as well as how to pursue the ones that one chooses not to beg, requires some principle or set of principles that must, on pain of circularity, be extraphilosophical, for it is trivially true that if philosophy is bounded at all (which it surely is), then it is bounded by something other than philosophy. There is a great deal of territory that lies beyond philosophy, any portion of which might serve to guide one's trajectory through the endlessly ramifying decision space. One might, for example, use neuroscience as a guide to philosophical enquiry, as neurophilosophers have advocated, or adopt computer science, as many functionalists have done. Or one might be guided by one's cognitive biases, semantically dignified by philosophers as "intuitions."[5] Doing philosophy requires, paradoxically enough, a kind of creative blinkering, a closing down of options, a filtering of possibilities. Yoking philosophy to biology is one way to do this. That is, roughly speaking, how biology *shapes* philosophy.

---

[4] Philosophers' fondness for using chess analogies may say something about the class background and intellectual pretensions of professional philosophy. After all, most of the same points that philosophers use the chess analogy to make could just as well be made using the examples of basketball, tennis, or boxing.
The game of chess is conventionally associated with the solitary exercise of pure intellect. It's a Cartesian game. See also Dennett (2006).

[5] I do not mean to suggest that intuitions are without epistemic value any more than I mean to suggest that cognitive biases are without epistemic value.

Biology's role in shaping philosophy does not involve interdisciplinarity as it is often conceived – that is, as a sort of melding of two disciplines or the incorporation of the elements of one discipline into another. Biophilosophy does not work like this because philosophy is not a discipline in the sense that biology is a discipline. Of course, there is a perfectly good sense in which philosophy is a discipline. There are departments of philosophy in universities, philosophy conferences, and learned journals. Philosophers employ a specialized language that is opaque to outsiders, make use of certain distinctively philosophical communicative and inferential practices, and reward certain kinds of expertise. In contrast, biology is individuated both by its domain and by the body of knowledge that it has accumulated about that domain through implementing research conducted in accord with certain methodological norms. Philosophy, however, does not have a proper domain – or, to put the point differently, philosophy addresses *every* domain. It is distinguished by the kinds of questions that it asks, the manner in which it goes about answering them, and the norms governing what answers count as acceptable rather than by the subject area toward which those questions are addressed.

It's the logical relation between philosophy and biology that delimits what biophilosophy is and thereby determines what it isn't. As I have pointed out, it is not a mixture of biological and philosophical claims (although biological claims can serve as premises in biophilosophical arguments) and it is not an entailment of philosophical claims from strictly biological premises. It is not a reduction of philosophical claims to biological ones either (which would involve the commission of a category error). The relations that obtain between biology and philosophy are considerably looser but no less significant than the alternatives canvassed earlier.

Speaking very generally, biophilosophers use biology to constrain, guide, and inspire philosophical theorizing. They use it to constrain philosophy by closing off certain conceptual options. In doing so, they use it to carve out a pathway through conceptual decision space. And they use it as a source of inspiration by drawing on biological models in the service of philosophical ends.[6]

---

[6] Millikan's (1984) *Language, Thought, and Other Biological Categories: New Foundations for Realism* is a paradigmatic example of the philosophical use of a biological model.

This brings me to this volume's subtitle: "New Foundations for Naturalism." "Naturalism" is an elastic idea. Most contemporary philosophers consider themselves to be naturalists, but this seeming consensus encompasses a wide variety of views, and it will serve no good purpose to attempt to itemize them here. Very generally, naturalisms are grouped into ontological and methodological varieties. Ontological naturalism concerns the *kinds* of things that exist. According to this view, everything that exists is either (numerically) identical to or constituted by physical things. So "ontological naturalism" is for the most part just another name for physicalism, or anti-supernaturalism, and is compatible with various finer-grained positions of reductionism, antireductionism, and eliminativism.

Obviously, naturalism of *this* sort is only tangentially related to biophilosophy. Biological items are physical items. But if physicalism is true, then everything else is physical too, so metaphysical naturalism, as it is commonly understood, does not have any special connection to the biological realm. However, one might distinguish metaphysical naturalism per se from *biological metaphysical naturalism*, which has it that nonparadigmatically biological attributes of organisms are identical to or constituted by paradigmatically biological items. This, too, might be understood from a reductive, antireductive, or eliminativist perspective – but in each of these cases, biology is used as a touchstone for metaphysical credibility.

Methodological naturalism is considerably more difficult to pin down and is probably best thought of as a philosophical sensibility rather than a commitment to a set of propositions. Most characteristically, methodological naturalists conceive of philosophy as in some sense *continuous* with science. From this perspective, the border between philosophy and science is a blurry, if not entirely fictional one. Methodological naturalists tend to make a deflationary assessment of "pure" philosophy. They tend, on the whole, to favor a posteriori claims over *a priori* ones, to pursue synthetic rather than analytic truths, to value contingency as much as necessity, to be suspicious of conceptual analysis, and to be wary of thought experiments set in exotic possible worlds. In short, they are not afraid of getting their hands dirty by grappling with the empirical domain, and they privilege those investigative procedures that reliably deliver knowledge about that domain.[7] Biophilosophy is clearly – perhaps paradigmatically – methodologically naturalistic.

Biophilosophy provides *foundations* for naturalism in biology in much the same way that any scientific discipline provides foundations for naturalism. Put a bit more explicitly, biophilosophy provides one way of grounding the conceptual apparatus of philosophy in the extra-philosophical world – the "real" world, that is, the world that we deploy our concepts and metaconcepts to make sense of – the world of plants and porcupines, genes and proteins, neurons and muscle, the world that makes it possible for us to do philosophy and on which all of the philosophy that we do depends.

---

[7] As I mentioned earlier, naturalisms come in many flavors. For a more nuanced look at the varieties of naturalism and the arguments offered on their behalf, see P. S. Kitcher (1992b), Rosenberg (1996), Flanagan (2006), Papineau (1993), Almeder (1998), and the useful collection of papers in de Caro and Macarthur (2004).

1 | *Darwin and the Overdue Demise of Essentialism*

DANIEL C. DENNETT

Now that Darwinian thinking has replaced essentialist thinking in biology, should "essence" be considered a dirty word, banished from the working vocabulary of philosophers in all but historical contexts? The term gets some protective coloration from its innocent, nontechnical use by nonphilosophers. For instance, Douglas Hofstadter, queried by me about its use in the title of his recently co-authored book, *Surfaces and Essences: Analogy as the Fuel and Fire of Thinking* (with Emmanuel Sander, New York: Basic Books, 2013) responded

I don't react negatively or fearfully when I hear the word "essence", because in my mind the word has not been tainted by a wearying set of arcane debates in philosophical circles. For me, it's just an informal, everyday word, not a technical term. I doubt that you and I have any disagreement about what the word "essence" means when it's used informally, as a synonym for "gist", "crux", "core", etc. And that's how it's used in the book. (Personal correspondence, January 3, 2014)

He is right that I have no disagreement with him regarding his (familiar, nontechnical) sense of the term. I worry, however, that the excellent insights he reaps from his celebration of essences in *this* sense will lend false respectability to the philosophers' "arcane debates" when they use the term – and, more recently, when they use thinly veiled substitutes (euphemisms, in effect), now that its credentials have been challenged.

Ever since Socrates pioneered the demand to know what all *F*s have in common, in virtue of which they are *F*s, the ideal of clear, sharp boundaries has been one of the founding principles of philosophy. Plato's *forms* begat Aristotle's *essences*, which begat a host of ways of asking for *necessary and sufficient conditions*, which begat *natural kinds*, which begat *difference-makers* and other ways of tidying up the borders of all the sets of things in the world. When Darwin came

Thanks to Diana Raffman for illuminating discussion and advice on an earlier draft.

along with the revolutionary discovery that the sets of *living* things were not eternal, hard-edged, in-or-out classes but historical populations with fuzzy boundaries, islands historically connected to other islands by vanishing isthmuses, the main reactions of philosophers were to either ignore this hard-to-deny fact or treat it as a challenge: Now how should we impose our cookie-cutter set theory on this vague and meandering portion of reality?

"Define your terms!" is a frequent preamble to discussions in philosophy, and in some quarters it counts as Step One in all serious investigations. It is not hard to see why. The techniques of argumentation inaugurated by Socrates and Plato and first systematized by Aristotle are not just intuitively satisfying ("self-evident" on reflection) but demonstrably powerful tools of discovery, indispensable for answering difficult questions and resolving contentious disagreements, often with an undeniable finality. Shouldn't the goal of all inquiry be the triumphant coda "*Quad erat demonstrandum*, which was to be demonstrated"? Euclid's plane geometry was the first parade case, with its crisp isolation of definitions and axioms, inference rules, and theorems. If only all topics could be tamed as thoroughly as Euclid had tamed geometry! The hope of distilling everything down to the purity of Euclid has motivated many philosophical enterprises over the years, different attempts to *euclidify* all the topics and thereby impose classical logic on the world. These attempts continue to this day and have often proceeded as if Darwin never existed. Philip Kitcher points to a glaring example:

Consider, for example, what Ernst Mayr has called Darwin's replacement of "typological thinking" by "population thinking." Darwin's recognition of a vast amount of intraspecific variation often goes unappreciated today in philosophical discussions, even though it has been uncontroversial for well over a century. Recent discussions of natural kinds, prompted by the seminal ideas of Saul Kripke and Hilary Putnam, often assume that one can revive essentialism. Yet, if species are natural kinds, no such revival is in prospect. Kripke and Putnam largely restricted their discussions to the cases of elements and compounds, and with good reason, for given the insights of neo-Darwinism, it is clear that the search for some analog of the microstructural essences can't be found. No genetic or karyotypic property will play for species the role that atomic number does for the elements. (Kitcher 2009)

It was Quine (1969) who reintroduced the term "natural kinds" to philosophy, and he, at least, appreciated that only a few of the kinds found in nature are natural kinds considered as modern-day essences, and he probably regretted the way his imprimatur was interpreted as a naturalist's blessing for some kind of return to carefree essentialism. "Green things, or at least green emeralds, are a kind," Quine observed (p. 116), manifesting his own appreciation of the fact that while emeralds may be a natural kind, green things are not. Colors are not natural kinds precisely because they are a product of biological evolution, which has a tolerance for sloppy boundaries when making categories that would horrify any philosopher bent on achieving good, clean definitions. If some creature's life depended on lumping together the moon, blue cheese, and bicycles, you can be pretty sure that Mother Nature would find a way for it to "see" these as "intuitively just the same kind of thing."[1]

The common unspoken presumption that somehow essentialism can be made to work outside the abstract realm of mathematics is, I suggest, a methodological, not metaphysical, prejudice. Spelled out, the presumption is this: we have this wonderful tool, classical bivalent logic, and we've invested our lives in mastering its use. Without sharp boundaries, as sharp as those of Euclid's classes of geometric elements, it is disabled, so we will take on as a working assumption that there is some way of euclidifying all the vagueness and fuzziness out of our terms. In this way, we can go back to business as usual, tolerating Darwinian population thinking among those with a taste for such practices but denying its application to our chosen topics.

The prejudice is widespread among philosophers, and not without reason. For instance, one of the most popular practices threatened by the Darwinian perspective is the tactic of confronting opponents with disjunction-elimination arguments of what might be called the fish-or-cut-bait variety. First, let me demonstrate why it is so often favored by philosophers and then show why they are wise to foreswear it in many – almost all – naturalistic contexts.

---

[1] This paragraph is adapted from Dennett, *Consciousness Explained* (1991, p. 381, n. 2). The heterogeneous set could mark an idiosyncratic kind for any organism that had three detector/sensor systems yoked with OR gates that were impenetrable to introspective analysis, a case of radical synesthesia. It could be lumpy with crisp edges or lumpy and fuzzy at the same time. What counts as a bicycle or blue cheese is negotiable in our world.

Consider the following simple proof that there are irrational numbers *A* and *B* such that *A* to the *B* power is rational. It depends on the assumption that every real number is either rational or it isn't.

Let *A* be $\sqrt{2}$.
Let *B* be $\sqrt{2}$.
Then what about *A* to the *B* power? Is $(\sqrt{2})^{\sqrt{2}}$ rational? *I don't know.*

But I do know that either it is or it isn't:

$((\sqrt{2})^{\sqrt{2}}$ is rational$)$ v. $\sim((\sqrt{2})^{\sqrt{2}}$ is rational$)$.
If it is rational, QED.
If it isn't, then keeping *B* as $\sqrt{2}$, let *A* be $(\sqrt{2})^{\sqrt{2}}$.
Now is $((\sqrt{2})^{\sqrt{2}})^{\sqrt{2}}$ rational? Yes, because it is $(\sqrt{2})^2$, which is 2.

One way or another – and I don't have any idea which way it is – there are such a pair of numbers.[2]

The great benefit of this form of argument is that it permits you to finesse your ignorance about difficult matters and still achieve a demonstration. But you really can't use this delicious form of argument when the topic is dogs, say, instead of numbers. Is it true that every animal either is a dog or isn't a dog? What about coydogs and wolf hybrids? The boundaries of the *dog* concept are vague, and so are the boundaries of *coyote* and *wolf* and many other important concepts. These undeniable borderline cases are not just a nuisance to anyone intent on framing a fish-or-cut-bait argument; they typically disable the argument form altogether.[3]

---

[2] It is often noted that this is an example of an intuitively satisfying proof that proponents of intuitionist logic such as Brouwer cannot accept – a stiff price for intuitionism.

[3] In a challenge to the standard conception of borderline cases, Raffman (2005, 2014) argues that a borderline case for a vague term Φ lies between Φ and an incompatible (contrary) category Ψ but is neither Φ nor Ψ; for example, a borderline case of a dog lies between a dog and a coyote but is neither a dog nor a coyote or between a dog and a wolf but is neither a dog nor a wolf, etc. Thus bivalence and excluded middle are safe: the sentence "*x* is Φ" is false, and the sentence "*x* is not Φ" is true in a borderline case. (*Aside:* There are no higher-order borderline cases between Φ and borderline Φ because borderline Φ items are not Φ and borderline cases are not defined between contradictories – only between incompatibles.) For present purposes, the important point is that on Raffman's view, the class of not-Φ things *includes the borderline cases*; the nondogs include the coydogs. Hence, if she is right, my criticism of the fish-or-cut-bait strategy will require reformulation (but only reformulation): the problem will be that the negated disjunct, now covering a heterogeneous class

An argument that exposes the impact of Darwinian thinking is David Sanford's (1975) nice "proof" that there aren't any mammals:

1. Every mammal has a mammal for a mother.
2. If there have been any mammals at all, there have been only a finite number of mammals.
3. But if there has been even one mammal, then by (1), there have been an infinity of mammals, which contradicts (2), so there can't have been any mammals. It's a contradiction in terms.

Because we know perfectly well that there are mammals, we take this argument seriously only as a challenge to discover what fallacy is lurking within it. And we know, in a general way, what has to give: if you go back far enough in the family tree of any mammal, you will eventually get to the therapsids, those strange, extinct bridge species between the reptiles and the mammals. (Technically, mammals are also classified as therapsids, the only surviving therapsids, but usually the term is used to refer to the premammalian nonreptilian species from which mammals descended.) A gradual transition occurred over millions of years from clear reptiles to clear mammals, with a lot of intermediaries filling in the gaps. What should we do about drawing the lines across this spectrum of gradual change? Can we identify a mammal, the Prime Mammal, that didn't have a mammal for a mother, thus negating premise (1)? On what grounds? Whatever the grounds are, they will compete with the grounds we could use to support the verdict that that animal was *not* a mammal – after all, its mother was a therapsid. What could be a better test of therapsid-hood than that? Suppose that we list ten major differences used to distinguish therapsids from mammals and declare that having five or more of the mammal marks makes an animal a mammal. Aside from being arbitrary – why ten instead of six or twenty, and shouldn't they be ordered in importance? – any such dividing line will generate lots of unwanted verdicts because during the long, long period of transition between obvious therapsids and obvious mammals there will be plenty of instances in which mammals (by our five+ rule) mated with therapsids

containing the borderline cases as well as the polar opposites – the coydogs as well as the coyotes, the wolf hybrids as well as the wolves – does not generally support the sort of conclusion that euclideans wish to draw. (Raffman's approach does not entail sharp boundaries for vague words; see Raffman [2014], especially chaps. 2 and 4.)

(fewer than five mammal marks) and had offspring that were therapsids born of mammals, mammals born of therapsids born of mammals, and so forth! Of course, we would need a "time machine" to see all these "anomalies" because the details are undetectable after all those millions of years. Just as well, since the details don't really matter in the long run. What should we do? We should quell our desire to draw lines. We can live with the quite unshocking and unmysterious fact that, you see, there were all these gradual changes that accumulated over many millions of years and eventually produced undeniable mammals.

The insistence that there *must* be a Prime Mammal, even if we can never know when and where it existed, is an example of *hysterical realism*. It invites us to reflect that if we just knew enough, we'd see – we'd *have* to see – that there is a special property of mammal-hood – the *essence* of mammal-hood – that defines mammals once and for all. To deny that there is such an essence, philosophers sometimes say, is to confuse metaphysics with epistemology: the study of what there (really) *is* with the study of what we can *know* about what there is. I reply that there *may* be occasions when thinkers do go off the rails by confusing a metaphysical question with a (merely) epistemological question, but this must be shown, not just asserted.[4] In this instance, the charge of confusing metaphysics with epistemology is just a question-begging way of clinging to one's crypto essentialism in the face of difficulties.

Richard Dawkins, in his recent essay recommending the retirement of the concept of *essence* (2014), writes

Paleontologists will argue passionately about whether a particular fossil is, say, *Australopithecus* or *Homo*. But any evolutionist knows there must have existed individuals who were exactly intermediate. It's essentialist folly to insist on the necessity of shoehorning your fossil into one genus or the other. There never was an *Australopithecus* mother who gave birth to a *Homo* child, for every child ever born belonged to the same species as its mother. The whole system of labelling species with discontinuous names is geared to a time slice, the present, in which ancestors have been conveniently expunged from our awareness (and "ring species" tactfully ignored). If by some miracle every ancestor were preserved as a fossil,

---

[4] Passages in the last few paragraphs have been drawn, with minor revisions, from Dennett (2013, pp. 240–3).

discontinuous naming would be impossible. Creationists are misguidedly fond of citing "gaps" as embarrassing for evolutionists, but gaps are a fortuitous boon for taxonomists who, with good reason, want to give species discrete names. Quarrelling about whether a fossil is "really" *Australopithecus* or *Homo* is like quarrelling over whether George should be called "tall". He's five foot ten, doesn't that tell you what you need to know?

So it isn't just philosophers who have trouble breaking the habit of presupposing essences. As Dawkins notes, there are good reasons for having tidy, "discrete" names for lineages, agreed-upon landmarks to work with, but then we mustn't mistake our convenient agreements for discoveries. Plato unforgettably recommends we carve nature at its joints, but there just aren't enough real, objective joints to suit our communicative purposes. We don't *need* to draw lines, but we *may* draw lines, arbitrarily, in the interest of practical taxonomy. Even if we make this move, disjunction elimination is pretty much disabled as a tool for demonstrating anything because wherever we've drawn our line, we are left with variations on one or both sides of our line that defy the sorts of generalizations that are needed to run the elimination arguments. But that is a good thing because the conclusions typically drawn from such arguments are apt to mislead us away from important truths, as we shall see.

In particular, the demand for essences with sharp boundaries blinds thinkers to the prospect of gradualist theories of complex phenomena, such as life, intentions, natural selection itself, moral responsibility, and consciousness.

If you hold that there can be no borderline cases of being alive (such as, perhaps, viruses or even viroids or motor proteins), you are more than halfway to *élan vital* before you start thinking about it. If no *proper part* of a bacterium, say, is alive, what "truth maker" gets added that tips the balance in favor of the bacterium's being alive? The three more or less standard candidates are having a metabolism, the capacity to reproduce, and a protective membrane, but since each of these phenomena, in turn, has apparent borderline cases, the need for an arbitrary cutoff doesn't evaporate. And if single-celled "organisms" (if they deserve to be called that!) aren't alive, how could two single-celled entities yoked together with no other ingredients be alive? And if not two, what would be special about a three-cell coalition? And so forth.

If, as Fodor (2008) insists, the frog either does or does not have the intention to catch a fly, you end up claiming that natural selection cannot account for adaptations:

I suppose it is likewise plausible that frogs catch flies with the intention of doing so. (If you are unprepared to swallow the attribution of intentions to frogs, please feel free to proceed up the phylogenetic ladder until you find a kind of creature to which such attributions are, in your view, permissible.) Now, intentions-to-act have intentional objects, which may serve to distinguish among them. A frog's intention to catch a fly, for example, is an intention to catch a fly, and is ipso facto distinct from, say, the frog's intention to sun itself on the leaf of a lily. (p. 2)

Now the intention to catch a fly is distinct from the intention to catch an "ambient black nuisance" even if, in the selective environment "fly" and "ambient black nuisance" are coextensive. Because natural selection "can't, as it were, 'see' the difference between intentional states that are extensionally equivalent" (p. 4), it cannot select *for* one intention rather than the other. This conclusion soon leads, by a cascade of disjunctions, to a killer disjunction: either natural selection has a mind or it doesn't. And, surprise, surprise, it doesn't. And because it doesn't, it cannot explain adaptations. The dubiety of this conclusion vies for top honors with the mythic aerodynamic "proof" that bumblebees can't fly.[5]

Fodor is well aware that he's legislating from an essentialist position by insisting that we all stick to "literal" readings of every term and either confirm or deny each proposition:

Surely, you may say, nobody could *really* hold that genes are *literally* concerned to replicate themselves? Or that natural selection literally has goals in mind when it selects what it does? Or that it's literally run by an intentional system? Maybe. Admittedly, the tactic of resorting to scare quotes when push comes to shove (as in 'what natural selection "prefers "', 'what Mother Nature "designs"', 'what the selfish genes "want"' and so forth) can make it hard to tell just what is being claimed in some of the canonical texts. Still, there are plenty of apparently unequivocal passages. Thus Pinker (1997, p. 93):

---

[5] The story appears to have some foundation in fact, though it has been transformed through retelling, a meme with quite a distinguished history, dating back to the 1930s, when August Magnan, a famous French entomologist, and his lab assistant, M. Saint-Lague, did the engineering calculations, as reported in Magnan's book, *Les Vols des Insects* (1934). Of course, Magnan realized that this was a *reductio* of current thinking in aeronautical engineering. See also John McMasters (1989). This footnote is drawn from Dennett and Plantinga (2011).

Was the human mind ultimately designed to create beauty? To discover truth? To love and to work? To harmonize with other human beings and with nature? The logic of natural selection gives the answer. The ultimate goal that the mind was designed to attain is maximizing the number of copies of the genes that created it. Natural selection cares only about the long-term fate of entities that replicate . . .

Fiddlesticks. The human mind wasn't created, and it wasn't designed and there is nothing that natural selection cares about; it just happens. This isn't Kansas, Toto. (p. 7, n. 12)

And if, like John Searle, you deny that there is any room for gradations of consciousness, or gradations of understanding, you end up declaring that "Strong AI" is impossible, or that consciousness is inexplicable, or both.

What we need to break through these self-imposed straitjackets of theoretical imagination is an appreciation of what I call the *sorta* operator, and a good way to see it in action is by putting Turing's revolutionary idea about computation in juxtaposition with Darwin's revolutionary idea about evolution. The pre-Darwinian world was held together not by science but by tradition: all things in the universe, from the most exalted ("man") to the most humble (the ant, the pebble, the raindrop), were the creations of a still more exalted thing, God, an omnipotent and omniscient intelligent creator – who bore a striking resemblance to the second-most exalted thing. Call this the trickle-down theory of creation. Darwin replaced it with the bubble-up theory of creation. One of Darwin's nineteenth-century critics put it vividly:

In the theory with which we have to deal, Absolute Ignorance is the artificer; so that we may enunciate as the fundamental principle of the whole system, that, IN ORDER TO MAKE A PERFECT AND BEAUTIFUL MACHINE, IT IS NOT REQUISITE TO KNOW HOW TO MAKE IT. This proposition will be found, on careful examination, to express, in condensed form, the essential purport of the Theory, and to express in a few words all Mr. Darwin's meaning; who, by a strange inversion of reasoning, seems to think Absolute Ignorance fully qualified to take the place of Absolute Wisdom in all the achievements of creative skill. (MacKenzie 1868)

It was, indeed, a strange inversion of reasoning. To this day, many people cannot get their heads around the unsettling idea that a purposeless, mindless process can crank away through the eons, generating ever more subtle, efficient, and complex organisms without having the slightest whiff of understanding of what it is doing.

Turing's idea was a similar – in fact, remarkably similar – strange inversion of reasoning. The pre-Turing world was one in which computers were people, who had to understand mathematics in order to do their jobs. Turing realized that this was just not necessary: you could take the tasks they performed and squeeze out the last tiny smidgens of understanding, leaving nothing but brute, mechanical actions. IN ORDER TO BE A PERFECT AND BEAUTIFUL COMPUTING MACHINE, IT IS NOT REQUISITE TO KNOW WHAT ARITHMETIC IS.

What Darwin and Turing had both discovered, in their different ways, was the existence of competence without comprehension (Dennett 2009, from which material in the preceding paragraphs has been drawn, with revisions). This inverted the deeply plausible assumption that comprehension is in fact the source of all advanced competence. Why, after all, do we insist on sending our children to school, and why do we frown on the old-fashioned methods of rote learning? We expect our children's growing competence to flow from their growing comprehension; the motto of modern education might be "Comprehend in order to be competent." And for us members of *Homo sapiens*, this is almost always the right way to look at, and strive for, competence. I suspect that this much-loved principle of education is one of the primary motivators of skepticism about both evolution and its cousin in Turing's world, artificial intelligence. The very idea that mindless mechanicity can generate human-level – or divine-level! – competence strikes many as philistine, repugnant, an insult to our minds and to the mind of God.

Turing, like Darwin, broke down the mystery of intelligence (or intelligent design) into what we might call atomic steps of dumb happenstance, which, when accumulated by the millions, added up to a sort of pseudointelligence. The central processing unit of a computer doesn't *really* know what arithmetic is or understand what addition is, but it "understands" the "command" to add two numbers and put their sum in a register – in the minimal sense that it reliably adds when thus called on to add and puts the sum in the right place. Let's say it *sorta* understands addition. A few levels higher, the operating system doesn't *really* understand that it is checking for errors of transmission and fixing them, but it *sorta* understands this and reliably does this work when called on to do so. A few further levels higher, when the building blocks are stacked up by the billions and trillions, the chess-playing program doesn't *really* understand that its queen is in jeopardy, but it *sorta* understands this, and IBM's Watson on *Jeopardy sorta* understands the questions it answers.

Why indulge in this *sorta* talk? Because when we analyze – or synthesize – this stack of ever more competent levels, we need to keep track of two facts about each level: what it *is* and what it *does*. What it *is* can be described in terms of the structural organization of the parts from which it is made – so long as we can assume that the parts function as they are supposed to function. What it *does* is some (cognitive) function that it (*sorta*) performs – well enough so that at the next level up we can make the assumption that we have in our inventory a smarter building block that performs just that function – *sorta* good enough to use. This is the key to breaking the back of the mind-bogglingly complex question of how a mind could ever be composed of material mechanisms. The *sorta* operator is, in cognitive science, the parallel of Darwin's gradualism in evolutionary processes. Before there were bacteria, there were *sorta* bacteria, and before there were mammals, there were *sorta* mammals, and before there were dogs, there were *sorta* dogs, and so forth. We need Darwin's gradualism to explain the huge difference between an ape and an apple, and we need Turing's gradualism to explain the huge difference between a humanoid robot and hand calculator. The ape and the apple are made of the same basic ingredients, differently structured and exploited in a many-level cascade of different functional competences. There is no principled dividing line between a *sorta* ape and an ape. The humanoid robot and the hand calculator are both made of the same basic, unthinking, unfeeling Turing bricks, but as we compose them into larger, more competent structures, which then become the elements of still more competent structures at higher levels, we eventually arrive at parts so (*sorta*) intelligent that they can be assembled into competences that deserve to be called *comprehending*. We use the intentional stance (Dennett, 1971, 1987) to keep track of the beliefs and desires (or "beliefs" and "desires" or *sorta* beliefs and *sorta* desires) of the (*sorta*)rational agents at every level from the simplest bacterium through all the discriminating, signaling, comparing, remembering circuits that compose the brains of animals from starfish to astronomers. There is no principled line above which true comprehension is to be found – even in our own case. The small child *sorta* understands her own sentence, "Daddy is a doctor," and I *sorta* understand "$E = mc^2$." Some philosophers resist this antiessentialism: either you believe that snow is white or you don't; either you are conscious or you aren't; nothing counts as an approximation of any mental phenomenon – it's all or nothing. And to such thinkers, the powers of minds are insoluble mysteries because they are "perfect" and perfectly unlike anything to be found in mere material mechanisms.

When we turn to moral responsibility, consider the influential argument by Galen Strawson (2010):

1. You do what you do, in any given situation, because of the way you are.
2. So in order to be ultimately responsible for what you do, you have to be ultimately responsible for the way you are – at least in certain crucial mental respects.
3. But you cannot be ultimately responsible for the way you are in any respect at all.
4. So you cannot be ultimately responsible for what you do.

The first premise is undeniable: "the way you are" is meant to include your total state at the time, however you got into it. Whatever state it is, your action flows from it nonmiraculously. The second premise observes that you couldn't be "ultimately" responsible for what you do unless you were "ultimately" responsible for getting yourself into that state – at least in some regards. But according to step 3, this is impossible.

So step 4, the conclusion, does seem to follow logically. But let's look more closely at step 3. Why can't you be (ultimately) responsible for *some* respects, at least, of the way you are? In everyday life we make exactly this distinction, and it matters morally. Suppose that you design and build a robot and send it out into the world unattended and unsupervised and knowing full well the sorts of activities it might engage in, and suppose that it seriously injures somebody. Aren't you responsible for this, at least in some respects? Most people would say so. You made it; you should have foreseen the dangers – indeed, you *did* foresee some of the dangers – and now you are to blame, at least in part, for the damage done. Few would have any sympathy for you if you insisted that you weren't responsible *at all* for the harm done by your robot.

Now consider a slightly different case: you design and build a person (yourself at a later time) and send yourself out into the risky world knowing full well the possible dangers you would encounter. You get yourself drunk in a bar and then get in your car and drive off. Aren't you responsible, at least in part, for the "way you were" when you crashed into a school bus? Common sense says of course. (The bartender or your compliant host may share the responsibility.) But how could this be, in the face of Strawson's knockdown argument? Well, remember that Strawson says that you can't be *absolutely* responsible for the way you are. But so what? Who would think it was important to be *absolutely* responsible? Here is what Strawson (2010) says:

To be absolutely responsible for what one does, one would have to be *causa sui*, the cause of oneself, and this is impossible (it certainly wouldn't be more possible if we had immaterial souls rather than being wholly material).

The burden falls on Strawson and others to show why we ought to care about ultimate or absolute responsibility. I think it is just as obvious that people can gradually become morally responsible – *sorta* responsible – during their passage from infancy to adulthood as it is that lineages of reptiles and then therapsids can gradually become a lineage of mammals over the eons. You don't have to be an *absolute mammal* to be a mammal, and you don't have to be *absolutely responsible* to be responsible. So the constructive way of reading Strawson's argument is that, like Sanford's argument that there are no mammals, it is a *reductio ad absurdum* of the concept of absolute responsibility. The law may oblige us to draw a line (like the line for minimal age for a driver's license or for voting) but will understand that it is arbitrary, an imposed boundary, not a discovered joint in nature.

There are other philosophical puzzles that can benefit, I suspect, from exploring the no-longer-forbidden territory opened up by Darwin's critique of essentialism. Might there be important precursor grades of *semi-quasi-proto-sorta*-altruism, from which we could get a better vantage point to look at "real" or "pure" altruism? Are there interesting epistemic states that are *almost* genuine knowledge? Once we give up essentialism for good, we can perhaps begin to reconstruct the most elevated philosophical concepts from more modest ingredients.

2 | *Darwinism as Philosophy*
Can the Universal Acid Be Contained?

ALEXANDER ROSENBERG

The history of science has a broad pattern. Each science, including mathematics, began its life as a subdiscipline of philosophy, or at least as among the concerns of philosophers. Mathematics – at first mainly the science of space – separated itself from philosophy in the time of Plato and Euclid, physics in the period from Galileo to Newton, chemistry in a process that mainly took place during the lifetimes of figures from Boyle to Lavoisier, and biology from 1859, when the "Newton of the blade of grass" was compelled to publish *On the Origin of Species*.

As each of these disciplines separated itself from philosophy, it left questions to philosophy that it didn't need to answer or was unable to answer, questions that looked like they should be addressed by the science that relegated them to "mere" philosophy. Two obvious examples: mathematicians never seemed to need to answer the question, "What is a number?" Physicists have for the most part steered clear of addressing the question, "What is time?" The agenda of philosophy is replete with questions the sciences (and mathematics) can't answer yet, may never be able to answer, and don't need to answer. In addition to this first set of questions the sciences cannot (yet or perhaps ever) answer or don't need to answer, there are the second-order questions about why the sciences can't (yet) or don't need to answer the first set of questions.

This pattern in the history of science was finally broken by Darwin. Instead of leaving questions to philosophy, his breakthrough enabled the sciences, in particular, biology, to begin to take on questions that from Aristotle's time onward had been the exclusive preserve of philosophy. It took more than a century of repeated forays by biologists and philosophers inspired by Darwin to convince the disciplines – biology and philosophy – that the former could deal with the questions of the latter and then to shape the answers biology provides to a host of perennial questions in philosophy. The prominence of "naturalism" in metaphysics, epistemology, the philosophy of mind, the philosophy of language, and moral philosophy is evidence of this achievement. Nowadays, philosophical "naturalism" pretty much means philosophy driven by mainly insights from Darwin.

In my view, Darwinian theory is particularly salient for answering questions philosophers and, even more persistently, nonphilosophers pose about the purpose of life, the meaning of human existence, the trajectory of our species' history, individual free will, and personal identity. In this chapter, however, I focus on the two most consequential issues where Darwinian theory has the greatest impact: moral philosophy and metaethics and the philosophy of mind/philosophy of psychology. In these two domains, the impact of the theory is even more revolutionary than is supposed by most naturalists who adopt it. It undermines most answers to normative questions about moral values. It raises fundamental questions about the nature of thought, the meaning of language, and the possibility of rational argument.

Of course, Darwinian theory does not do this by itself. It requires some auxiliary assumptions that should be innocuous to naturalists: first, that science is our best guide to the nature of reality; second, that its methods are the most reliable means to secure knowledge; and third, the corollary that philosophical theories need to be compatible with the most well-established theories in the natural sciences. Without these auxiliary assumptions, naturalism can't rule out special pleading about some special domain in which intuition, common sense, or revelation guides us to truths that trump naturalism.

The first section of this chapter explains why Darwin's theory must have the gravest repercussions beyond biology. The second section sketches its corrosive impact on moral philosophy. The third section shows how a Darwinian solution to the mind/body problem subverts all but the most radical philosophies of mind. The final section identifies the fundamental unsolved philosophical problem of justification that Darwinism poses to naturalists.

Throughout this chapter, readers will note how large Daniel Dennett's work bulks. Dennett is surely the contemporary philosopher who earliest, most persistently, and with greatest influence has made Darwin central to philosophy. This chapter argues only that he has not gone quite far enough in tracing its revolutionary impact on our discipline.

Blind Variation and Environmental Filtration: Darwin's Universal Acid

It's worth emphasizing that random variation and natural selection are blind variation and environmental filtration. Indeed, it would be advisable to substitute the second description for the first. Blind is better than random because it draws attention to the fact that the cause of variation is never need, benefit, advantage, or suitability to the local environment. Darwinian processes don't really even need variation to be random in a sense of "indeterministic," though random variations will perforce be blind in the requisite sense. Passive filtration is better than natural selection. "Selection" carries the suggestion of intentional choice, and "natural" encompasses more than the local factors that impinge on the lineage and traits that evolve. Filtration is the passive work of sieves, and it's the local environment that does the sifting.

Daniel Dennett said it best in a book title: *Darwin's Dangerous Idea* (1995). In fact, it is *universal acid*: "[I]t eats through just about every traditional concept ... Darwin's idea is a universal solvent, capable of cutting right to the heart of everything in sight. The question is what does it leave behind?" (p. 521). Even less than Dennett thinks, this chapter shall argue.

Take anything you like that reflects an economy of means to ends, especially anything that seems to show creativity, ingenuity, wisdom, forethought, considered design, and clever execution, and it will turn out to just be another instance of a long and boring process that has none of these features, a process that is in fact, again as Dennett so tellingly describes it, an algorithm – a substrate-neutral, mindless, mechanical procedure. Substrate neutral: it can work its magic on properties of almost any composition of matter as an input: aspects of macromolecules one by one, features of complexes of them, traits of larger agglomerations of matter, monadic properties, relational ones, and properties of spatially distributed objects of many different kinds. The process can be implemented on many different kinds of inputs and by many different kinds of procedures. Mindless: all these procedures operating on different substrates will have to have some things in common to count as Darwinian. They can't work by magic or any other process that requires ingenuity, creativity, advanced planning, foresight, design, judgment, discretion, or, God forbid, wisdom.

The process of natural selection is as mindless as a Turing machine, as mechanical as a spring-driven pocket watch.

Every case of means/ends economy in nature is the result of the operation of this mindless, substrate-neutral algorithm that Darwin discovered. Adaptations such as the eye-spots on the moth's wing or the greater oxygen affinity of fetal hemoglobin over adult hemoglobin or the bat's echolocation equipment are its results. But, in addition, and this is an even more radical claim, processes that track "moving targets" and attain or maintain outcomes or end products, including human thought, also proceed by the same mindless, substrate-neutral algorithmic process Darwin discovered. Blind variation and natural selection are not just mechanisms that put adaptations in place, they are also operating in real time whenever behavior shows environmentally appropriate plasticity, and that includes when humans act for what they describe as purposes. How can we be sure of this? Because we know there are no purposes.

That there are no purposes in nature is something that has been increasingly vouchsafed to us since Newton. It was Descartes who first decried the appeal to purposes, but Newton who substantiated Descartes' rejection of any role for them in physical dynamics. Kant endorsed this view but famously drew the line at biology: "There will never be a Newton for the blade of grass." The "Newton of the blade of grass" was born to the Darwin family in Shropshire twenty years after this pronouncement. Though Darwin provided a positive account of how the appearance of purpose results from purely physical, causal, nontelological processes, well before he did so corpuscularians, and anyone who took mechanics seriously would have had trouble even granting purposes a role in the world. Spinoza recognized this earliest (*Ethics*, Appendix, p. 59). Purposes acting from the future to bring about the past can be ruled out because the as yet nonexistent future cannot bring about existent occurrences and events from their past. Aristotelian entelechies could be ruled out on grounds from mechanics – they violate conservation laws, on evidential grounds, and because they beg the question of how mindless mechanical causes can bring about purposes. This leaves the designs of a benevolent deity as the only available explanation of the appearance of purpose in the universe. On this theory, the apparent purpose reflected in all the means/ends economies of nature are real but derived because, like human artifacts, they result from the original purposes of a (mainly) benevolent and omnipotent or at least very powerful agent.

Darwin's theory has so much more evidential support than this theory and so much consilience with purely mechanistic theories in physical science that there is no serious dispute about its standing as a better, indeed the best, explanation of the appearance of purpose. The Darwinian process is mandated by the second law of thermodynamics. It provides the only way the appearance of purpose could have emerged in a world driven by physics (see Rosenberg 2014). The reasons are obvious. First, whatever process causally produces adaptations – the appearance of purpose in the universe – must start with zero adaptation. Helping one's self to the merest sliver of an adaptation as the starting point of evolution begs the question of where this sliver came from. Second, an acceptable explanation of adaptations must show how they emerge from nonadaptations. Because all the basement-level laws in nature except for the second law of thermodynamics are time-symmetrical, a time-asymmetrical process such as adaptational evolution must harness the second law, producing local order at the expense of greater global disorder. Darwinian natural selection can begin when the thermodynamics of the chemical soup somewhere randomly results in a molecule that combines limited bonding stability with limited (probably catalytic) self (or similar)–replication. Third, it carries on from there algorithmically but irreversibly, building local adaptations and newer adaptations on top of older ones. And it does so only at the expense of global entropic increases. The result is the sequence of quick and dirty solutions to "design problems" with us to this day. What is more, anything else that honors the two requirements – start with zero adaptations, harness the second law – will turn out to be just a faster or slower version of the same Darwinian process that produced us.

Many philosophers who agree that Darwin's theory offers the best or the only acceptable theory of how the appearance of purpose arises in nature also hold the view that Darwin's discovery somehow naturalizes, tames, purpose, makes it safe for science. It admits that there really are purposes and shows how purposes arise and how they work in nature – via the mechanism of blind variation and environmental filtration. On this view, when we say that the heart beats *in order to, for the sake of, with the purpose of, so that* the blood is circulated, we are uttering a truth that is in no way incompatible with the *denial* of future causation, immanent teleology, the ordinances of a benevolent deity. The claim is a true one, and its truth conditions are a set of facts about variation and selection in the past.

One might suppose that whether Darwin successfully banished purpose from nature or somehow tamed it for science is a verbal dispute. This cannot be right, however. It would be a version of Orwellian "Newspeak" where the meaning of the word is changed from one that invokes future causation, eminent or immanent design, to one of passive algorithmic mechanical processes driven by the second law of thermodynamics. Try it out: "War is Peace," "Freedom is slavery," "Ignorance is strength," "Purpose is blind variation/environmental filtration," indeed.

If Darwin had simply naturalized purpose, he would have effectively vindicated the Aristotelian conception of nature, at least done so for the domains beyond physical science. We have all been taught that the scientific revolution of the seventeenth century overthrew Aristotelian scholasticism, purging nature of "natural place" teleology. To think that Darwin restored Aristotelian thinking to a portion of the field would certainly be a surprise to Thomists and others who still defend an Aristotelian conception of nature or some part of it. If, along with most historians of science, you hold that Darwin hammered the last nails into the coffin of the Aristotelian word view, you cannot credit him with taming purpose and making it safe for science.

Here is still another way to see that Darwin banished purpose from nature. Caloric theory, developed by Lavoisier and before him by Joseph Black, treated heat as a substance, "caloric," a weightless but incompressible fluid that moved, like water to equalize its level in adjacent containers between which it could travel. Using caloric theory, Black was able to calculate specific heats accurately, and Lavoisier was able to develop the calorimeter to measure the heat generated by chemical reactions. Chemistry still makes use of Black's tables of specific heat and employs successors to Lavoisier's calorimeter, but no one supposes that Kelvin showed that caloric – the incompressible weightless fluid Black and Lavoisier hypothesized – is really molecular motion. They are too different from each other for anyone to treat molecular motion as vindicating the existence of caloric. What Kelvin showed was that there is no such thing as caloric. *Mutatis mutandis* for Darwin and purpose.

Here is still another way to see that Darwin disposed of purpose. From the time of Newton, the most problematic concept in physics was gravity: a force that is transmitted at infinite speed, through total vacuums, that nothing can be shielded from, even by the thickest insulation. Given the rest of physics' commitment to contact forces that operate through causal chains, gravity was not just a fly in the ointment but the major embarrassment of the discipline. Newton famously replied to demands that he explain how it operates by saying, *Hypotheses non fingo*.[1] Now what exactly was the achievement of Einstein's general theory of relativity? No one supposes that he resolved Newton's embarrassment by providing an explanation of how this force operates. No. What Einstein did was show that there is no such thing and that the curvature of space-time produces the accelerations that Newton explained by appeal to his mysterious and unaccountable force. Curved space-time is too different from gravitation for anyone to suppose that the former provides the causal mechanism whereby the latter operates. Einstein showed that there is no such force as gravity. *Mutatis mutandis*, Darwin should that there is no such thing as purpose. Blind variation/environmental filtration is as different from purpose as molecular motion is from caloric or space-time curvature is from gravity.

In physics, chemistry, and biology, it proved convenient to use terminology redolent of the foregone notions: chemists still employ calorimeters, for example. In physics, there is "gravitational lensing." But it's just a convenient shorthand for the way the curvature of space-time affects photons; in biology, it's the concept of "function" that misleadingly suggests purpose despite biologists' best efforts (more on this later).

Why is this issue important? Because the "manifest image," common sense, ordinary life, and parts of philosophy proceed by employment of concepts that require real purposes, the sort Aristotle invoked. These concepts are immune to the naturalistic project of reconciling the manifest image and the scientific one. The attempt to convert them to concepts that are nothing like them in their very natures is a version of Orwellian "Newspeak."

---

[1] "I do not feign hypotheses."

Darwinian Genealogy of Morals Imposes a Transvaluation of Values

The impact of Darwin on serious moral philosophy was long delayed but ultimately profound. It seemed at first to give rise to a moral theory every sensible person should have repudiated. But social Darwinism should never have been so called. The thesis that goes by that name should have been labeled "social Spencerism" to record the fact that it was Herbert Spencer who held that the morally best outcome or the morally right course of action is the one that maximizes biological fitness.

In fact, Darwin was famously puzzled about how selection for maximization of biological fitness could even have produced people committed to the moral codes familiar to him. His only attempt at explanation involved an appeal to group selection that was long afterwards repudiated among biologists. T. H. Huxley, Darwin's bulldog, famously held that our moral norms were in conflict with those traits that evolved by natural selection. Both he and Darwin certainly read enough Hume to have come across his arguments against inferring from "is" to "ought." These arguments were even more well known to philosophers, and in the early part of the century, they were reinforced by G. E. Moore's "open question argument" against what he called "the naturalistic fallacy" of identifying a moral property with a natural or descriptive one. "Naturalistic fallacy" came to label Hume's objection to arguments from "is" to "ought" as much as Moore's. Naturalism of any kind, and especially moral naturalism, became anathema in moral philosophy and metaethics. Then, in the last quarter of the last century, things changed. The growth and success of naturalism in epistemology and metaphysics encouraged others to take it seriously in moral philosophy. The process was accelerated by a breakthrough in biology. Starting from the work of W. D. Hamilton in the 1960s and 1970s, it became apparent first that core morality is compatible with Darwinian processes. Operating through genes and culture, it could after all have created and shaped a core morality shared by almost all people everywhere throughout human history. Biology, evolutionary game theory, paleoanthropology, and experimental economics began to provide the detailed genealogy of morals.

Core morality is a set of norms shared by humans everywhere. The norms are difficult to articulate. Some are so obvious that stating them seems pointless ("Don't cause your infant child gratuitous pain"). When we try to express others, they bear large numbers of qualifications, hedges, and exceptions ("Thou shalt not kill"). Some pairs of the norms in the moral core infrequently enjoin incompatible actions. Besides the moral core, there are also moral norms that differ across cultures (e.g. honor killing). Interestingly, these nonuniversal norms, repugnant to many, appear often to be adaptative, fitness-enhancing in their respective, and quite different local ecologies.

Darwinian theory and a great deal of data from paleoanthropology now show that core morality is not just an adaptation but the only solution to a suite of problems that faced our ancestors when they found themselves exiled from the shrinking rain forests of Africa, facing the challenges of survival at the bottom of the food chain on the African savanna. The "design problem" facing *Homo erectus* was posed by predatory megafauna both competing with and preying on humans in an ecology with little else to support life except for animal protein and fats. And the problem was not simply warding off predators. Humans had already begun to exacerbate their problem in three ways: producing many more offspring over the life course than other primates, producing them much closer together that other primates, and producing offspring that required long childhood dependence. The long dependence was required because in humans most brain development had to take place after birth, not before it. The human birth canal is too narrow to allow for much prenatal neural development. Unless Mother Nature found a way to turn large populations of young children with long periods of dependence into an adaptation, these three traits were bound to carry us to extinction.

At least initially the only steady source of food on the savanna was scavenging protein and fats from whatever the top predators might leave. But *Homo* came out of the rain forest with three advantages: (1) use of stone tools, an adaptation shared with chimps and other primates, that we learned quickly to apply to break into marrow and brain inaccessible to predators; (2) a theory of mind – or rather a capacity to predict the behavior of conspecifics – which we also shared with other primates; and (3) a trait primates lacked but we shared with a few other species – dogs, tamarins, dolphins, and elephants – a tendency to cooperative child rearing. It is impossible to say why we acquired this trait when other primates did not. It may be enough for a Darwinian account of human evolution simply to note that it was an available variation hit on by several species independently, but only by one among the primate species.

Theory of mind and cooperative child rearing soon synergized to encourage the division of labor, hunting, gathering, and child rearing; a long childhood and a large brain can be exploited for teaching and the labor specializations and norms required for increasingly complicated cooperative projects. The result is a coevolutionary cycle that selects for further improvements in those traits – improved theory of mind and greater inclination to cooperate – until we arrive at core morality. It is worth noting that this scenario is robust. It can be varied significantly and still have the same result. In particular, we don't have to think of increased fertility, closer child spacing, and a long period of postnatal dependency constituting prior problems to be solved when our ancestors arrived on the savanna. Instead, we can identify an increase in protein and fat consumption as the cause of increased fertility and reduced birth spacing, along with postnatal brain growth, all of which enabled humans to climb to the top of the food chain on the savanna. In fact, the right account is almost certainly a co-evolutionary feedback loop between both scenarios: successful scavenging required increased brain size, scavenged proteins and fats build brain size, and increased brain size increases scavenging success and then hunting payoffs. To this scenario may be added an evolutionary process that drives the selection of packages of emotions and norms that strengthens enforcement of cooperative institutions and produces internal moral motivation that increases the cost of defection and free riding. The result is the evolution of core morality as a ubiquitous adaptation across humans everywhere, along with local moralities selected for under differing ecological conditions as well.

This Darwinian explanation of the evolution of morality begins with two assumptions that it seems difficult to deny:

1. All cultures, and almost everyone in them, endorse most of the same core moral principles as binding on everyone.
2. The core moral principles have significant consequences – good or bad – for humans' biological fitness – for our survival and reproduction.

So much for the genealogy of morals.

Now suppose that we philosophers set out to underwrite, ground, justify, or support the claim that core morality, or some significant component of it – utilitarianism, duties (perfect and imperfect), respect for (natural) rights, a set of virtues – as the right morality, the correct or "true" ethical theory, or if not "truth-apt" then at any rate the one that should be endorsed by all rational agents. Starting with core morality and carving out a particular moral norm or set of them to underwrite as foundational is certainly a central task of moral philosophy. Many naturalistic philosophers aspired to accomplish this central task or at least to show that naturalism had the resources to do so. One motivation to account for moral knowledge naturalistically is obvious. If we can't do so, and we must grant there is moral knowledge, we make it easier for nonnaturalists to argue that there are other sorts of knowledge – including knowledge of nonnatural, supernatural, and, for that matter, divine things.

But naturalists seeking to certify morality face a very serious problem: the two assumptions of the Darwinian explanation of the emergence of core morality severely constrain possible arguments for the rightness of core morality or some part of it. The only way all or most normal humans could have come to share a core morality is through selection on alternative moral codes or systems, a winnowing process that resulted in some set of core norms being selected for in the evolutionary competition and becoming "fixed" in the population. But note, if our universally shared moral core were both the one selected for and also the right moral core, then the correlation of being right and being selected for couldn't be a coincidence. Here we need to add another auxiliary premise naturalists share: science doesn't tolerate cosmic coincidences. The correlation needs to be explained. Two alternatives immediately suggest themselves: either core morality is the right morality because it evolved through a Darwinian process, or core morality evolved through a Darwinian process because it is the right morality.

Alas, there are fatal objections to either of these alternatives. On the one hand, Darwinian processes are not particularly good at driving human beliefs to true ones. On the other hand, just showing that a practice or the norm that governs it enhances reproductive success provides no normative justification of it whatsoever.

Darwinian forces drive humans and other creatures to outcomes – including belief dispositions and occurrent beliefs – that are adaptive in their environments even when those beliefs are false. Consider folk physics, folk biology, and folk psychology; consider religious beliefs; beliefs about strangers, foreigners, women, the mentally ill; beliefs about probabilities. The list of beliefs heuristically useful and manifestly false but fostered on us by their adaptive payoff is endless, revealing a set of norms that Darwinian pedigree has no tendency to underwrite, warrant, justify, or ground their rightness, correctness, or truth.

Could the causal relationship operate in the reverse direction, from fitness to moral rightness? There is nothing morally right about having more offspring rather than less or many offspring as opposed to none. No one supposes Genghis Khan to be credited with much moral standing despite the fact that his genes are the most widely represented among all human beings (1 of every 300 males is a direct descendant).

Can we explain the coincidence that our moral core is the right morality and that it was the result of Darwinian selection by identifying a joint cause of both its correctness and its Darwinian pedigree? Naturalists really can't seek two independent causal processes, one for core morality's fitness and the other for its rightness. That would be to accept the coincidence. There has to be at most one such process. But the only "prior" cause available for the emergence of adaptations through natural selection is the operation of the second law of thermodynamics on local conditions that obtained on the Earth (and might obtain elsewhere for all we know). We naturalists are unlikely (to put it mildly) to find a joint cause for core morality's rightness and its Darwinian pedigree in the second law.

Here is another logically possible but bizarre alternative: true morality was the overdetermined result of Darwinian selection and some other process, which both justifies and brings it about. Ethical intuitionists might be able to contrive a scenario that combines natural selection with the independent emergence of an epistemic apparatus that enables us to see moral truths. But naturalism cannot accommodate that possibility. Note that natural selection will result in humans at least sometimes being able to discern the heuristically useful truth that core morality enhances cooperation, which, in turn, increases biological fitness. But this instrumental justification is far from what we need if we are to show that core morality is the right morality.

The solution to the coincidence problem is glaringly obvious, of course. It's just extremely unpalatable. Simply deny that our moral core is the right, correct, true morality, that it has anything like the sort of justification philosophers have traditionally sought for it. This move solves the coincidence problem by taking seriously core morality's Darwinian pedigree. Identifying the strong selective forces that have foisted core morality on us doesn't just explain its emergence, it explains the strong feelings we share in favor of core morality, feelings that led to its treatment as objectively right and thus to the persistent search for the grounds of its objectivity, truth, correctness, rightness. These facts about how core morality was selected for also make denying the rightness of our moral core unpalatable.

The repugnant conclusion forced on the naturalist is that core morality, including its components, spin-offs, and presumptive foundations in utilitarianism, or a theory of duties (perfect or imperfect) or (natural) rights, or a set of virtues, lacks a justification. This conclusion may be mitigated by the recognition that there is no other different set of moral norms that is better justified or justified at all, including the denial of the norms we actually embrace. The one thing naturalists cannot do is seek another source of justification beyond science that could or does underwrite core morality or some component of it or a moral theory that formalizes it. To suppose otherwise is to surrender naturalism altogether.

It is easy to illustrate the problem raised for a naturalistic justification of ethical norms in the strategy common to several naturalists – Patricia Churchland, Sam Harris, and Daniel Dennett – of grounding a set of norms that enjoin us to take steps to enhance human flourishing. What enhances human flourishing can be learned from science, in particular, biological science, and more especially cognitive neuroscience and evolutionary psychology suitably qualified. These scientific findings guide the selection of the moral norms that promote human flourishing and, perhaps more important, enable us to combat norms that don't. Naturalists are well aware of Hume's injunction against inference from "is" to "ought," from "conducive to good health" to "the morally right thing to do." To Hume's challenge, some reply, with Dennett, "From what can 'ought' be derived? The most compelling answer is this: ethics must be *somehow* based on an appreciation of human nature – on a sense of what a human being is or might be, and on what a human being might want to have or want to be. If *that* is naturalism, then naturalism is no fallacy. No one could seriously deny that ethics is responsive to such facts about human nature" (Dennett 1995, p. 468). To this it seems perfectly reasonable to respond, "What's so good about satisfying a nature and a set of wants just because they enhance the probability of creatures like us having more offspring?" Notice that this question immediately gives rise to a version of G. E. Moore's "open question" argument: what's so good about Darwinian fitness?

As the preceding passage from Dennett suggests ("If that is naturalism, naturalism is no fallacy"), naturalists are well aware of the naturalistic fallacy (as a label for Hume's problem and for Moore's distinct and different objection). Lacking a reply to Hume (and perhaps also to Moore), and unwilling to accept a nihilism about moral norms, at least some naturalists have sought to change the subject of ethics altogether. The project of justifying core morality in the way at least two centuries of moral philosophy has sought to do as unconditionally rationally compelling to the individual must be surrendered. Naturalism recognizes that its Darwinian genealogy has no reply to Hume's claim that "Tis not unreasonable for me to prefer the destruction of the whole world to the scratching of my finger." Core morality will have no rational grip on an individual who does not share in a cooperative enterprise that enhances the number of his offspring. But it will have instrumental value for a group and for its individual members *as long as they* wish individually to reap the benefits of cooperation for them and their posterity. Instead of viewing moral philosophy as the search for intrinsic values, categorical imperatives, or timeless virtues, some naturalists have begun to treat it as a compartment of political theory, as a component of institution-design: given a set of ends, goals that humans have (owing to the Darwinian process that shaped them), how do we best achieve these ends? The Darwinian genealogy of morals obviously gives a great deal of the answer to this question and constrains the rest of the answer to this question: how can we improve on the norms of core morality to attain human ends? Kim Sterelny gives voice to this strategy of changing the subject of moral philosophy:

A natural notion of moral truth falls out of the picture that moral belief evolved (in part) is to recognize, respond to, promote and expand the practices that make stable cooperation possible. For there are objective facts about the conditions which make cooperation profitable, and about the individual capacities and social environments which make those profits more or less difficult to realize ... No doubt there is no single set of optimal norms: the best normative packages for a group will depend on its size, heterogeneity, and way of life. But ... a natural notion of moral truth seems to emerge from the idea that normative thought has evolved to mediate stable cooperation ... The moral truths are those maxims which are members of all or most near-optimal normative packages; sets of norms that if adopted, would help generate high levels of appropriately distributed and hence stable cooperation profits. ("Evolution and Moral Realism," Ben Fraser and Kim Sterelny, draft of November 2013, p. 3)

Moral truth is optimality in providing the fruits of cooperation. "Thou shalt not kill" is on a par with conventions like "Everyone should drive on the left or the right exclusively." The question remains whether as a subject ethics can tolerate this much of a change in its subject matter.

Darwinism in the Philosophy of Mind

The mind/body problem has been with philosophy in pretty much its present shape since Descartes. What began as a proof, from the nature of cognition and sensation, that the mind is not identical to the brain became for most naturalists in the late twentieth century a puzzle about how the mind could be the brain.

The problem is that thought has content; it is always thought *about* things and their properties. But, as Leibniz noted, no arrangement of matter in the brain or elsewhere can have this property of being about some other clump of matter, still less being about nonexistent clumps of matter or properties they don't have. He invites us to conduct a thought experiment:

Imagine there were a machine [i.e. the brain] whose structure produced thought, feeling, and perception; we can conceive of its being enlarged while maintaining the same relative proportion among its parts, so that we could walk into it as we can walk into a mill. Suppose we do walk into it; all we would find there are cogs and levers and so on pushing one another, and never anything to account for a perception [or a cognition, for that matter]. So perception must be sought in simple [nonmaterial] substances, not in composite things like machines. (*Mondadology*, Section 17, Bennett translation[2])

It's easy to update Leibniz' thought experiment: substitute for the machine's cogs and levers, the brain's neural network with electric charges passing along neurons to synapses, where neurotransmitter molecules change concentrations and shapes, shifting gradients of calcium, potassium, and chlorine ions. No aboutness, no *intentionality*,

---

[2] www.earlymoderntexts.com/pdf/leibmona.pdf.

no matter how many neurons, no matter how cleverly arranged, and not even the ghost of an idea of how they could realize intentionality and thus cognition.

One chunk of matter just can't just by itself be "about," be "directed at," mean another chunk of matter. If I am thinking about Paris, there is no neural circuitry in my brain that just by virtue of its physical configuration is "about" Paris. This is the mind/body problem.

It was probably in Charles Taylor's *Explanation of Behavior* (1964) that the problem of the intentionality of thought was first clearly recognized as one of teleology. Daniel Dennett certainly twigged to it in *Content and Consciousness* (1969), and Jonathan Bennett made it explicit in *Linguistic Behavior* (1976).[3]

The realization emerges from the diagnosis of behaviorism's inability to eliminate behavior's appearance of purposiveness, especially its plasticity and persistence. But these apparently teleological features of behavior were "inherited" from the thoughts, about environment and target state, that drive behavior. Behavior is purposive because thought is. Its intentionality, aboutness, content, is really a matter of fine-grained teleology. To a first approximation, what makes a bit of the brain, a neural circuit, a desire is the way it can combine with beliefs to bring about some end, goal, or purpose. What makes a neural circuit a belief is the way it can combine with desires to bring about some end, goal, or purpose. The content of the desire is a description of the end; the content of the belief is a description of the means – the facts about the circumstances relevant to attainment of the desire.

The intentionality of thought is the teleology of the neural circuitry. Therefore, it ought to succumb to a Darwinian analysis. So was borne the program of *teleosemantics*, the strategy of showing how, as a result of a Darwinian etiology, a brain state can be said to be about, to contain information regarding something beyond itself, and thus reveal how thought can be physical. This program is not just inevitable for naturalism, it is the only one that can hope to succeed once it is

---

[3] It was clear as far back as Dennett (1969) that consciousness can't be by itself the original source of intentionality. Perform a Humean thought experiment: look into yourself. What's intrinsically originally intentional about tokens moving through consciousness? Their appearance of intentionality is just a matter of the pattern of succession and association of these tokens passing through consciousness. Think of the tokens passing through the conscious states of the newborn infant. What are they "about"? Consciousness just obscures the fact that the ground of intentionality is in goal-directedness purpose.

recognized that the essence of intentionality is its teleology. For there is only one way that the appearance of teleology arises: the way Darwin discovered. No wonder Leibniz wouldn't have been able to detect thought in the machinery of the brain. You can't see past Darwinian processes in present structures.

The approach of teleosemantics begins with a prior insight: the taxonomy of common sense and science is largely "functional" and not "structural" in the sense that most things are classified in terms of their characteristic causes and effects, usually their effects. Consider the noun "chair." There is almost nothing in the meaning of the word that restricts its material composition (a chair could be made of dry ice), size or shape (doll house chairs are chairs), number of legs (a three-legged stool is a chair; so is a one-legged shooting stick), whether it has arms (bar stools are chairs), and so on. A chair is by definition what can be sat in, that is, something with effects in maintaining a certain posture. The same must be said for psychological states such as beliefs, desires, emotional states, memories, and perceptual states. They are, physicalists hold, all brain states, but their natures are given by their functional roles. Their roles as bearers of content, as being about something, their representational functions are their *biological functions*, what they and their ancestors were shaped by, selected for, through a Darwinian process of blind variation and natural selection. The classical illustration of how this thesis works is the frog snapping at flies. When the frog snaps its tongue at a fly located at x, y, z, at time t, somewhere in the neurology of the frog there is a set of neurons whose firing in the presence of the fly caused the snapping. This set of neurons has the content "fly at x, y, z, t" if and only if there is a Darwinian etiology, a history of selection on ancestors of this frog that shaped their neurology so that when the frog requires nutrition and flies are snappable, its tongue snaps at them. We might even loosely say the firing of the neurons *means* fly at x, y, z, t. Developing a theory of psychological content out of this insight has been a forty-year program in the philosophy of mind.

As noted earlier, something like this approach to content, aboutness, intentionality must be correct because it is mandated by the recognition that the appearance of purpose is always the result of blind variation and natural selection. But this is too programmatical a consideration to have much sway when it comes to solving the mind/body problem. So, for forty years, working out the details of exactly how the Darwinian

insight illuminates the mind has been the joint project of two generations of philosophers of mind. The programmatic consideration – that
this approach has to be right – should, like the gallows, focus our
attention. The stakes are high. For instance, if it pans out, we can
have a naturalistic epistemology without changing the subject, as in
the case of ethics. If it does not pan out, the alternatives are radical
indeed: dualism – the mind must really be some sort of spooky spiritual
substance – or eliminativism – content, aboutness, intentionality are
illusions.

One strong nonprogrammatical argument for teleosemantics begins
with the fact that cognitive states can misfire; we can be wrong, commit
errors or mistakes, and have false beliefs. This is a feature of thought
that seems to cry out for a Darwinian approach. The idea is that when
the frog snaps at a lead beebee at x, y, z, t, the content of the relevant
neurons is "fly at x, y, z, t" because that is what its biological function
is, and it is mistaken on this occasion. The biological function, the
"proper function" (a term that Millikan [1984] introduced), the
normal function (Neander's [2012] term) is given by the Darwinian
etiology, in which the presence of flies, not beebees, shaped the
neurology of frogs. False content should turn out to be a matter of
malfunction. A lead beebee can trick the frog's neuron's into "fly at x,
y, z, t," a falsehood that produces tongue snapping at a nonnutritional
object that is in fact harmful to the frog.

Another consideration strongly suggesting the rightness of this
approach is the way Darwin's insight encompasses learning, both
operant and classical. The latter was selected for much earlier than
the former. It's crucial to survival of the sea slug. We know its
molecular biology, thanks to Eric Kandel. Operant conditioning has
long been recognized as a species of Darwinian selection operating by
nongenetic transmission within the organism's lifetime (Dennett 1975):
at a certain level of neurological evolution, Darwinian processes hit on
operant mechanisms, which were then selected for owing to the fact
that the environment was changing too fast to be successfully tracked
by genetic variation in neural structures. As these mechanisms improve,
animals show behavioral plasticity/persistence sensitive to more rapid
environmental changes. At this point, teleosemantics provides the
resources to attribute content to new neural states that are not
hardwired or built by previous reinforcement (selection) to produce
behavior appropriate to the environment. Dennett later developed the

thesis that there is a hierarchy of Darwinian mechanisms producing increasingly seemingly purposive behavior that is fine-grained and environmentally appropriate, that is, adapted behaviors in four stages: first, there are Darwinian creatures, whose behavior is hardwired, by presumably hardwiring content in their neurology or what passes for it. Consider amoebae, who are built to detect nutrient gradients and whose motion-directing equipment can be accorded the content "More sugar to the left of here" and is somehow about the sugar. Second, Skinnerian creatures, whose neurology is sensitive to operant conditioning, can acquire new content; pecking the blue key will be followed by a food pellet appearing below the key. Third, there are Popperian creatures, so called in honor of Popper's observation that thinking consists of generating variants and testing them in imagination – permitting "our hypotheses to die in our steads." Popperian creatures that that show even greater apparent purposiveness in behavior, that is, even fine, more environmentally appropriate behavior than Skinnerian creatures. They must do it by an inward Darwinian process: there must be an inner environment that represents a great deal about the actual environment and that selects among possible behaviors. Birds that can spontaneously figure out how to use a tool to fashion another tool to reach insects they feed on provide examples. Somewhere in the bird's brain there is a set of neurons with content "Bending this stick will enable me to reach a longer stick that will reach down into the insect mound." Finally, there are what Dennett calls "Gregorian creatures," after the cognitive neuroscientist Richard Gregory, who first articulated the role of concepts in perception. Gregorian creatures, like us, are capable of stocking their Popperian inner environment with words, silent versions of external signs that extend memory and inference vastly beyond the powers of Popperian creatures.

These broad brush strokes are suggestive, but they don't substitute for careful analysis that shows how patterns of variation and filtration shape specific representations – neural states – with content we can pin down and be confident the neural circuitry actually bears. To illustrate the point, the frog's neural circuitry can't really contain the thought "fly at x, y, z, t," for to do so, it would presumably have to have the concept "fly," and it is either doubtful that the frog has such a concept or pointless to pursue experiments to decide whether it does so. "Fly at x, y, z, t" is just a placeholder we use to identify the content, whatever it is, of those neurons. Teleosemantics is going to have to do better than placeholders, at least for Popperian and Gregorian creatures, or give a good reason why, despite its inability to do so, it should not be deemed a failure, one that brings us face to face with dualism or eliminativism.

It is worth following out some of the ways the program of teleosemantics proceeds: it's the usual method of definitions, counterexample, refinement. Let's begin with the problem of distal content. When the neurons tell the frog to snap its tongue in a certain direction, why is their content a fly or, if not a fly (because it lacks the concept of "fly"), something vaguer, such as "a dark object a centimeter or two in front of the tongue"? Why couldn't the content just be "incoming photons distributed in a pattern previously associated with nutrition" or "small black retinal excitation associated with pleasing taste" or even just "pattern of action potentials coming from visual cortex previously reinforced when associated with tongue snapping." These contents less distant (distal) than "fly" need to be ruled out, don't they? So also should more distal content be excluded, content that includes too much, for example, "fly in the presence of photons that reflect off of the fly." A neat solution to this distal content problem suggested by Millikan (1984) also adds important detail to a teleosemantic theory: consider the downstream "consumers" of the neural state representing "fly at x, y, z, t." These "consumers" include the neural system and eventually the digestive system of the frog, with the function of converting food into energy. The selection process that brought about these downstream digestive functions narrows the neural state's content to being more like "food at x, y, z, t" than "fly at x, y, z, t." And they rule out at least some other things as content, such as retinal images or the photons that reflect off the fly, because there is no distinctive historical pattern of variation and selection for responding to retinal images or photons *qua* food. Moreover, the neural signals from the retina through the visual cortex to the neural circuits that do have the content "food at x, y, z, t" were selected for, but only because they were the means by which the frog responded to flies and not vice versa. Thus teleosemantics employs the notion of a selective etiology to zero in on the "right" distal object – food – as the content of the neural state. Still it seems unreasonable to attribute the concept "food" or "chow" or "tasty stuff" or any other concept describable in human vocabulary to the frog, even as we recognize the fine-grained environmental appropriateness, the adaptiveness of, the neural states of the frog. Moreover, the selection history that shaped these neurons to output signals to the tongue from inputs caused by flies in fact selected for such inputs only when caused by healthy flies, in the absence of predatory birds, etc., where the "etc." needs to be unpacked by indefinitely many factors that might reduce frog fitness. Yet the neural content cannot reasonably include all these qualifications on "fly" or "food" or "yummy stuff at x, y, z, t."

Of course, frogs are very limited in their abilities to fine-tune their behavior, including their tongue-snapping behavior, even through operant conditioning. But mammals, especially primates and, more to the point, humans, engage in complex learning. This higher learning leads to such fine-grained behavioral discriminations that attributing concept to them is uncontroversial. What we learn, for example, linguistic behavior, shows the systematicity of the thought that produces it. We therefore need to have an idea of how teleosemantics deals with the systematicity of human thought and that of other primates perhaps. It was Chomsky who first emphasized this feature, especially as it reflects itself in language. Teleosemantics certainly can accommodate inborn, innate, or hardwired recursive computational algorithms selected for to provide adaptive synaptic and grammatical structure to thought and through it to speech (especially once cooperation became required for survival [see earlier]). It can quite happily build complex thought out of stored neural states each of which is shaped by separate selective histories, whether Darwinian in the case of innate concepts, if any, or Skinnerian, Popperian, or Gregorian learning conditioning (all operated by versions of "long-term potentiation" – the mechanism of information storage in the brain) in the language of the neuroscientist.

But the problem of carving out the right distal object as the "topic" in Dretske's terms (Dretske 1989) of the neural content is the tip of a large iceberg. There is also the problem of specifying the "comment" in Dretske's terms, what the thought "says" about its object: can a teleosemantic approach specify the content of neural states enough to match up with content ascriptions that common sense ascribes or that cognitive science requires or that solve the physicalists' mind/body problem? These may not all be the same, but it seems unlikely that teleosemantics can do them all. The problem it faces is identified by Fodor (1990): the disjunction problem. Owing to a long history of selection for neural circuitry that leads the frog's tongue to snap at flies, the frog's tongue also snaps at occasional lead beebees. When it does so, why is the content of the neural circuitry the falsehood "fly at x, y, z, t" instead of the truth "fly or beebee at x, y, z, t"? What in principle is the difference between misrepresentation and disjunctive representation? Since adding a disjunct that converts a false content into a true one is always logically permissible in a content attribution, the question emerges, "How can teleosemantics distinguish any malfunctions, such as false beliefs from well-functioning disjunctive beliefs?" The problem gets worse, according to Fodor. Consider biologically significant properties that are always instantiated together, such as being warm-blooded and bearing live young. These distinct properties are universally coinstantiated in the environment. Two properties always coinstantiated can't make separable contributions to the selection of neural circuits by the environment. Accordingly, neural circuits that are *about* the instantiation of two such properties cannot differ in teleosemantic, that is, Darwinian, that is. biologically identifiable, content.

The right conclusion to draw is that the content of neural circuitry really is indeterminate, or at least far less determinate than the content of natural language in the hands of clever philosophers can construct. Well before teleosemantics hit its stride, Quine had warned us about the epistemic underdetermination of content by behavior, and now teleosemantics reveals that content is in fact indeterminate. Dennett put it poignantly way back in 1969, invoking the dog's interest in a particular T-bone instead of the frog's interest in flies: "What the dog recognizes this object as is something for which there is no English word, which should not surprise us – why should the differentiations of a dog's brain match the differentiations of dictionary English?" (p. 85). But what about the differentiations of a competent English speaker's brain? Should they line up with the *Oxford English Dictionary*? What if they don't?

The intentionality, the content, the aboutness, the meaning of written inscriptions and spoken sounds derives from the neural circuits that produce them. The specific way in which neural circuits endow marks and noises with meaning was first begun to be made explicit by Grice (1957). To Searle (1980) we owe the distinction between original intentionality – what the neural circuits have – and derived intentionality – what speech, writing, and symbols have owing to their neural causes. But, if neural circuitry is indeterminate in its content, then so is the content, the meaning of all speech and writing. No matter how precisely one wants to express oneself, there is no precision to be had. Take any sentence token, no matter how simple, "The cat is on the mat," or how specific, "2 + 2 = 4," and there is no unique proposition the inscription expresses because the neural state that gives it its intentionality has no unique propositional content – and that is the only place content can come from. In other words, the differentiations of the English speakers' brains don't match the differentiations of dictionary English or Serbo-Croatian, Mongolian, and so on. Is that serious? Couldn't it turn out that every one speaks an *idiolect*, the product of the unique Darwinian consequence-etiology that trained up each of us to speak our natural languages? With enough overlap, there wouldn't really be a breakdown in communication anyone would notice. But this isn't the conclusion we have to draw. A much more radical outcome faces us. If our neural circuits all lack unique propositional content, then they lack all propositional content. Why? Well, there will be no finite disjunction of propositions that constitutes their propositional contents either because the disjunction would be a unique proposition too. To say that a finitely large (in fact, very small) neural circuit contains, is about, means, an *infinitely* long proposition (even one we can "identify" by some fancy recursion) is a subtle means of acknowledging the conclusion that it's not about any proposition at all.

This conclusion should surprise and disturb us, for it turns out not that everything we say or write is ineliminatively vague, without specific content, not just that the content of what we think cannot with any precision be expressed by any natural language. It turns out that there really isn't any propositional content at all in the neural circuits despite their fine-grained control of apparently purposive behavior, including speech and writing.

The attribution of specific contents to utterance and inscription is, as Dennett observed long ago, "just" a stance, an instrument, a heuristic technique, indispensible for creatures such as us to make our ways in the world. And now we understand the Darwinian processes through which it emerged and why it did so – to move us from the bottom of the savanna food chain to the top. There is no independent fact of the matter about what humans or any other sentient creatures think, that is, what their thoughts are *about*. Thoroughgoing Darwinians will understand what is going on here. Recall that Darwin's breakthrough was to banish purpose from the world, not make it safe for science. Teleosemantics does the same thing for content: it ends up forcing us to be as eliminativist about intentional content as its parent, Darwinian theory, ends up forcing us to be eliminativist about purposes. Just as the appearance of purpose turns out to be the reality of blind variation/environmental filtration, so the appearance of content turns out to be the same thing. If the essence of intentionality is teleology, of course, this conclusion was fated long ago.

But if content is an illusion foisted on us, like purpose, what can we make of speech and writing? Darwin's universal acid has eaten through too much. It has made a mockery even of the claim that there is no content, meaning, intentionality, derived or original, for that matter. The inscription you have just read turns out to have no content because there is no such thing.

Eliminativist materialism has always been criticized as self-refuting: "I believe that there are no beliefs," says the eliminative materialist. The claim that there is no original intentionality has no derived intentionality. Accordingly, it has no truth value, or if it does, then it has all the semantic virtues of the sentence, "This sentence is false." So much the worse for eliminative materialism. Now it seems, however, that Darwin's insight threatens to drive everyone who embraces it into the eliminativists' cul-de-sac.

Propositions are true or false. Sentences express propositions and derive their truth values from the truth values of the propositions they express. The intentionality – the propositions sentences that "express," "contain," are "about" – is supposed to be derived from the intentionality of thoughts *directed at these propositions*. If neural circuitry doesn't have these propositions as content, then neither do the sentences we utter and write. Darwinian acid has eaten right through the meaning of the statements that express it.

The challenge to the Darwinian naturalist is nothing less than the development of a workable alternative to all the machinery required for semantic evaluability – syntax, semantics, a substitute for truth and falsity, and a naturalistic account of justification, argument, reason – for without a causal account of justification, then no matter what substitute for truth or falsity of thoughts a Darwinian approach provides, it won't be able coherently to argue for it.

Can Naturalism Avoid the Burn of Darwinian Acid?

What are the prospects for a naturalistic account of justification and reason? Dennett has explicitly addressed this challenge in some of his most recent work ("The Evolution of Reasons" [2013]; page references in what follows are all to this essay). An examination of his approach to reason in a Darwinian world shows how hard it will be to make the case. He starts on familiar ground, noting that we have an "unsuppressible" proclivity for reading meaning and purpose into things, an instinct we share with other animals, one "as much in need of a biological account as the distraction display of birds" (p. 59). The account will not only explain but it will also vindicate reason, meaning, design, and purpose in human thought by revealing their existence in nature so that all we have to do is read them off our own experience of the environment:

The biosphere is utterly saturated with design, with purpose, with reasons. What I call the design stance predicts and explains features throughout the living world using exactly the same assumptions that work so well when we reverse engineer artifacts made by (somewhat) intelligent human designers. Evolution by natural selection is a set of processes that "find" and "track" reasons for things to be arranged one way rather than another. (p. 49)

For the process of natural selection to find and to track reasons, at least some reasons have to obtain prior to the process that finds them and tracks them; reasons need to play a nonteleological, purely causal role in nature. That is, naturalism will have to provide a purely causal analysis of justification. "Wherever there are reasons, there is room for, and a need for, some kind of *justification* and the possibility of *correction*" (p. 51). We need to show how "[e]volution by natural selection starts with *how come* and arrives at *what for*." As Dennett recognizes, "We start with a lifeless world in which there are lots of causes but no reasons, no purposes at all. There are just processes that happen to generate other processes until at some 'point' (but don't look for a bright line) we find it appropriate to describe the *reasons* why some things are arranged as they now are" (pp. 50–1). The appropriateness of this finding, however, can't just be a matter of a stance that is convenient for our survival. The emergence of reasons from causes has to be a fact about nature. Let's consider whether Dennett's recipe does this.

There are, he notes, two kinds of norms and modes of correction: *Pittsburgh normativity* and *Consumer Reports normativity*. The former are created only once humans begin to communicate. *Consumer Reports* normativity emerged long before humans, indeed long before metazoans. It is the instrumental normativity of hypothetical imperatives, "quality control or efficiency, the norms of engineering" (p. 51). "Wherever there are *What for* reasons an implicit [instrumental] norm may be invoked: real reasons supposed always to be good reasons, reasons that justify the feature in question. No demand for justification is implied by any *How come* question" (p. 51).

How do the *What for* reasons and their accompanying norms emerge? Start with the prebiotic realm of molecules in motion, combining and breaking up in accordance with the stoichiometric equations of chemistry and the laws of thermodynamics. Given world enough and time, say, 10 billion years or so, these mindless processes will produce some local chemical equilibria – hydrophobic lipid bilayers are an example familiar from every coffee cup with a bit of cream in it.

Imagine we are back in the early days of this process where persistence is on the verge of turning into multiplication and we see a proliferation of the same type of item where before there was none and ask, "Why are we seeing these here?" The question is becoming equivocal. For now there is both a narrative answer, how come, and, for the first time, a justification, what for. We are confronting a situation in which some chemical structures are present while chemically possible structures are absent, and what we are looking at are things that are better at persisting/reproducing in the local circumstances than the alternatives … In other words there are reasons why the parts are shaped and ordered as they are. (p. 53)

Is it supposed to be obvious that when thermodynamic noise produces some molecules that are more stable than others, the causes become reasons, the *How comes* turn into *What fors*? Are there really reasons, justifications for the presence of molecules that combine stability and replicability?

Consider a scenario that differs only in the order of events but is equally thermodynamically possible: through random chemical reactions, several molecule configurations come into existence that all replicate, either because they serve as templates or catalyze the synthesis of other copies of themselves or one another or otherwise make the conditions for their synthesis thermodynamically favorable. How come? Thermodynamic churning. What for? No reason at all. If just one of them is more stable than the others, lasts longer before succumbing to random shocks that break it up, its numbers will begin to increase. There are answers to the *How come* question, "How did this particular distribution of these particular molecules come about?" But the question "*What* did these particular distributions come about *for?*" seems entirely unmotivated. They didn't come about for anything. In describing one of these molecules, it's a stretch to ask what is the "function" of the covalent bond here or the methyl group there. It looks like a job for Cummins' (1975) "causal role" kind of function, but at most it's going to need for the selected-effects Darwinian kind of function several rounds of replication. No *What fors*, just several *How comes*. But Dennett sees this empty glass half full:

We can reverse any reproducing entity, determining its good and bad, and saying *why* it is good or bad. This is the birth of reasons, and it is satisfying to note that it is a case of Darwinism about Darwinism: we see a proto-Darwinian algorithm morphing into a Darwinian algorithm, the gradual emergence of the species of reasons out of the species of mere causes, *what fors* out of *how comes*, with no essential dividing lines between them. (p. 54)

Here, as in the case of the naturalization of ethical values, one wants to ask what's so special about replication, about copying, about reproduction? The question is not just rhetorical because depending on what counts as replication, there will be many cases of it that engender no temptation to pose *What for* questions. For example, consider a line of computer code that takes as input lines of code and produces copies of them. Start with a data set of random inputs, including one copy of the line of code itself. When this program takes itself in as a line of code, it produces a copy of itself. The other lines of code do nothing to inputs to them. Given world enough and time, the number of copies of the self-replicating code proliferates exponentially. There is no temptation to ask *What for: what* is the replicating code replicating *for, what* is any part of the replicating code doing *for* it.

Once he has reasons, Dennett is prepared to ascribe them promiscuously: "Sponges do things for reasons, bacteria do things for reasons. But they don't have the reasons. They don't need to have the reasons" (p. 56). One can't help but be reminded of the confusion caused by Leibniz's misnamed "Principle of Sufficient Reason."

We use the word "reason" somewhat casually when what we mean is "cause." In these two cases of sponges and bacteria, it's not just easy to replace "reasons" with "causes." It is mandatory, on pain of committing a version of what Ruskin (1856) called "the pathetic fallacy."

The sequence in Dennett's narrative doesn't take us from *How come* to *What for*. At most, it takes us from the physical reality to the design stance, from causal processes entirely free of teleology to descriptions of them that impose a heuristically useful teleological overlay. And Dennett's own peroration reveals as much:

> We can have our cake and eat it too. We can use the intentional stance to discover and articulate the reasons evolution (mother nature) has mindlessly unearthed ... We can use the intentional stance with a clear conscience, but only because Darwin has shown us how to cash out the intentional language in suitably austere talk about algorithmic processes of design generation and refinement. Darwin showed us how to get to what for from how come. Darwinian *what for* explanation coexists with its obligatory *how come* backing ... we use *what for* speculative hypotheses to help us frame testable *how come* hypotheses to test. (p. 61)

If Darwin had in fact shown us how to cash out intentional language, we wouldn't need to treat it as a stance. We could help ourselves to it as "real," not just a device creatures like us use to guide the formulation of purely causal hypotheses about *how come* things are arranged the way they are. Still, the passage indicates that Dennett recognizes the "stakes" in this matter. The intentional stance needs backing. The only backing that will work is provided by Darwin's analysis of the only way the appearance of purpose can arise in a physical world, a way that shows that the appearance isn't real. So, in the end, Darwin doesn't make the world any safer for reasons than he does for purposes.

This should be no surprise because, as Dennett's arguments show, the concept of reason, as in "reason for," is built on real purposes, not their Darwinian nonpurposive replacements.

For all its achievements in solving, dissolving, and otherwise settling a raft of perennial philosophical problems, naturalism's greatest challenges lie before it. If it can surmount them, the philosopher who hitches his or her wagon to the sciences will be able at last to rest easier about the agenda of our discipline's problems. Meanwhile, we philosophers inspired by Darwin will have to take Hume's counsel to speak with the vulgar while we try hard to think with the learned.

3 Animal Evolution and the Origins of Experience

PETER GODFREY-SMITH

Introduction

How can we find a way to understand the simplest and most basic forms of subjective experience? What is the set of living organisms for which it *feels like something to be* one of those organisms? When did this phenomenon begin, and what was its earliest form?

The intrinsic interest of these questions is obvious, I take it, and they are important in at least two other ways. Progress here should help in other areas of philosophy of mind, including the most basic debates about how mental and physical are related. To this you might say: attempts to answer my questions won't *help* with the mind/body problem itself, but rather would *be* helped by resolving (if we can) the more fundamental questions. However, it may be that if we better understand the relations between simpler and more complex forms, this will help us to see how subjective experience can have a basis in the material. I think that the shape of an eventual theory will be one that relates the material to the living, the living to the cognitive, and subjective experience to the kind of cognitive operations that living systems engage in. In a companion paper to this chapter (forthcoming), I address the first couple of these relationships in detail. In this chapter, I discuss the later ones, looking especially at the evolution of animal life and how stages in animal evolution might be related to subjective experience.

These issues are also pressing in a more practical way. Here I have in mind ethical questions about the treatment of animals in farming, experimentation, and elsewhere. Questions about how we should treat animals

Earlier versions of this chapter were presented at the 2014 Harvard University Symposium "Animal Consciousness: Evidence and Implications," organized by Dale Peterson, Irene Pepperberg, and Richard Wrangham and at the 2015 Macquarie University Conference "Understanding Complex Animal Cognition," organized by Rachael Brown. I am grateful to members of the audiences at both talks for helpful comments, and also to Diana Reiss, Rosa Cao, and Jane Sheldon.

of various kinds are closely connected to questions about subjective experience, especially questions about suffering. A plausible view is that when we are dealing with an organism with no subjective experience at all, there are few ethical concerns about ways we might treat it. Or perhaps there are some concerns, but they are different from those that apply in the case of an animal that has the capacity to suffer – questions in environmental ethics may still apply, for example. For animals that do have subjective experience, especially of a negative kind (pain and suffering), there is a strong initial case that this should be factored into decisions about how we treat them. In those discussions, the important questions are not evolutionary per se, but about the distribution of subjective experience among present animals. Does a fish or crab have any subjective experience at all? But evolutionary questions are connected to these questions.

The next section sets up the topics of this chapter in more detail and sketches some features of evolution *before* animal life. I then look at the early history of animals, focusing on stages that seem likely to have some relation to the evolution of subjective experience. The last section looks at the relation between the evolutionary history and recent work in neurobiology and philosophy of mind.

Subjective Experience and Early Evolution

I said that the aim is to understand subjective experience. How does this relate to questions about consciousness? It is common now to use the word "consciousness" in a broad way, to cover all kinds of subjective experience. Distinctions might then be made between different kinds of consciousness, some more complicated than others. I don't think there is any error in setting things up this way, but it's not the best. An earlier way of framing the issues, seen more often in the 1980s, distinguished three main problems for philosophy of mind: "qualia," "consciousness," and "intentionality." The problem of "qualia" was seen as the problem of explaining the first-person feel of the mental, "intentionality" involves semantic content or "aboutness," and "consciousness" was seen as a sophisticated kind of mentality, with special features on both the cognitive and qualitative (subjectively felt) side.

Now "qualia" and "consciousness" are often seen as amounting to the same thing, not because of an argument for reduction of one to the other, but because there is only one phenomenon to consider. If there is something it feels like to be a system, then the system is said to be conscious or have some kind or degree of consciousness

(Nagel 1974).[1] I prefer the earlier setup and think the difference is not merely verbal. "Qualia" was a very unattractive term, but it fit quite naturally with the idea that some organisms might undergo very simple forms of experience that are distinct from anything we would usually call consciousness. I wonder whether squid feel pain, for example, but I don't think of this as wondering whether they are conscious beings. "Sentient" is a better adjective for the more general property, and some people do use that term, though it is not a common one in philosophy, and many would probably say that consciousness can be understood very broadly, and to wonder whether a squid feels pain is to wonder whether it is *phenomenally conscious*.

In this chapter, the phrase "subjective experience" will be used more broadly than "consciousness"; a system undergoes subjective experience when there is something it feels like to be that system. "Qualitative" (in a sense derived from "qualia") will be used as an adjective for the felt features of those mental states that are subjectively experienced. "Consciousness" is something beyond mere subjective experience, something richer or more sophisticated, though it is hard to say how this is best understood, and several different kinds of sophistication might be relevant – a topic for another chapter. I'll use the term "cognitive" in a broad way for the processes in organisms that manage sensory input, establish and access memories, control behavior, and so on. I don't assume that an information-processing or computational view of all these processes is the right view (though it might be). I want a general term for the side of the mental that involves behavioral control and intelligence.

The next section will work through some of the history of animal life. Before that, I will describe some of the evolutionary setting *before* animals, especially because it's important to appreciate that a substantial amount of cognitive or proto-cognitive capacity was in place before animals evolved.

Suppose that we approach the history of life from the point of view of functionalism in philosophy of mind, looking for the initial evolution of perception, memory, and behavioral control – the things functionalism tells us are important in giving a physical system psychological properties. All these capacities evolved well before animals did, and some are seen in quite sophisticated forms even in single-celled organisms, including prokaryotes (bacteria and archaea). Bacteria, for example, can track and respond to desirable and undesirable chemicals in

---

[1] Chalmers (1996) also frames things this way.

their environments in very effective ways. *Escherichia coli* bacteria control their swimming with a form of short-term memory. At each time step, the bacterium compares the chemicals it is presently sensing with those encountered a few seconds before. If conditions are better now than they were a moment ago, the cell continues along the line it has been following. If conditions are getting worse, it randomly "tumbles." This system is much more sophisticated than the usual philosophical example of bacterial behavior, magnetotaxis, especially because it involves something beyond the simplest relations between input and output.[2] The present stimulus has a significance that depends on the preceding time step.

Bacteria are prokaryotes, cells with no nucleus that also lack further internal structures that other single-celled organisms have. An important event before the evolution of animals was the evolution of *eukaryotic* cells, which are larger and more complicated and whose initial evolution features the engulfing of one prokaryote (a bacterium) by another (an archaean) something like 1.5 billion years ago. One feature of eukaryotic cells that is especially important to the evolution of behavior is the *cytoskeleton*. This is a skeleton-like internal collection of fibers whose movements can be chemically controlled. In particular, they can contract. This makes possible changes in the cells' overall shape and is the beginning of nontrivial manipulation of objects and new kinds of locomotion. Single-celled eukaryotes also evolved richer forms of sensing, such as detecting the direction of light.[3]

Transitions in Animal Life

From a world of the more complex single-celled organisms just described, the evolution of multicellularity occurred perhaps a dozen times, independently, with different results. One of these gave rise to animals.

---

[2] For the *Escherichia coli* system, see Baker et al. (2006). For magnetotaxis, see Dretske (1986) but also O'Malley (2014).

[3] See Spang et al. (2015) for an important bridge between prokaryotes and eukaryotes with respect to quasi-behavioral capacities in unicellular organisms. For the evolution of light sensing and vision in unicellular organisms, see Jékely (2009).

Multicellularity

Animals are a branch of multicellular organisms originating perhaps 800–900 million years ago. There is much uncertainty about the dates and the pattern of the first branchings in this part of evolutionary tree of life. I'll work provisionally in this chapter within a fairly traditional view of the history of animals. This view has been challenged, but debates about the first events do not have too many consequences for the principles central to this chapter.

Though sensing and the control of behavior were not animal inventions, multicellularity made possible great shifts in the evolution of these capacities because it enabled a specialization of sensing and acting parts within the larger unit. This division of labor requires interaction between parts – some sort of effect of one cell on another in real time. There are various ways to do this, and some of it can be achieved without a nervous system, but only a small range of extant animals do not have nervous systems – sponges, placozoa, and a few reduced oddities whose ancestors had and lost them.[4] So that is the next step to consider.

Nervous Systems

Nervous systems arose perhaps 700 million years ago. There is ongoing debate about whether they arose once or several times, but they certainly evolved early and are present in nearly all animals. For someone interested in the evolution of subjective experience, this might look like *the* transition, the landmark. And so it may be, but I said all that without addressing the question of what a nervous system *is*. This is not a question with a straightforward answer.

Nervous systems enable interactions between cells with respect to their electrical properties. Cells can "depolarize" – the usual charge difference across a cell membrane can be lost and quickly restored. Nervous systems induce patterns of these changes in collections of cells. Those features, however, are also seen outside animals – depolarization of cells and effects of one cell on another's electrical properties. Suppose that the category *neuron* were to be understood purely in terms of excitability and cell-cell interaction; perhaps any cell is

---

[4] See Jékely et al. (2015).

a neuron if it is electrically excitable and can influence another cell's electrical excitation by means of chemical intermediaries or more direct effect. If this is what a neuron is, then various organisms that are usually called "nonneural" do have neurons, including some plants.[5] This broad functional definition of a neuron is quite a reasonable one, though it is at odds with many habits of description in biology. What might a reasonable, narrower definition look like? In a paper about early nervous systems that I co-wrote with Gáspár Jékely and Fred Keijzer (2015), we opted for a definition that combines functional and morphologic elements: a neuron is an electrically excitable cell that influences another cell by means of electrical or secretory mechanisms and whose morphology includes specialized projections. Neurons in this sense are restricted to animals, as far as we know. A nervous *system* can be understood as a system made up, in part, of cells of this kind.

There might appear to be an element of arbitrariness in adding this morphologic criterion about projections to narrow down what counts as a neuron. As I said, I think there's nothing wrong with the broader definition. But the combination of excitability, chemical signaling, and a morphology with projections is an important one; it enables nervous systems to achieve specific patterns of cell-cell interaction, especially interactions that are tightly targeted, even over long distances. An important and almost exceptionless generalization can also be stated: all and only organisms with neurons (in the narrower sense just defined) also have muscle cells. Muscle and neurons seem to have coevolved.[6]

Did neurons evolve once or several times? This has been debated at length over recent years due to changes in our understanding of the early evolutionary branchings in animal history.[7] A traditional view has it that sponges are the "sister group" to all other animals: there is an evolutionary branching deep in the past that goes on one side to sponges and on the other to all other animals. Another possibility, based on genetic evidence, is that comb jellies, or *ctenophores*, are the sister group to all other animals, including sponges. In other words, there is an early evolutionary branching that goes on one side to ctenophores and on the other to other animals including sponges. Ctenophores used to be grouped with

[5] For a review, see Volkov and Markin (2014).
[6] For the exceptions, see Jékely et al. (2015).
[7] For this debate, see Moroz (2015) and Jékely et al. (2015).

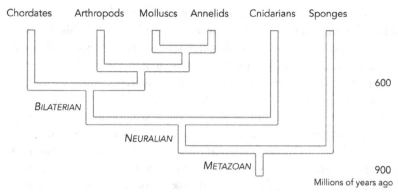

Figure 1 A representation of some early branchings in the animal part of the tree of life with dates (in millions of years, not to scale) tentatively associated with some events. Names along the top designate some of the main groups within animals. The italicized and capitalized labels show the initial appearance of broad kinds of organisms important to this chapter. First is the evolution of *metazoa*, or animals; then *neuralia*, animals with nervous systems (or perhaps a subset of these animals [see the main text]); then *bilaterians*, animals with bilaterally symmetrical bodies, including ourselves.

jellyfish – and some researchers still think this is correct – but a number of recent papers have made the case that ctenophores are more distant from us than any other living animal. The debate is important because sponges do not have nervous systems, while ctenophores do. If ctenophores are the sister group to all other animals, then either nervous systems evolved at least twice (once for ctenophores and once for everyone else) or the ancestors of sponges had nervous systems and lost them.

Figure 1 represents some animal groups and their evolutionary relationships. To keep things simple, in this figure I leave out ctenophores altogether. The term "neuralian" was introduced by Claus Nielsen (2008) for all animals with nervous systems. When he introduced the term, Nielsen assumed that the neuralia are a single branch of the tree – a "clade." (Any portion of an evolutionary tree such as that in Figure 1 is a clade if it can be generated by picking a point on the tree and including all organisms downstream of it.) Nielsen assumed that ctenophores were located somewhere internal to the neuralian clade. If ctenophores belong outside sponges in Figure 1, then animals with nervous systems do not form a clade. It would be a bit unusual to keep the term "neuralian" if

that is how things turn out, but I am going to set this issue aside and
allow that there might also be neuralia outside Figure 1.[8] This does not
affect the points made later in this chapter because my focus will be on
evolution in the organisms that are represented in the figure.

Suppose for a moment that nervous systems were a one-time animal
invention. Again, this might look like *the* landmark for the early
evolution of the mind. But what were the first nervous systems doing
for their owners? One natural assumption is that these early nervous
systems played a simpler version of the same sort of role – coordinating
perception with action – that is seen now in us. In bacteria, in early
animals, and in ourselves, a crucial task is coordinating what is perceived with what is done, and nervous systems evolved in animals to
enable this in an especially complex way.

Perhaps this is right, but we should not simply assume it. First, many
things present-day nervous systems do aren't a matter of controlling
behavior, and these may have been important in the early stages;
nervous systems often control aspects of development and
physiology.[9] And even within behavior, there is a possible discontinuity
between then and now. The ideas I'll sketch next draw on a paper by
Fred Keijzer, Marc van Duijn, and Pam Lyon (2013), building on
earlier work by Carl Pantin (1956).

When people imagine the role of early nervous systems, they often
picture a flowchart starting with perception and terminating with
behavior. Behavior itself is taken for granted: something is done. But
how is it done? In the case of a multicellular animal, it is a substantial
task to perform a coherent behavior at all, coordinating the microacts
of cells into a useful macroact by the whole organism. There's an
important *internal coordination* role that nervous systems play,
which is distinct from their role in coordinating perception with action.

I noted earlier the coevolution of nervous systems and muscle.
Without muscle, an animal can't do much. Motion then must be
achieved with cilia (little hairs), whose powers are limited. Keijzer
and his co-authors argue that it was the demands of coordinating
muscle action into useful behavior that first gave rise to the patterns

---

[8] As well as sponges, *Placozoa* are animals without a nervous system. I've not
marked them on the tree. They are thought to have branched off later than
sponges but earlier than cnidarians.

[9] These roles are discussed in more detail in Jékely, Keijzer, and Godfrey-Smith
(2015).

of interaction between cells associated with nervous systems. Guidance from the senses in simple animals can be done nonneurally, at least in large part. They suspect that the first nervous system evolved as a way to control a complex new effector system – muscle – in something that might have looked like an early cnidarian (present-day cnidarians include jellyfish, corals, and anemones). In some of the preceding passages I assumed what Keijzer and his co-authors call an "input-output" role for nervous systems – the emphasis was on a division of labor between some cells specializing in sensing, others in acting. Keijzer and colleagues want to challenge this assumption. Early nervous systems might have had a lot to do with just *pulling the animal together*.

So far these are points of principle, regarding possibilities for early nervous systems and their function. Is there any way to make claims about how things actually went? I will raise some possibilities (which depart now from the views of Keijzer, van Duijn, and Lyon).

Before 600 million years ago or so, we have no idea what the lives of animals were like. The only evidence that animals existed at all, and had nervous systems, is genetic evidence. The branching point that connects humans and cnidarians, for example, was probably earlier than 650 million years ago.[10] Then we reach a period now called the "Ediacaran" (635–540 million years ago), from which some soft-bodied animals are preserved as fossils. Once we find animals whose lives we can say something about, we see something of philosophical interest. Many Ediacaran animals seem to have lived on the sea floor, grazing on microbes or filter feeding. Some appear to have been mobile, and some probably had nervous systems. What were they doing with them? We can make some defeasible inferences from their bodies. Ediacaran animals have no legs, no antennae, no sign of complicated eyes, no shells, no spines, and no claws. They had none of the bodily tools of complex interaction between animals and none of the obvious tools of complex real-time behavior at all. There appears to have been little or no predation – there are no fossils of half-eaten individuals.[11] In Mark McMenamin's apt term (1998), it seems to have been "The Garden of Ediacara."

---

[10] Here I draw on Petersen et al. (2008).
[11] There is just one possible known exception, some *Cloudina* fossils from the late Ediacaran.

If we employ the distinction introduced earlier between *internal coordination* and *input-output* roles for early nervous systems, then this feature of the fossil record suggests that in the Ediacaran, a lot of what nervous systems did was internal coordination. What was going on that needed *reacting* to? Not much; lives appear to have been quite self-contained. Nervous systems in the Ediacaran may have functioned mostly in "pulling the animal" together, as I put it earlier, enabling simple locomotion and feeding and controlling physiology and development without complex real-time sensorimotor arcs being present at all.

Another piece of evidence may push in a different direction, however. This comes from the evolution of associative learning. Standard frameworks in learning theory distinguish between "classical" and "instrumental" conditioning. Classical conditioning, exemplified by Pavlov's dogs, is a means by which correlations between stimuli can be tracked – a behavior apt as a response to A comes to be produced in response also to B when B is a predictor of A. Instrumental conditioning is learning to produce (or avoid) behaviors that have been previously followed by good (or bad) consequences (or such consequences in specific situations). The origins of associative learning are unclear, but classical conditioning is very widespread across bilaterian animals – animals with left-right symmetry, such as ourselves.[12] Within this group, classical conditioning is seen in animals as simple as nematodes, which have only 302 neurons. Instrumental conditioning, in contrast, has been seen (so far) only in invertebrates with larger nervous systems, such as crabs, various insects, and some mollusks.[13] Classical conditioning still may have evolved independently in bilaterians a number of times, but suppose, for the sake of argument, that it evolved once and was passed down many lines from there. If so, it probably evolved something like 600 million years ago, either in the Ediacaran or before it. The function of classical conditioning seems tied very much to an *input-output* role for nervous systems; it is a tool for dealing with external patterns, not a tool for internal coordination. (Instrumental conditioning, in contrast, has both roles.) If classical

---

[12] See Perry et al. (2013) for a review. They also accept a single finding of classical conditioning in an anemone, an animal outside the bilaterians. This would push a single origin for classical conditioning further back in time or else be an independent origination of the ability.

[13] See Perry et al. (2013) again.

conditioning evolved in or around the Ediacaran, it makes it less likely that nervous systems at this time were concerned mostly with internal coordination.

Neither of these historical arguments is strong, though both point toward directions from which further evidence might come.

Sensorimotor Complexity and CABs

At the end of the Ediacaran, we reach the Cambrian "explosion," when many new kinds of animal appear in the fossil record. From these bodies we can again make inferences – stronger ones this time – about lifestyles. From the early Cambrian, we *do* see legs, antennae, complicated eyes, shells, spines, and claws. There is much controversy and rampant speculation about the Cambrian, but a family of main-stream views has particular importance here.[14] These views hold that at least one important thing that happened in the Cambrian was a process of "feedback" that linked the evolution of behavior and bodies in many groups. This shift may have first taken place in arthropods (which now include insects and back then also included trilobites). Whether arthro-pods were first or not, the evolution of more complex behavior in some animals seems to have made life more complicated for others. In the early Cambrian, predation arose – seen clearly in the fossils – and with predation a series of "arms races" appear to have followed, improving the senses and the means for bodily action. The evolution of rapid and fine-grained behavior in one animal makes the choices of others more acute. Parker (2003) has argued that a crucial event in this process was the evolution of image-forming eyes. Another possibility is that eyes were part of a suite of important features that evolved together. Either way, the *details* of what was going on around an animal came to matter to its life and prospects. This rather obvious feature of present-day animal life may not have been in place at all before the Cambrian.

The coevolutionary processes of this time linked senses, behaviors, and bodies. Michael Trestman, in a useful categorization (2013, p. 81), marks out what he calls *complex active bodies*:

This is a cluster of related properties including: (1) articulated and differentiated appendages; (2) many degrees of freedom of controlled motion; (3) distal senses (e.g. "true" eyes); (4) anatomical capability for

---

[14] See Marshall (2006) and Budd and Jensen (2015).

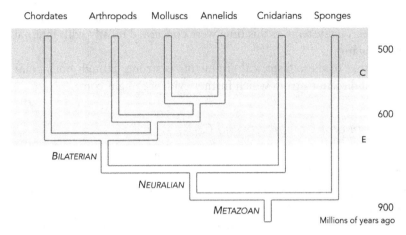

Figure 2 Further events in animal evolution. Many groups are not included. The lower shaded band marks the Ediacaran (E), and the upper band marks the Cambrian (C). Genetic evidence suggests that various familiar animal groups branched off from each other in the Ediacaran, though we have little fossil record of them there.

active, distal-sense-guided mobility (fins, legs, jet propulsion, etc.); and (5) anatomical capability for active object manipulation (e.g. chelipeds, hands, tentacles, mouth-parts with fine-motor control).

These are bodies that can manipulate objects, sense things at a distance, and react to them. CABs originated in the Cambrian, and as Trestman has it, only three groups of animals have given rise to bodies of this kind: vertebrates, arthropods, and a small group of mollusks, the cephalopods. With these bodies, the role for nervous systems that we are familiar with – the fine-grained linking of perception and action – becomes prominent. There's an opening up of senses to the world and, through new capacities for behavior, tighter *loops* between perception and action. Not only does what you do come to depend in a finer-grained way on what you see, but what you *do now* affects what you *see next*.

Figure 2 summarizes the steps described in the last few pages. As in Figure 1, we start from the evolution of animals (metazoa) and nervous systems (neuralia), at least on the nonctenophore line. A shaded band marks the Ediacaran. Around this time we see the evolution of bilaterian animals. As the figure shows, genetic evidence suggests that many of the major animal groups had already diverged at this early stage, without much morphologic fanfare. Then we reach the Cambrian, a band shaded differently, a time of rapid evolution of bodies and behavior.

If this is right in broad outline, we reach the following picture: the first nervous systems may have done rather little of what we now see nervous systems as enabling – behavior in real time, the fine-grained processing of what the senses tell us. Eventually, these did become central to animal life, in a process that began perhaps in the Cambrian. From that point on, the mind evolved in response to *other* minds – in response to demands that the speeding-up of behavior, more complex senses, and an ecology of individual-on-individual interaction placed on each organism. Further, new *bodies* evolved in response to other minds. Bodies that would not have been advantageous before these new behavioral regimes now became essential. The ecology in which new bodies evolved was an ecology of behavior.

The stage we've reached is well before we get to any of the animals that people usually think of as having subjective experience. We are in a world in which the behaviorally significant animals are arthropods, simple fish, and (more so a little after the Cambrian) some mollusks. With respect to the senses, behavior, and the nervous connections between them, though, some plausible basics are now in place. From this point, some animals evolve *more* neurons, more complex modes of interaction between them, and consequently more complex patterns of behavior. Other animals remain, or become, simpler.

Latecomer and Transformation Theories

I'll now start to bring this historical material into closer contact with the philosophy of mind. I'll organize this discussion with another diagram, Figure 3.

Figure 3 has more than one interpretation. First, it is a picture of part of the animal tree in which the shading within branches represents the sensorimotor and cognitive complexity seen in some species within particular groups. It is also an attempt to represent complexity that is indicative of subjective experience, but I will reach that second interpretation in a moment. For now, set subjective experience aside and think of sensorimotor and cognitive complexity only. The figure doesn't represent overall values for each group but *high* values within each group. The figure mixes taxonomic levels and leaves a great deal

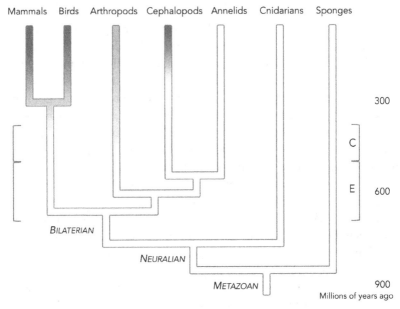

Figure 3 Part of the animal branch of the tree of life, with the shading within branches on the left showing the location of high levels of sensorimotor and cognitive complexity within some groups in those lineages. (Other groups are omitted to simplify the figure.) The brackets mark the Ediacaran (E) and the Cambrian (C).

out; I'm very aware of the limitations of this kind of representation, but at a coarse grain, I think it's informative. The drawing shows a partially *parallel* development in these features, beginning especially in the Cambrian. I've replaced "mollusks" in Figure 2 with cephalopods, a small group within the mollusks. Within arthropods, in contrast, complex behavior is seen across several different groups – bees, spiders, hermit crabs, mantis shrimp, and so on. Cephalopods of the relevant kind appear later than the other animal groups with complex behavior, but once they arise, they evolve very large nervous systems, especially octopuses.[15]

---

[15] See Darmaillacq et al. (2014). Some groups are omitted to keep the figure simple, including fish and nonavian reptiles. As discussed later, I regard fish as likely to undergo some forms of subjective experience.

Stepping back for a moment, the overall picture looks like this: animals are a branch of the tree of life – one way of being a multicellular organism – unified even across nonneural groups by shared patterns of cell-cell signaling. Nervous systems arose quite early in this branch, though, making use of preexisting electrical capacities of cells, and created a new kind of control system. Bilaterian bodies, like ours, also seem to have evolved well before any regime of complex behavior. Such behavior evolved in the Cambrian, and it evolved in parallel in several groups, not as a radiation from a single source. This process was genealogically parallel but also coevolutionary, with one animal's changes making life more complicated for others. The graded nature of the shading within some branches is supposed to suggest that a lot of what happened along these lines was quantitative change, in a rough sense of that term. The behavioral advance is not one that can be measured on a single scale of complexity but shows many forms – different kinds of sensory sophistication, different means for locomotion, and different ways of acting on the world.

Now let's look at another interpretation of the figure. It can be seen as a hypothesized map of the evolution of subjective experience. To offer any such interpretation, we need to make further assumptions. I'll set up these next steps with a distinction between two general views. The first asserts a kind of *proportionality* – though that is not quite the right word – between the cognitive side of the mind and subjective experience. This can be seen as an application of a simple kind of functionalism. What we call the "qualitative" side of the mind is just the first-person point of view, the *subject's* point of view, on cognitive processes at work in the system. The qualitative is in no sense an *extra* feature of the mind, something that might in principle be absent; it's an insider's point of view on the cognitive.

This view gives us a way of thinking about the origins of subjective experience, a way that emphasizes gradients and differences of degree. It might be hard for us to imagine simple and minimal kinds of subjective experience because we can't give ourselves the point of view of an animal very different from ourselves. But, on the cognitive side, we can probably understand the gray areas quite well and get a reasonable grip on the differences between a minimal scrap of mind and none at all. The qualitative will then exhibit a gradient of the same general shape. A failure of imagination is encountered on the qualitative side, but that's just a limitation in us. There can still *be* a gradient on the qualitative side that maps to a gradient on the cognitive side.

This view does lead to surprises. On the cognitive side, as I've emphasized, there is a gradient in complexity that stretches well past animals, all the way to unicellular life. In Figure 3, I started my shading in the Cambrian, but why not much before, in earlier animals or single-celled life, very faintly? Isn't that the message of the analysis? A view like that initially looks absurdly generous, but it need not be; after all, gradients reach very low values.

Clearly, there is much uncertainty about how the details of this first view would go, but the overall picture is one that makes sense. The second view I'll discuss has developed in more recent work – the literature has seen a shift, I think, from the first to the second view. This view rejects any sort of "proportionality" assumption about the richness of the cognitive and qualitative. *Divergences* between these are now emphasized, and a large body of work charts the apparently quirky manner in which some of the cognitive activity going on inside humans has a subjective feel, along with much that does not. This I take to be the theme of much recent neurobiology (Dehaene 2014). There's a distinctive kind of cognitive processing that brings with it subjective experience, embedded in much that does not. Work of this kind can motivate a view in which subjective experience is an evolutionary *latecomer*. The small fraction of what's going on in humans that has a subjective feel seems to be indicative of a particular way of organizing perception and cognition, a late-evolving way that features the achievement of forms of cognitive unification that many nonhuman animals probably do not have.

I'll take a closer look at these arguments. First, it is uncontroversial that there is a lot of sophisticated processing going on in our brains that we do not subjectively experience. The initial stages of visual processing and the processing of the syntax of sentences we hear are standard examples. These, however, might have limited force as arguments against the first view. Perhaps such kinds of processing (the second, in particular) are just like doing sums or accounts in the background, very different from the sensorimotor capacities we might associate with simple forms of subjectivity in animals. Other work, though, shows that there really is a problem here. An example is the work of Dale Milner and Melvyn Goodale (2005) on vision. They argue that there are two "streams" of visual processing in our brains. Only one, the "ventral stream," leads to visual *experience*. This stream is concerned with tasks such as the categorization of objects. The "dorsal" stream, however, handles tasks related to basic navigation, and dorsal stream vision feels like nothing – or perhaps like *something*, but very different from vision. Here, allegedly, we have a sensorimotor arc guiding biologically important behavior in a way that does not give rise to sensory experience. Or if some faint subjective experience is present in this sort of perception, there is still a surprising divide between the cognitive and qualitative sides. Dehaene (2014) surveys a wide range of work that shows further divergences between complex cognitive processing and subjective experience; we do a huge amount of sensing and thinking in a way that feels like nothing at all.

It is sometimes unclear how work of this kind relates to the ideas developed earlier in this chapter. Much of the neuroscientific work is presented as an investigation of "consciousness." Some of the scientific writers may be using, tacitly, a framework similar to mine, in which a theory of consciousness is *not* a theory of subjective experience in the broadest sense. But other parts of this work do seem committed to the idea that recently evolved sophistications are necessary for an animal to have any subjective experience at all.

What marks the difference between processes we experience and those we don't? A range of views is being defended. According to one family of theories, what we are conscious of is information made available in a "global workspace" that integrates information from various sources (Baars 1988; Dehaene 2014). This machinery of integration is something that many animals probably do not have because it is machinery linked in specific ways to memory, attention, and executive control. Views of consciousness that give a special role to "working memory," such as Jesse Prinz's AIR theory (2000) and Peter Carruthers's view (2015), have a similar character. All this work shares the following picture: a lot of cognitive activity goes on in us that has no felt side, and we need to work out which are the special pieces that do have this feature. Once we find those special cognitive activities, and – better still – their neural correlates, we know what other animals need to have. It remains imaginable that animals without brains of the right kind do have subjective experience; those inner structures might only be necessary in our case, not in everything. But why should we believe this? To entertain this possibility threatens to remove the study of consciousness from science once again. It is better to conclude that when these features first evolved, so did subjective experience, and not before; vague talk of "gradients" in this area does not take seriously what we have been learning.

If so, subjective experience is not something that arises in all animals with complex sensorimotor capacities, but only in those with a particular kind of organization. There is a difficult question of what an animal has to be like to have machinery that is *close enough* to what enables subjective experience in us. Prinz (2000) thinks that this question is probably unanswerable. But according to this second family of views, there is no reason to regard Figure 3 as telling us much about subjective experience. Instead, we should probably shade a narrow band for subjective experience at the top of the mammal branch (or perhaps the mammal and bird branches) and leave the rest blank. The evolution of a significant amount of sensorimotor complexity in other parts of the tree is beside the point because an animal can have a lot of sensorimotor complexity and no subjective experience associated with it – we know this from our own case.

I'll now offer a reply to these ideas. I agree that some earlier work assumed too simple a mapping between cognitive and qualitative. A latecomer view is not the only response to what recent research has taught us, though. Another possibility is what I will call a "transformation" view. According to this view, the forms of processing studied in recent work on consciousness may have substantially affected subjective experience but did not bring it into being. Instead, they made it richer, perhaps, and brought it into different kinds of contact with memory and verbal report. Basic forms of subjective experience were present earlier and require less, and in us, these have been transformed.

What argument can be given for this view? Is it a vague plea for retention of a more generous attitude and no more? The best argument I can offer at the moment is based on the role of what seem like old forms of subjective experience that seem to appear alongside, and often intrude into, more unified kinds of processing. Examples include pain and what Derek Denton calls the "primordial emotions" – bodily feelings that register important metabolic states and deficiencies, such as thirst and the feeling of not having enough air. As Denton says, these bodily feelings have an "imperious" role when they are present: they press themselves into experience and can't be easily ignored (Denton et al. 2009). Do you think that those things (pain, shortness of breath, etc.) *only feel like something* because of sophisticated cognitive processing in mammals that has arisen late in evolution? I doubt it.

I will focus on the case of pain and evidence for pain in animals that are unlikely to pass the tests for consciousness that people like Dehaene and Prinz would impose. Making this argument is not straightforward. One might initially say that it's obvious that even simple animals respond to pain in a way that indicates that they feel it. But many responses to bodily damage that might initially appear to involve pain and distress probably do not. For example, rats with a severed spinal cord, and hence no channel from body damage to the brain, can exhibit some of what looks like "pain behavior" and can also respond in quite sophisticated ways to the damage.[16] Given this, it is significant that other experimental work has shown that more complex pain-related behaviors are present in animals far from us on the evolutionary tree, including some invertebrates. What I see as important in this work is its indication that these animals respond to damage with more than reflexes, with modifications to their behavior that are flexible, sensitive to novelty, and balanced by other cost-benefit considerations.

The clearest results are in vertebrates, though some invertebrates also have shown this pattern. In one study, prior testing was used to work out which of two environments (empty or enriched) was preferred by some zebra fish. After injection with a chemical believed to cause pain, the fish then preferred the normally less-favored environment when it had painkiller dissolved in it and not otherwise: "the fish were willing to pay the cost of being in an unpreferred environment to obtain analgesia, and thus it can be inferred that these fish must have obtained some reward possibly in terms of pain relief such that the pain was reduced" (Sneddon 2011). Similarly, in a study in chickens, birds with damaged bodies chose a food that would usually be less preferred, provided that it contained analgesic: "lame birds selected significantly more drugged feed than sound birds, and ... as the severity of the lameness increased, lame birds consumed a significantly higher proportion of the drugged feed" (Danbury et al. 2000). Finally, Robert Elwood reports that hermit crabs could be induced to leave their shell

---

[16] "The spinal cord distinguishes noxious stimuli from other stimuli, and adaptive changes in behavior result ... [L]earning about noxious stimuli can occur in the absence of conscious awareness of pain" (Allen 2004). For recent work on animal pain, see also Key (2015), who defends a latecomer view, and Jones (2013).

by a shock, but they were more reluctant to leave a higher-quality shell or to leave when the odor of a predator was around: "hermit crabs trade off competing demands in their responses to electric shock in a way that cannot be explained by a nociceptive reflex response" (Elwood 2012, p. 26).

It is important, also, that other animals appear to fail these tests. Crabs may be very different from their fellow arthropods, the insects. An older review, but not one that has been superseded as far as I know, says: "No example is known to us of an insect showing protective behavior towards injured body parts, such as by limping after leg injury or declining to feed or mate because of general abdominal injuries. On the contrary, our experience has been that insects will continue with normal activities even after severe injury or removal of body parts" (Eisemann et al. 1984).

These results do provide some support for a view of pain as a basic and fairly widespread form of subjective experience, one unlikely to be dependent on late-arriving mechanisms of working memory, integration of information, and so on.

One response to this argument is to say that it suggests that many more animals than we realized *have* the complex features that enable subjective experience in us, including fish and hermit crabs. It would require further empirical work to assess this view. Another possibility, one that surely becomes vivid once these results are on the table, is that there are forms of subjective experience that are simpler and older than the form of consciousness that recent neurobiological work on humans has been investigating. If so, there is something it's like to be a fish or hermit crab, even if (as I would put it) they are not conscious.

If the arguments offered over the last few pages are right, the transformation view may well be correct, and the latecomer view is not as well supported as it might have appeared. A case then can be made for some sort of the separation of categories that have recently been conflated. There's the evolutionary origin of a subjective *feel* to life, in a very broad sense, and this was later shaped eventually into something with the familiar features of *consciousness*. In these pages I've not attempted to say much about the relation between these two things. My aim has been to say something about the evolution of subjective experience in a broad sense. The uncertainties in this area are enormous. The shape of the tree of life around the time that nervous systems first evolved is not yet clear, and there are puzzles about the relation between genetic and fossil evidence. But the idea of parallel evolution of sensorimotor and cognitive complexity from the Cambrian onward is better supported. As I emphasized, this process was genealogically parallel but also coevolutionary, with one animal responding to behavioral evolution in another. Any mapping between behavioral complexity and subjective experience will also be controversial, at least for now, and many assessments of particular cases may change. The point of Figure 3 lies not in its details but in contrasts with very different charts that might be drawn – charts that present subjective experience as a latecomer, for example, and charts marking a single origin of experience in one lineage, with radiation from there. With all this in mind, I think that Figure 3 might be a reasonable rough map of the history of subjective experience.

4 | *Neurophilosophy*

PATRICIA CHURCHLAND

Introduction: What Is Neurophilosophy?

"Neurophilosophy" explores the impact of discoveries in neuroscience on a range of traditional philosophical questions about the nature of the mind. This subfield aims to move forward on questions such as the nature of knowledge and learning, decision making and choice, and self-control and habits by drawing on data from the relevant sciences – not only neuroscience and clinical neurology but also evolutionary biology, experimental psychology, behavioral economics, anthropology, and genetics. It draws also on lessons from the history of philosophy and the history of science, which saw mysteries about the nature of the blood or fire or infectious disease become less mysterious as experimental science began to provide new observations and tested explanations (Thagard 2014).

The massive accumulation of neurobiological data from many levels of brain organization and many species of nervous systems is a recent development because neuroscience did not really reach full steam until about the 1970s. Why was the development of neuroscience delayed until recently?

Although clinical observations had long implicated the brain in mental functions, understanding exactly *why* lesions affected mental functions remained out of reach. This was so because essentially until very recently nothing was known about the microstructure of brains – about neurons and how neurons worked, about how the brain was organized into networks and systems, and about how neurochemicals mediated interactions between neurons. Notice that detailed drawings of nerve cells were produced by Camillo Golgi and Ramón y Cajal only in the latter part of the nineteenth century. How neurons *interacted*

Particular thanks to Paul Churchland and Joshua Brown for clear-headed discussion and to David Livingstone Smith for wise advice.

with each other to yield effects such as a behavior was still a profound mystery.

Chemistry, by contrast, was a vastly more mature science in the early nineteenth century, strengthened by basic organizing principles of atomic theory, as outlined by Dalton in 1805 and a clear appreciation of the fundamental elements – no longer deemed to be earth, air, fire, and water. Instead, the elements were characterized by Mendeleyev in the 1880s in the periodic table – things such as oxygen, hydrogen, tin, and gold. As for neuroscience, it is perhaps surprising to realize that the existence of *inhibitory* connections between nerve cells was demonstrated by John Eccles and colleagues only in the 1950s. Physics, far more mature in terms of theory and explanation by that time, had begun to investigate the inner structure of the atom.

To get a perspective here, note that effective brain imaging techniques came into their own only in the last two decades of the twentieth century. At the micro level, many details regarding the synapse and how neurons communicate are not completely understood even now, nor are the functions and dynamics of neural *networks*. Neuroscience is a young science.

Because the brain's basic units work by changes in voltage across the cell membrane and by chemicals that regulate such changes, and because the units are not visible to the naked eye, development has depended on a theoretically and experimentally rich physics and chemistry. Specifically, neuroscience depends on tools and devices that exploit the knowledge of physics and chemistry, for example, the electron microscope, microelectrodes, nuclear magnetic resonance, monoclonal antibodies, and most recently, optogenetics. It is noteworthy that understanding how neurons work required knowledge of electricity, and that knowledge was not in hand until Michael Faraday's discoveries in the first half of the nineteenth century.

Some philosophers take it as dead obvious that the enduring existence of many puzzles in neuroscience entails that neuroscience can never, *ever* discover much in the way of mechanisms of cognitive function. One major reason for this conclusion is that they have generally failed to appreciate the clear historical point that the sciences of the nervous system are very young indeed.

The Relation Between Mind and Brain

The words "mind" and "brain" are distinct. Even so, that linguistic fact leaves it open whether mental processes are in fact processes of the physical brain. (Remember: water and H_2O are different words, but they do name the very same stuff.) A favored theory in philosophical thought, championed by Plato, developed by Descartes, and even now defended by Thomas Nagel (2012), holds that just as the words are distinct, so too are the processes. This approach is known as "dualism" – a "two stuffs" theory embracing physical stuff and the utterly different soul stuff. Thinking, seeing, and choosing, according to dualism, are processes of the nonphysical mind or soul. For dualists, the mind/body problem is the problem of how a physical state of the brain can interact with a totally nonphysical state of the soul. By contrast, according to an equally venerable if less popular tradition, there is only the brain; mental processes are processes of the physical brain whose exact nature remains to be discovered. This is known as "physicalism" and found adherents in Hippocrates, Hobbes, Hume, and Helmholtz. Physicalists realize that there is no problem about how the mind and body *interact* inasmuch as there are not two things, but only one thing: the brain. The mind is what the brain does. For them, the important problem concerns how the brain learns and remembers, how the brain enables us to see and hear and think, and how it enables us to move our eyes, legs, and whole body. Their problem concerns the nature of the brain mechanisms that support mental phenomena. Interestingly, dualists also have a closely related set of problems: how does soul stuff work such that we learn and remember, see and hear and think, and so forth. Whereas in neuroscience physicalists have a vibrant research program to address their questions, dualists have no comparable program. No one has the slightest idea how soul stuff does *anything*.

Neurophilosophy as a research program has poor prospects unless mental processes such as remembering and attending are processes of the brain. Otherwise, we should just study the stuff that *does* perform attending and remembering and find out how *that* works, stuff such as the "soul stuff" postulated by Descartes. At this stage in the sciences, the evidence overwhelmingly indicates that all mental events and processes, including visual or auditory perception, learning, memory, language use, and decision making, are in fact events and processes of the physical brain. It is not that there is one single experiment that decisively shows this. Rather, the evidence has steadily accumulated over countless observations and experiments, and no counterevidence raises doubts.[1] Even though we may not understand in detail the mechanisms whereby we recall an event that occurred in childhood, we are reasonably sure that such a recollection is a brain process. This is not unlike Michael Faraday's realization that electricity was not an occult phenomenon but rather a natural physical phenomenon, even though he did not understand in precise detail the nature of electromagnetism.

One of the most dramatic observations of mind/brain dependency came from the split-brain studies published in the late 1960s. These studies involved patients whose cerebral hemispheres were surgically separated in order to treat drug-resistant epilepsy. The nerve sheet connecting the two hemispheres – the corpus callosum – was the structure that was cut, thereby disconnecting the cortex of the right and left hemispheres. The aim was to aid the patient by preventing a seizure from traveling from its origin in one hemisphere to the other hemisphere. Astonishingly, tests of "split brain" subjects showed that the mental life of the two hemispheres was also disconnected: the right hemisphere might have knowledge the left did not or see something or decide something that the left did not, for example (Gazzaniga and LeDoux 1978). The implications for the mind/body problem were obvious: if mental states were *not* brain states, why would cutting the corpus callosum allow knowledge and experience to be confined to activity in *one* hemisphere? Although a defiant dualist might invent some story to accommodate the facts (and a diehard few did this), the best and most reasonable explanation for the disconnection effects was simply that a *physical* pathway was interrupted, a pathway essential for mental unity, and that soul stuff was just not in the game. As Michael Gazzaniga (2015), one of the leading split-brain researchers puts it, consciousness can be split.

The many observations made by clinical neurologists of patients who suffered focal brain damage also weighed in. Focal brain damage could result in highly specific losses of cognitive function, such as the loss of

---

[1] See P. M. Churchland (1996a), Frith (2007), and P. S. Churchland (2002) and excellent textbooks such as Baars and Gage (2007).

the capacity to recognize familiar faces, loss of recognition of a limb as one's own, and loss of the capacity to perform an action on command, such as saluting or waving hello. The Damasios, Hanna and Antonio, launched a huge project at the University of Iowa Medical College to systematically document as many cases as possible involving similarly located lesions to test whether there were similar functional effects. This important project elevated brain lesion studies beyond the single case study to a more systematic understanding of the outcome of focal brain lesions and their effect on capacities.[2]

Studies of a few patients who had suffered bilateral damage to the hippocampus (a small curved structure beneath the cerebral cortex) showed them to be severely impaired in learning new things (anterograde amnesia). This finding initiated a massive research program to understand the relation between learning and memory and the hippocampal structures (Squire, Stark, and Clark 2004). Memory losses associated with dementing diseases also linked memory with neural loss and further suggested the tight link between the mental and the neural. Important also are studies of attention using brain imaging along with single neuron physiology. These varied studies suggest that at least three anatomic networks, connected but somewhat independent of the other, are involved in different aspects of attention: alerting, orienting, and executive control. Moreover, each of these functions has been the target of detailed further study, indicating, for example, that there are strong associations between these functions and awareness, especially between detection of a target (consequent on orienting) and awareness (Petersen and Posner 2012).

Developments in psychology, especially visual psychology, also implicated neural networks in mental functions, and this work tended to dovetail well with the neuroscientific findings on the visual system. Explanations of color vision, for example, depended on the retina's three cone types and on opponent processing by neurons in cortical areas. It was well appreciated that much in the world – such as ultraviolet and radio waves – could not be detected by our visual system because of its physical organization.[3] Perception of visual motion was linked to the behavior of single neurons in a visually sensitive area of

---

[2] For a simple account, see Grens (2014).

[3] See Solomon and Lennie (2007), pp. 276–86, and also chapters 9 and 10 in P. M. Churchland (2007).

cortex known as MT (middle temporal). Visual hallucinations were known to be caused by physical substances such as LSD or ketamine, and consciousness could be obliterated by drugs such as ether, as well as by other substances employed by anesthesiologists, such as propofol. No evidence linked these drugs to soul stuff. On the contrary, many anesthetics appear to work by altering the normal balance of excitation and inhibition of neurons in circuits.

Short-term memory can be transiently blocked by a blow to the head or by a drug such as scopolamine; emotions and moods can be affected by Prozac and by alcohol; decision making can be affected by hunger, fear, sleeplessness, and cocaine; elevated levels of cortisol cause anxiety. Very specific changes in whole-brain activity corresponding to periods of sleep versus dreaming versus being awake have been documented, and explanations for the neuronal signature typifying these three states have made considerable progress (Pace-Schott and Hobson 2002). In aggregate, these findings weighed in favor of the hypothesis that mental functions are a subset of functions of the physical brain, not of some spooky "soul stuff."

Evolutionary biology encouraged us to dwell on the fact that nervous systems are the product of evolution and that the human nervous system is no exception. Comparisons of anatomy, between human and nonhuman nervous systems, have revealed that the functional organization, at both macro and micro levels, has been highly conserved over hundreds of millions of years (Allman 1999). Although human brains are larger than the brains of other land mammals, we share all the same structures, pathways, innervation patterns, neuronal types, and neurochemicals. Neurons in a fruit fly work essentially the same way as neurons in the human brain. Molecular biology revealed that the genetic differences between humans and our nearest relatives, chimpanzees (*Pan troglodytes*) and bonobos (*Pan paniscus*), are very small (Striedter et al. 2014).

These evolutionary relationships imply that either no mammals have nonphysical souls or all do. Now questions flood in: if humans *alone* do have a soul, where do human souls come from, and why does the soul suddenly appear, some 4 million years after the *Homo* species branched off from our common ancestor with chimpanzees? Did extinct *Homo* species such as *Homo erectus* and *Homo neanderthalensis* have souls too? Based on cranial measurements, anthropologists believe that the brains of *Homo neanderthalensis* were typically larger than our brains.

Neanderthals probably had some form of acoustic communication even though they may not have been able to make all the vocalizations of which humans are capable (Lieberman 2013). Moreover, genetic data reveal that they did interbreed with *Homo sapiens* (Pääbo 2014). What about *their* souls? Still other questions challenge the idea that the human soul, not the human brain, is the repository of all that makes us clever. How can ravens and rats and monkeys solve complex problems – how can they sleep, dream, pay attention, and so forth – if a soul is needed for such functions?

By the 1980s, there was impressive, if cautious, agreement among scientists as well as philosophers that the existence of a nonphysical soul that feels, decides, sees, and reasons was improbable. Where disagreement flourished unabated, however, concerned whether neuroscience could *explain* those functions, physical though they may be. Neuroscientists tended to expect that with new techniques and more experiments, progress would continue to be made. How far we shall get, time and research effort will tell.

Some philosophers, by contrast, confidently predicted that neuroscience would never explain cognitive functions, a view particularly associated with Jerry Fodor (1975, 1980, 1998) and his colleagues but widely espoused within the subdiscipline of philosophy of mind. This view tended to be known as the "autonomy of psychology" – autonomous with respect to other sciences, especially neuroscience. It is important to understand that this claim about the limits of neuroscience was just a *prediction*, and it was supported by philosophical speculation, not scientific evidence. Although highly popular until about 1990, the idea has slowly and systematically been undercut by actual progress in the neurosciences, especially by increasingly suggestive links between data at the behavioral, whole-brain, and neural levels. Embarrassingly for the philosophical prediction, convergent studies on functions such as decision making (Glimcher and Fehr 2013), attention (Petersen and Posner, 2012), and spatial representation (Moser et al. 2014), for example, have revealed much more about mechanisms than some skeptical philosophers thought was remotely conceivable (Fodor 2000).

One further reason for ignoring much of neuroscience arose from a misguided analogy. The idea was that cognition is like running software on a computer, where the brain is analogous the computer hardware.[4] Just as you need not know anything about a computer's hardware to understand an application such as PowerPoint, so you need not understand anything about the brain to understand cognition, or so the argument went. To anyone who looks at all closely at the brain, the disanalogies between brains and conventional computers are so numerous and so profound that the brain/hardware analogy was not taken seriously in neuroscience or bioengineering. Not least among the differences are that brains are parallel not serial processors, that storage and processing in brains are not done by separate modules but by the same structures, and that brains change their structure as they develop from gestation to adulthood and at all stages as they learn (Churchland and Sejnowski 1992). The actual nature of the brain's anatomy and physiology became an inspiration for developing unconventional computers that are more brainlike (Hinton 2013; Yu et al. 2013).

The point where influential philosophers are still confident that the mysteries permanently have the upper hand concerns conscious experience. Typically, there are two distinct arguments to support this conviction. The first argument makes a straightforward prediction about where science will go in the future. It is based on current intuitions about the tractability of the problem of explaining consciousness in neurobiological terms. With great confidence it will be claimed that consciousness is so completely and utterly and thoroughly mysterious, it will *never* be explained at all, period (McGinn 2012, 2014). By way of illustration, it may be suggested that expecting any science to explain how conscious experience emerges from the activity of neurons is like expecting a rat to understand differential equations. Despite its chest-pounding confidence, this prediction should be taken with ample doses of caution because predicting where science will go and what will be discovered is really a rather risky business, to put it politely.

The second and more influential argument rests on the dualist's belief that although nonconscious events such as memory consolidation and preprocessing in vision are brain events, *conscious* events such as feeling nauseous are not brain events. Hence neuroscience cannot explain them. Thus, when I am aware of a pain in my tooth or a decision to kick off my shoes, some philosophers, such as

---

[4] Dennett (1987) was especially fond of this analogy and appears still reluctant to abandon it.

David Chalmers (1996) and Thomas Nagel (2012), consider those conscious events to be extraphysical, merely running parallel to the physical events.

A methodologic point may be pertinent in regard to the dualist's argument: however large and systematic the mass of empirical evidence supporting the empirical hypothesis that consciousness is a brain function, it is always a logically *consistent* option to be stubborn and to insist otherwise, as do Chalmers and Nagel. Here is the way to think of this: identities – such as that temperature really *is* mean molecular kinetic energy, for example – are not directly observable. They are underwritten by inferences that best account for the mass of data and the appreciation that no explanatory competitor is as successful. One could, if determined, dig in one's heels and say, "temperature is *not* mean molecular kinetic KE, but rather an occult phenomenon that merely runs parallel to KE" (Churchland 1996b). It is a logically consistent position, even if it is not a reasonable position.

In a similar vein, causality, as Scottish philosopher David Hume famously noticed, is not directly observable. It involves an inference to the best explanation available.[5] I cannot literally observe the causal relation between a mosquito on my arm and the itch that follows its departure. But my causal inference is based on strong background knowledge. For another example, despite the powerful evidence that human immunodeficiency virus (HIV) is the major cause of AIDS, some still insist, without contradiction, though perhaps with much mischief, that the cause of AIDS lies elsewhere, such as God's punishment for bad behavior.

To be sure, caution concerning accepted theory does sometimes facilitate the emergence of new causal hypotheses that surpass the prevailing theory in predictive and explanatory power. Scientists, if they are not foolish, then upgrade their causal explanations. For example, it was widely believed that anxiety and poor diet were the major causal factors behind gastritis (inflammation of the stomach lining) until Barry Marshall and Robin Warren in the 1980s challenged that hypothesis experimentally. They discovered the more fundamental cause – a bacterium known as *Helicobacter pylori*. They did not merely vaguely wave in the direction of a *conceivable* different causal claim,

---

[5] For a new and quite possibly correct account of how causality is represented in the brain, see Danks (2014).

however. They showed experimentally that they had *discovered* a more powerful causal explanation. In the case of conscious experience, although philosophers such as Chalmers and Nagel express their reservations about the brain, the only thing they really do have are reservations. Moreover, their reservations are based on intuitions about how different experience seems to be from states occurring in the physical brain. They have neither competing experiments nor a competing hypothesis with any power or detail; in particular, they have no hypothesis that surpasses let alone competes seriously with the neuroscientific hypothesis.[6] For example, there is nothing that even begins to approach the richness of the neuroscientific literature on attentional mechanisms, for example, that alerting is different from orienting, which, in turn, is different from detection and from executive control. Surprisingly perhaps, with the appropriate intervention, these functions are dissociable, and they are supported by different neural networks.[7]

How do the dualists address the dependencies – the causal dependencies that suggest identification – between consciousness and brain activities? A favored strategy is to propose that conscious states just run parallel to brain states. This proposal may be embellished, perhaps by the idea that conscious states neither cause nor are caused by brain states – the two streams are causally isolated. A variation of this opts instead for a one-way causal street – brain states cause conscious states, but conscious states do not cause brain states. Traditionally, the view that mental states do not cause brain states is called "epiphenomenalism." Actual evidence is lacking for both hypotheses – both are merely empty denials of the idea that consciousness is a biological phenomenon.

Historically, the most renowned defender of two-way causal isolation was Gottfried Leibniz. Leibniz held this view because he thought that it was inconceivable that completely different substances could interact causally. If they shared no properties – not even spatial properties – how could they affect each other? Moreover, with the benefit of contemporary physics, we can see that the causal interaction between *nonphysical* stuff such as a soul with *physical* stuff such as electrons

---

[6] For discussion of a brain-based hypothesis, see P. S. Churchland (2013a) and Graziano (2013).
[7] See again Petersen and Posner (2012).

would be an anomaly relative to the current and rather well-established laws of physics. More exactly, it would affect the law of conservation of energy. If brains can cause changes external to the physical domain, there should be an anomaly with respect to conservation of energy. No such anomaly has ever been seen or measured. The absence of anomalous data suggests either that the hypothesis of a nonphysical conscious stream of states lacks credibility or that the conscious stream of conscious states does not interact with brain states at all.

When the neuroscientist Josef Parvizi used a tiny electrical stimulus to activate a very specific part of the brain (the middle cingulate gyrus) as part of the preparation of his human patient for surgery, his patient described the emergence of a conscious state consisting of the determination to muster courage to deal with a problem. When the stimulus was off, the feeling vanished (Parvisi et al. 2013; P. S. Churchland 2013b).[8] This experiential event was repeatable in that patient. Moreover, a very similar state was also reproducible in yet another patient stimulated in the same region. The reasonable conclusion is that the stimulus caused the change in conscious state. Some naysayers may wish to take the option that the brain events and the experienced event happen synchronously without causation: the experience stream and the brain stream are separate.

What keeps the two streams synchronized? That is the stunning puzzle that emerges from the epiphenomenal hypothesis. Here is how Leibniz dealt with the puzzle: God sets up and maintains a "pre-established harmony" to keep mental and physical states properly aligned. Needless to say, Leibniz' solution is completely ad hoc, cobbled together to in order to fill an embarrassing silence. Chalmers' does not appeal to God, but he does advert to a future physics that allegedly will explain the alignment between noninteracting streams of mental and brain events. A revolutionary new physics, according to Chalmers' (1996) conjecture, ultimately will explain the nature of consciousness as a nonbrain phenomenon. I have been unable to escape the feeling that this is really the old Leibniz solution suited up in the duds of a future physics instead of theology.

Granting that there are uncertainties in physics, is there a rationale within physics for claiming that a revolution provoked by the mysteries

---

[8] For a review article on drug-resistant surgery for epilepsy, see Ryvlin, Cross, and Rheims (2014).

of consciousness is in the cards? According to Chalmers, there will be, because nothing less will explain consciousness. Consciousness is so extraordinarily mysterious that only a revolution in physics will account for it.

My small sampling of physicists indicates that they do not wish to rush into investing heavily in a new physics just to address *consciousness*, especially when neuroscience has not by any means been stopped dead in its tracks. And especially when neuroscience has not yielded anomalies that challenge *particle* physics, but only puzzles that might possibly challenge neuroscience. Physicists acknowledge puzzles concerning the possibility of a new theory at the subatomic level to link strong forces, weak forces, and gravity, but these are phenomena in the range of 10^{17}, not in the range of milliseconds and micrometers (10^{-3}), where neurons exist and function. As physicist Steven Weinberg said, the puzzles in physics that motivate a possible revision to the standard model are at the wrong spatial and temporal scale to offer even the barest hint of a solution to the matter of explaining consciousness.[9] Have the philosophers themselves proposed anything substantive by way of a new physics to replace existing physical theory? No. There is nothing substantive – nothing even weakly semisubstantive.

If you are a dualist, either you can pretend that the huge accumulation of dependency evidence in neuroscience is not really there (not a realistic option), or you can say something substantial to address them. Rationally, something must be done insofar as this accumulation appears strongly to favor the hypothesis that conscious states are brain states. A novel strategy, tendered by Chalmers, claims that neuroscientific data are actually *neutral*, as between his parallel-stream hypothesis and the hypothesis that mental states are states of the physical brain.[10]

To assess the figures of merit of this "neural data neutral" strategy, try it elsewhere in science and see what results. Consider the nature of light as understood within contemporary physics: light is electromagnetic radiation (EMR) – light visible by humans is just one part of

---

[9] This was Weinberg's answer to a question at Gustavus Adolphus College, October 8, 2014. See also Weinberg (2015).

[10] This is a view Chalmers has made explicit only in conversation, though he acknowledges that it is implicit in his earlier writing, even in *The Conscious Mind*.

a larger spectrum that includes x-rays, microwaves, and so forth. Here is what the "neutral strategy" could say about light: "actually, the physical evidence is neutral between the hypothesis that light *is* EMR and that light is not EMR but a spooky thing. That is, light and EMR run in parallel streams, whose synchrony will be explained by a revolution in physics."

Here is what the "neutral strategy" says about life: "all of cell biology is neutral between the hypothesis that life is an occult force (vitalism) and the hypothesis that life is the outcome of the biological structure and organization – cells, membranes, genes, ribosomes, mitochondria, and so forth."

Scientifically, these "data neutral" proposals look counterproductive and more elaborate that the facts require. Silly though they may be, they are not, however, internally incoherent hypotheses. One bizarre claim that oddly appeals to various philosophers of mind is that if the "parallel stream" hypotheses are not internally contradictory, they are as reasonable as established scientific theories. Notice that it is not internally contradictory to say that the Earth is only one hour old, but it would be strange to say that this is as reasonable as saying it is about 5 billion years old.

The twin predictions regarding mind and brain – that neuroscience will never account for conscious experience and that a revolution in physics will explain why – are generally motivated by emphasizing the difference between a neuron, on the one hand, and a feeling of tooth pain, on the other, for example. On reflection, it is argued, the differences appear to be so profound and so complete that surely, *surely* it is inconceivable that the pain in my tooth might really be the activity of neurons in the brain.

Striking though the touted differences are, it is sobering to recall that the history of science is full of discoveries in which seemingly very different phenomena turn out to be one and the same but were viewed from different perspectives (Thagard 2014; Churchland 1989). Breathlessly dramatizing the striking differences lacks the scientific heft to make the dual streams hypothesis compelling.

One problem with relying on what seems inconceivable is this: what is and is not conceivable is, after all, merely a psychological fact about us – about what we can and cannot imagine given our current beliefs and our capacity for imagination. It is not a metaphysical fact about the nature of the universe. In the opinion of some philosophers, however, trained philosophical intuition has special status and must be taken as revealing deep, "necessary" truths unavailable to untrained others – in particular, unavailable to those with only a scientifically educated intuition (McGinn 2014).[11]

An issue that spells trouble for a nonbrain theory of consciousness concerns the fact that the division between awareness and lack of awareness is typically blurry and often fluid. One place this really shows up is in the automatization of behavior as a skill is acquired, a commonplace phenomenon. As a child learns to read, she ceases to be aware of a word's individual letters; this is also demonstrated in the "word superiority" effect, whereby it is easier for an accomplished reader to read a *word* than to read individual *letters*, as measured by reaction time and errors. Another simple case: I can ride a bike without being aware of my feet working the pedals as I zoom along and think about my upcoming swim. Not so at the beginning of learning to ride a bike, where I had to pay attention to every aspect of riding. Here is the issue: are the many behavioral decisions of which I am unaware just mental brain events that blink out of the mental experience stream until an emergency arises and I must pay attention? Ditto for skating, driving a car, lots of speech and conversation, and, in my case, recently learning to be proficient at standing on my head. And here is related issue: are you aware of body position when you are concentrating on pitching a tent? Sort of and sort of not. Moreover, the neurobiological research on attention helps us to see why the answer is not simple. Apart from automatization of skills, what about shifts of attention, for example, where I cease to hear the speaker and reflect on what I will order for dinner? When I lose awareness of what the speaker is saying, does that just snap out of the consciousness stream and then snap back in? How does that work? What orchestrates and coordinates the snapping? And what *is* snapping?

This raises a second issue. Are our short-lived conscious experiences properties of a "substance"? Or are they just events, properties of "no thing" in the experience "stream"? What maintains the stream as *one* stream? Compared to the serious research in neuroscience on the mechanisms of sleep, attention, visual perception, coma, anesthesia,

---

[11] See my reply to McGinn (2014) in the *New York Review of Books*, June 19, 2014, p. 65.

and so forth, the naysayers seem to have a totally threadbare alternative, with very little in the way of a substantive explanatory framework.

Why do some philosophers of mind oppose so strenuously the two
hypotheses: (1) mental states are states of the brain and (2) probably
neuroscience can at least outline the mechanisms of cognitive functions? A range of reasons contributes, but as the frontiers of the
behavioral and brain sciences push ever forward into what seems like
a thicket of unapproachable mysteries, questions about turf and territory inevitably emerge. A strong assumption in the philosophy of
mind is that philosophers are uniquely equipped to set the boundaries
of what we can know and to outline the essential and enduring
features of concepts that scientists might apply. Philosophical intuition,
in this view, is a special trained capacity that can home in on those
necessary properties of a phenomenon that science must respect and
not challenge. In this way, philosophy sets the foundations for the
science. And if philosophers characterize necessary properties of the
mind that intuition and logic show cannot be explained by properties of
the brain, then that is the contribution of philosophy that science needs
to honor.

Thus some philosophers of mind believe that they own a problem
space that is concerned with conceptual necessities – necessary truths
about psychological states and processes, discovered by conceptual
analysis and so-called thought experiments.[12] A necessary truth cannot, according to this approach, be falsified by scientific data. Intuitions
trump data. Scientists, not surprisingly, are puzzled by where such *a
priori* knowledge might really come from, and they do not want to be
bamboozled by philosophical flimflam. After all, intuitions appear to
be just strongly held beliefs that are likely grounded in education and
reinforcement learning. Intuitions are not, by anyone's account, special
reports from Plato's heaven concerning Absolute Truths.

Philosophers are apt to defend their intuitions as supported by
thought experiments about what could obtain in any possible world.
Supposedly, the outcome of the "thought experiments" will identify
the *necessary* truths about, for example, the nature of knowledge. This
is a suspect strategy. Recall that Kant thought that he had shown by

---

[12] This view is not limited to a small minority but is widely espoused and widely
taught in philosophy courses. This is readily seen in entries in the online *Stanford
Encyclopedia of Philosophy*, which presumably represents the mainstream in
the field. See, for example, the entry under "Analysis of Knowledge."

thought experiments that space – the space our Earth and solar system inhabit – is necessarily Euclidean. Alas, the Euclidean claim is not even true, let alone necessarily true. Space is non-Euclidean. Thought experiments, for all the homage paid to them by philosophers, are not real experiments in any sense. Starting an inquiry with intuitions is fine if that is all you have to go on, but then experiment and observation should subject those intuitions to test, and other hypotheses should be considered. In this well-known fashion, experimental psychology and neuroscience have illuminated the nature of our knowledge of the world and the nature of learning, along with the broader question concerning the nature of how nervous systems of all mammals represent the external world (Squire et al. 2012).

How could our intuitions be misguided? Here is how: complex nervous systems are not mere reflex machines or simple conditioning machines; they build models of the external world that are deployed in navigating the world. But not all models are equally accurate to the world itself. A mouse's model of the spatial world may be sufficient to get it around its environs given its limited goals, but it will not be as accurate as *my* model of the spatial world or indeed that of a wolf. Brains also build models of the *causal* world – for example, that fire is hot and can burn us, that red raspberries are tasty, and so on. Regarding causality, too, models have different degrees of accuracy – my general causal model of the world is more accurate than that of my great grandmother or my dog, for example. Finally, the brain builds models of the *inner* world – the world of brain events, including processes we call emotions, drives, and attention. Here again, there are varying degrees of accuracy, and in particular, according to Michael Graziano (2013), the brain's ongoing model of attention can be inaccurate. In particular, it *will* be inaccurate if it embodies the idea that attention is a nonphysical, spooky phenomenon and hence that consciousness is also. Can this sense of "spookiness" be easily shed?

Probably not. By and large, our brains update our world models for us, but the control we have on the updating is limited. I might successfully update my causal model of the world as I come to realize that cholera is caused not by "bad air" but by bacteria. Somehow that information will modify and reshape my causal model of the world. However, a rainbow will still *look* like it has a location in space, even though I know full well that it does not. What about the model of attention and mental states generally? The model of mentality may persist in *seeming* to be spooky, even when I know "cognitively" that spooky is not accurate to the facts. This may be owed to deep biological features of the way the neural model works.

Here is a comparison: it is a deep biological feature of brains that we extend touch sensations to the end of the pencil or scalpel, to the digger end of the backhoe, and so forth. It seems that we can feel the end of the tool. We all know full well that we have no sensors at the end of the backhoe bucket, but our brain's model finds it very efficient to work that way anyhow – an evolutionary adaptation, no doubt. The point is that as we learn more about the brain, our scientific understanding of our model of attention may become more accurate, but the brain's model of conscious states we use on a moment-to-moment basis may itself be largely unmodified by such neuroscientific knowledge. Thus we may understand more about why it is so easy ("intuitive") to think that consciousness is a spooky phenomenon, even when we appreciate scientifically that consciousness is not spooky but brainy.[13] What is really interesting to me is that we can simultaneously hold both ideas – "spooky" and "brainy" – in our minds, albeit in different ways.

How Did Neurophilosophy Get Started?

Neurophilosophy was more or less inevitable, given the progress in neuroscience and the many links between higher functions and neural activities. Because I happened to be the first to publish using the word "neurophilosophy" (the title of my 1986 book bore that name), I will say a little about my own history.

In about 1978, I came to think that the arguments for an autonomous psychology – a science of the mind autonomous with respect to neuroscience – were too flimsy and self-serving to be taken seriously (e.g. Fodor 2000). If, as seems probable, there is no nonphysical soul but only the physical brain, then surely what is known in neuroscience cannot help but be relevant to understanding the nature of psychological phenomena, including vision, decision making, memory, and learning. Although I have always emphasized that understanding neuroscience was necessary to understand the mind, some philosophers read me as saying neuroscience is both necessary *and* sufficient. This

---

[13] I owe this point to Michael Graziano in conversation. But see also Graziano (2013).

was a poorly disguised straw man designed to make the project look extreme and unproductive (see McGinn 2014; Churchland 2014).

To appreciate more exactly the contribution neuroscience might make, I recognized that I needed to know as much as I could about neuroanatomy (structure) as well as about the developments in neurophysiology (function). I went to the head of the Neuroanatomy Department at the University of Manitoba Medical College and explained my need. To my everlasting gratitude, he warmly welcomed me and encouraged me to take courses alongside the medical students. The arrangement was informal because I was not enrolled as a medical student – I was, after all, still being paid to teach philosophy to undergraduates. Soon thereafter, I was invited to attend neurology rounds and neurosurgical rounds with the clinicians, a weekly event in which patients with neurologic conditions were presented, following which their cases were discussed in detail. After finishing all available courses, I then became associated with the spinal cord laboratory of Dr. Larry Jordan, which was focused on the neural circuitry that maintained rhythmic walking motions. In the lab, I began to dig much deeper into basic neuroscience.

Among other things, the experience in the Jordan lab taught me that understanding the available techniques is essential to evaluating an experimental article. Data will be unreliable if the technique is unreliable. It also taught me to remember that nervous systems, including our own, are the products of evolution. One of the deepest insights I learned from visiting neuroscientist Rodolfo Llinas was this: the fundamental function of nervous systems is to move the body so that the animal may survive and reproduce. Perception, emotions, and cognition are functions whose features were selected for insofar as they served behavior in the business of survival and reproduction. More exactly, perception and cognition serve prediction, and the capacity to make good predictions is a major driver of brain evolution. Commonplace thirty years later, Llinas's insight provoked me to see everything about cognition and perception in a fresh way.

Of course, my husband and philosophical colleague, Paul Churchland, was as fascinated as I by the adventures in the lab, and he too began to participate in experiments. He readily saw how his own ideas about weaknesses in parts of folk psychology fit with emerging data in the behavioral and brain sciences. Among my colleagues, Jeff Foss and Michael Stack also became hooked, and our daily lunches were effectively seminars batting around what we were all eagerly learning.

After Paul and I moved to the University of California San Diego, we encouraged our graduate students to have some laboratory exposure while engaged in philosophical research. Many of them did, and some, such as Elizabeth Buffalo, Adina Roskies, and Eric Thomson, eventually left philosophy to find their professional home in neuroscience. Others, such as Rick Grush and Brain Keeley, successfully straddled the two fields. In San Diego, the main neuroscience lab that I was associated with was run by Terry Sejnowski, whose lab was located in the Salk Institute. Francis Crick was also an associate of the lab and was an active participant on a daily basis. Terry's lab focused on a range of topics, including reinforcement learning and the question of what kinds of computations neurons and networks might be using.[14] We also frequently discussed the problem of consciousness and what experiments might help us to understand it as a brain-based phenomenon. Some of the most productive, broad-based, large-scale (one might say "philosophical") conversations took place over tea at that lab. Lab meetings and teatime continue even now to be a source of inspiration and reflection for me.

By and large, the reception of philosophers to the publication of *Neurophilosophy* in 1986 was anything but welcoming. Neuroscientists, by contrast, gave it a much warmer reception, something that seemed to further exasperate philosophers of mind.[15] Owing largely to the blossoming of the brain sciences, the book apparently facilitated the decision of many philosophy undergraduates to do graduate work in neuroscience rather than philosophy.

The hostility from philosophers that greeted neurophilosophy in its early days has mostly abated, and a small but enterprising cohort of younger philosophers has eagerly embraced its general intellectual attitude. They tend to be comfortably immersed in the neuroscience of psychology and philosophy with no sign of metaphysical angst.

---

[14] What emerged early on was the collaboration that resulted in Churchland and Sejnowski (1992).

[15] John Marshall, a well-known neuroscientist and frequent reviewer of books in Oxford, told me he was asked by the *New York Review of Books* in 1986 to review *Neurophilosophy*. Several years after he submitted his review, he gave me a typewritten copy of the gratifyingly positive review he had written. He explained that the *New York Review of Books* had declined to publish it. He vowed never to write for them again and did not.

Washing their hands of conceptual necessities seems to have left their creativity undiminished. Washington University in St. Louis was the first to set up a graduate program called "Philosophy, Neuroscience, and Psychology" (PNP), which has truly flourished, as has the coordinated undergraduate program. It set the benchmark for other similar programs. Duke University also saw a future in linking with psychology and neuroscience programs, and its programs also have flourished.

No one would call the shift to recognizing the relevance of scientific data a philosophical stampede, however. A quick look at the current graduate courses and syllabi from high-ranking schools in the United States reveals that conceptual analysis tends even now to dominate the philosophical agendas. Mainstream philosophical research on the mind/brain prides itself in being mainly about words, not things. Philosophers in other countries may be moving ahead more quickly. For example, Poland's prestigious Copernicus Center is at the forefront of research on such difficult problems as norms – what norms are; how they are learned, expressed and changed; and what data from psychology and neuroscience reveal about how they guide behavior.[16] Moscow's Center for Consciousness Studies likewise has a cohort of young researchers who are aiming to make progress on traditional problems about the nature of consciousness, knowledge, and representation by integrating data from many labs.[17]

Quine and the Conceptual Analysis Dogma

A powerful but oft-ignored lesson of Quine's (1960) discussion concerning naturalizing philosophical inquiry[18] is simple: clarifying a concept used to categorize the world can be very helpful in avoiding confusion in a seminar, but that clarification cannot itself tell us whether that concept truly applies to phenomena in the world, whether it should be revised in the light of facts, or even whether it possibly should be ditched altogether.

The applicability of a concept to phenomena in the actual world depends on science (broadly speaking) and discovery of the facts. This is obvious in the case of a concept such as "caloric," where we can be

---

[16] See, for example, Brozek (2013) and Heller, Brozek, and Kurek (2013).

[17] See the well-informed interviewers, Vadim Vasilyev and Dmitry Volkov discuss neurophilosophy with me at: https://youtu.be/GP8o-yjZePc.

[18] See also my Preface to the second edition (2013).

reasonably clear about what were believed to be the properties of caloric fluid, were it to exist; for example, it moves from hot things to cold things, hot things have more of it than cold things, it has no mass, and so forth. All that clarity notwithstanding, there is no such thing as caloric fluid. Differences in temperature are a matter of differences in mean molecular kinetic energy, not in volume of caloric fluid.

Consider now the case of a concept such as "soul," where we might have something like Descartes' idea of what we mean by the concept. A philosophical analysis of that concept tells us precisely nothing whatever about whether souls really exist or even whether they have the properties outlined in its analysis. The meaning of a word merely reflects current beliefs, and those beliefs may be misguided. Think of wholesale revisions to the concept of an "element," originally believed to comprise *earth, air, fire* and *water*, not one of which is now considered an element. The point extends more generally. In particular, it extends to words such as "knows," "believes," "rational," and "decides."

To elaborate, Quine's point was that what is *meant* by a word reflects what is *believed to be true* about the *things* the word denotes. Thus, meaning changes as knowledge expands. This point has been stoutly resisted by scientifically naive philosophers who supposed that if something is considered part of the very meaning of a word, then it is a necessary feature of the stuff denoted by that word. That the phenomenon has that meaning-linked feature is, allegedly, a necessary truth, and necessary truths are, needless to say, necessarily true regardless of what science discovers. Thus, these philosophers convince themselves that they can dope out the deep – *necessary* – features of the world by conceptual analysis.[19]

The argument sinks into the fallacious when it shifts from saying that something is part of the meaning of the *word* to saying what is a necessary feature of things in the *world*. Hence, even if, for nineteenth-century physicists, "is indivisible" is part of the very meaning of the word "atom," this does not make it necessarily true – or even true at all – that atoms are indivisible, that they have no substructure. Nevertheless, philosophers have been prone to make claims about what must be true

---

[19] See, for example, the entry in the online *Stanford Encyclopedia of Philosophy* under "Analysis of Knowledge." The authors, Jonathan Jenkins Ichikawa and Matthias Steup, state that a proper analysis of knowledge "should at least be a necessary truth."

about the mind based on their analyses of the meaning of words, words such as "knows" and "believes" and "conscious."

One quick further point about conceptual analysis: typically what is marketed under the banner of "conceptual analysis" is not actually a reflection of what a word means in its everyday use by ordinary folks (Schooler et al. 2014). Rather, it is a *theory*, albeit a camouflaged theory, about the nature of some phenomenon, such as consciousness or choice or knowledge. Consider, for example, the idea that beliefs require language because beliefs are states of mind standing in relation to a sentence. This idea is not based on what ordinary speakers of the language mean or even on what is implied by what they mean. Such claims go well beyond meaning. These are actually empirical hypotheses, *disguised* and *sold* as conceptual truths, based on scanty, or even no, empirical evidence.

Theorizing is an important undertaking in the effort to advance knowledge and understanding of the world, including the world of the mind/brain. Philosophers are as welcome into the theorizing tent as anyone else, and certainly some philosophers have made important contributions in this domain.[20] Clinging to outdated ideas concerning conceptual analysis and necessary truths impedes the progress that philosophers might otherwise make. In general, it is more rewarding to take account of existing data when trying to generate an explanatory theory of a phenomenon than to troll one's intuitions for "necessary truths," something the witty biologist Sir Peter Medawar (1979) suggested is the philosophical equivalent of "psychokinetic" spoon bending.

Concluding Remarks

As more is discovered about brain organization and the dynamics of neural networks and whole systems, our knowledge of mental functions also will expand, undoubtedly in unpredictable ways. Whether unsurmountable obstacles will be encountered is not known – certainly not known even by philosophers who insist that their well-trained intuitions have already spied such obstacles.

[20] For example, Eliasmith (2013), Craver (2009), Silva, Landreth, and Bickle (2014), Smith (2011), Danks (2014), Bickle (2013), Arstila and Lloyd (2014), P. M. Churchland (2013), and Glymour (2001).

In science, we typically cannot tell whether a problem is just not yet solved or absolutely unsolvable. You cannot tell just by looking – or just by using your intuition. Just as the Straits of Gibralter were once thought to mark the outer limits of the world, so it may be tempting to think that what we cannot now imagine marks the limits of what science can discover. This is a mistake, one that is rooted in philosophical complacency and a failure of intellectual courage. Of course, some problems are not problems for neuroscience or for philosophy – such as the problem of making a vaccine against the Ebola virus or sequencing the genome of an extinct species of humans such as *Homo erectus*. Some problems, as Sir Peter Medawar wisely reminded us, are political problems concerning the more effective way to address terrorism or whether to allow doctor-assisted suicide for the terminally ill. Some problems are personal problems about whether to change jobs.[21] But some problems *are* problems for science, and it is highly likely that the nature of consciousness is one of those problems. Whether we do actually solve it remains to be seen.

Young philosophers need to ask themselves a basic question: what is it that I really want to understand? Is it just what other philosophers *say* about a problem and how I might figure out a clever response within their framework of assumptions? Is it something about current English usage, such as what the word that names the problem usually *means*? Or is it the *nature* of the thing – how it works? These are quite different questions, using very different methods, and leading a researcher in very different directions.

---

[21] I too make this point, for example, in Brain-Wise (2002), yet philosophers such as Roger Scruton (2014) continue to wag their finger and warn that science cannot solve all problems.

5 | Teleosemantics

DAVID PAPINEAU

The Problem of Representation

"Teleosemantics" is a theory of representation. There are many different kinds of representations. Some representations are mental states: beliefs, perceptions, hopes, fears. Others are public, nonmental items: sentences, maps, diagrams, pictures.

What all representations have in common is "truth conditions." Any representation will portray the world as being a certain way. It will draw a line in logical space, dividing the possibilities into those that verify it and those that do not. When I assert that "Elvis Presley once visited Paris" or think the corresponding thought, my words, or my mental state, will be true if and only if Elvis did once go to Paris and false otherwise.

("A picture is worth a thousand words." It is not always easy to articulate what is being claimed by a perception or by a map or other pictorial means of representation. But this does not mean that these states lack truth conditions, just that they have dense and complex ones.)

Representation can seem puzzling. How *can* one state *stand for* another? When I say or write something, my message will be conveyed by sound waves or marks on paper, and when I believe or perceive something, the vehicle of my ideas will be some arrangement of neurones inside my head. What mysterious force gives these ordinary physical arrangements the power to reach out and lay claim to further possible states of affairs – often far removed in space and time, such as when I represent that Saturn has forty-five moons or that England won the World Cup in 1966?

A natural first thought is that the physical vehicles of representation gain truth conditions in virtue of being *interpreted* as having those truth conditions. The English sentence "Elvis Presley once visited Paris" means what it does, for instance, because speakers of English understand it a certain way. This is not a bad initial thought, and we shall return to it later, but without further elaboration, it can only get us so far. Interpreting a sentence as meaning a certain truth condition is most naturally understood in terms of its being associated with a certain type of mental state by speakers of the language – in our example, the state of thinking that Elvis once visited Paris. But this then leaves us with the question of what gives those mental states *their* truth conditions. And if the answer is that those mental states are interpreted with the help of yet further mental states, we are clearly off on a regress. We need some account of *original meaning* ("original intentionality," as it is often called) – we want to explain the kind of meaning that states can have in their own right and not in virtue of being interpreted with the help of other meaningful states.

Many philosophers think that original intentionality is a product of consciousness. As they see it, it is specifically conscious states that have the intrinsic power to represent the world to subjects.[1] They appeal to the way that perceptions and thoughts strike us introspectively. Suppose that you are currently seeing a tree. Isn't it built into the conscious nature of your sensory state, these philosophers urge, that it represents there to be a tree in your environment? Some take a similar line with thoughts, holding that it can be intrinsic to the conscious nature of a thought that it represents, say, that the stock market has fallen.

This line of thought is seductive but fundamentally misguided. Conscious states do represent, of course, but not in virtue of their conscious properties. States with just those conscious properties could, in principle, have represented different things or nothing at all. But this is not the place to argue these points in detail; I have done so elsewhere (Papineau, 2016). In this chapter, instead of tackling consciousness-based accounts of representation head-on, I shall instead explore an alternative approach to representation as depending on nonconscious properties and relations. This approach will explain the representational powers of conscious states in term of such properties and relations, rather than their conscious nature, and moreover will also allow that nonconscious states also can be representational in the fullest sense.

---

[1] This is, of course, the dominant tradition in the history of philosophy. For contemporary defenses of this approach, see the Introduction and essays in Kriegel (2013).

One last preliminary point before proceeding: not all representations are *categorical*, in the sense of being offered or entertained as saying what *is* the case, as opposed to what *might be* the case. For example, conjectures, imaginations, hopes, and fears are representations all right, but they are not categorical. They have truth conditions and so can turn out to be true or false, like other representations, but they aren't embraced as categorically conveying how things are, in the way that assertions, beliefs, and perceptions are. They are merely possibilities to be considered. In what follows, we shall be concerned specifically with categorical representations. Once these have been explained, then perhaps an account of noncategorical representation can be built on that basis.

Representation as a Biological Category

The key to understanding representation is to view it as a biological phenomenon. According to the teleosemantic program, representations are states whose biological function is to guide behavior in ways appropriate to such-and-such conditions. Those conditions are then the truth-conditional contents of the representational states. The representations are true if those conditions obtain but not otherwise.[2]

Let me illustrate. Vervet monkeys in Kenya have three distinct alarm calls – for leopards, eagles, and snakes respectively. These calls are designed to prompt specific behaviors in the monkeys. As Seyfarth, Cheyney, and Marler explain in their classic 1980 paper, the monkeys "respond to leopard alarms by running into trees, to eagle alarms by looking up, and to snake alarms by looking down." These responses determine what the calls represent, in that the truth condition of each call is that circumstance in which the monkeys' consequent behavior would be appropriate to its survival.

It will be helpful to analyze this case in terms made familiar by Ruth Millikan (1984). Distinguish the "producer" of the call, the signaler, from the "consumer," that is, the monkey that responds to the call. According to the teleosemantic analysis, it is the behavior of the consumers that determines the truth-conditional content of

---

[2] The first works developing this teleosemantic idea include Millikan (1984), Fodor (1984), Papineau (1984, 1987), and Dretske (1986, 1988).

the call, not the circumstances that prompt the producers. Suppose that the consumers respond to some call with behavior that is appropriate to an impending eagle, say. This then shows that the call means "eagle." And this remains the case even if the producers also regularly produce that call in response to fast-moving clouds, low-flying airplanes, and so on. The truth condition of the signal depends on how the consumers behave in response to it, not on what causes producers to emit it.

In this example, the producer is one organism, and the consumer another. But the story will work just the same if the producer and the consumer are inside the same individual. This then gives us a model for representation by mental states as well as public signals. Suppose, as seems plausible, that the monkeys also have three kinds of cerebral states, "produced" by their visual systems and "consumed" by their motor control systems. Then the story runs just the same. These brain states will represent leopards, eagles, and snakes respectively, in virtue of the fact that they are designed to lead the monkeys to behave in ways appropriate to just those threats.

So here we have a simple explanation of representation. It isn't magic. It is just a matter of certain states having the biological function of instigating behavior that is appropriate to such-and-such conditions.

There is a sense in which this account preserves the intuitive idea that the meaning of a representational state depends on how it is *interpreted*. Representation arises whenever a consumer *interprets* some state as signifying some circumstance, in the sense that it *acts* in a way appropriate to that circumstance. The crucial point is that the idea of interpretation now in play is not the idea of the representation prompting some further *mental* state in the agent. As we saw earlier, if we understand interpretation in this mental way, the appeal to interpretation is no good for explaining original intentionality because it inevitably degenerates into regress. However, the current proposal cuts through this regress. Interpretation is now a matter of acting, not thinking further thoughts. A representation is interpreted as having a certain truth if it leads a consumer to *act in a way appropriate to that condition*, not to *form a thought with that truth condition*. By this means, we explain representation without presupposing it.

Generalizing the Story

It will be useful to schematize more fully how the teleosemantic account explains representation in terms of biological functions. Suppose that we have some consumer system that responds to representation R with some behavior B and that the biological purpose of this consumer is to achieve some end E. Then the system that produces R will have the function of producing R when condition C obtains, where C is that condition that will ensure that B causes E. If all this is in place, then R will *represent* C: the producer is biologically supposed to produce R when C because this will then enable the consumer to serve its function of achieving E.

In the vervet monkey example, we took the relevant end E simply to be survival and reproduction. But we can also view biological ends in a more fine-grained way, as aimed at more specific results than survival and reproduction, and this will then allow the teleosemantic approach to deal appropriately with more complex kinds of representation.

To understand how biological items can have biological functions that are more specific than survival and reproduction, note that biological systems can be decomposed into nested structures of interlocking components. For example, the human body is composed of the brain, the temperature-regulating system, the cardiovascular system, and so on. The cardiovascular system can itself be decomposed into the heart, the lungs, and the blood vessels. Now all these components have the eventual biological function of fostering survival and reproduction. But they are all supposed to contribute to this in special ways: the brain by managing behavior and hormonal levels, the temperature-regulating system by maintaining a constant temperature, the cardiovascular system by circulating oxygen and nutrients and removing carbon dioxide and toxins, and, in turn, the heart, lungs, and blood vessels are supposed to do the latter by pumping, oxygenating, and transporting the blood, respectively.

Given this, we can identify the functions that are *specific* to the components in an overall biological system. For example, the specific function of the heart is to pump the blood. Of course, the heart also has the further functions of circulating oxygen and nutrients and removing carbon dioxide and toxins and thereby of fostering survival and reproduction. But these further functions are not specific to the heart, as is shown by the fact that their nonfulfillment does not mean that the heart isn't doing *its* job: if oxygen isn't circulated, this might be because the lungs aren't oxygenating the blood, not because the heart isn't pumping. In general, we can say that the specific function of some biological component is the most immediate effect it is supposed to produce at that level of decompositional analysis where it itself appears as an unanalyzed component. So, for example, pumping is the immediate effect attributed to heart once we decompose the cardiovascular system into its components.[3]

This means that the representational functions attributed to contentful states need not always be geared to survival and reproduction as such. If the producer-consumer system in which some representation R is a component itself has some specific end E, then the representational content of R will be the condition C that ensures resulting behavior will achieve E, whether or not survival or reproduction follows.

One obvious application of this idea will be to organisms that form and activate *desires* and other motivational states. These motivational states can be viewed as themselves consumer mechanisms whose specific biological purpose is to achieve certain proximal results, such as water, or sex, or social esteem – or whatever the motivation is aimed at. These motivational states are not aimed at such general results as survival and reproduction because it is not their fault, so to speak, if these further results do not follow once their specific ends have been achieved. The specific job of my desire for water, for example, is to get water into my body, and it will have fulfilled this aim even if my stomach is malfunctioning and this does not help my survival and reproduction.

Given this, representations that direct the selection of actions in pursuit of desires and other motivations will then represent circumstances relevant to achieving the specific ends of those motivations. The representational state that tells me what to drink when I am thirsty has the specific function of tracking water and fulfills this function even in cases where water will not aid my survival or reproduction.

Another important range of cases will be where the purpose of the relevant consumer is itself to produce further representations. For example, many producer mechanisms within our perceptual system have the purpose of detecting "features" (e.g. edges of physical objects)

---

[3] This account of specific functions is borrowed from Neander (1995).

on the basis of which further consumer mechanisms will construct representations of more complex phenomena (say, whole three-dimensional objects). In this kind of case, the producer (the edge detector) will represent *edges*, say, in virtue of the fact that its outputs are treated by the consumer (the object representer) in ways appropriate specifically to the presence of edges. (And this consumer will then have the purpose of representing *objects* in virtue of the fact that *its* outputs are in turn consumed by further mechanisms in ways appropriate, given their specific purposes, to the presence of objects.)

These examples show that the general teleosemantic approach will discern different kinds of representations in different components of different organisms depending on the details of their internal cognitive architectures. What the states in an organism's component mechanisms represent will depend on such things as its structure of motivational states, the computational structure underpinning its visual perception, and so on. This is not a weakness of teleosemantics. On the contrary, it shows that it is a powerful framework that can be applied to a wide range of cognitive architectures to identify the specific representationalist purposes served by their components.

Is Truth Functional?

The teleosemantic approach to representation hinges on the idea that the *truth* of a representation coincides with *fulfilling its biological function*. This opens the teleosemantic approach to a common objection – namely, that truth and biological function can come apart for representations. In particular, so the objection goes, there are plenty of cases where representations fulfill their biological functions even though they are false. Evolution doesn't care about truth, but just about practical biological success, object the critics. If we want to understand truth, they conclude, we need to look beyond the biological realm with its exclusive focus on practical results (Plantinga 1993; Burge 2010).

There are three different kinds of cases worth discussing here. First, there are representations for which the *biological expectation is falsity* rather than truth. Second, there are representations that luckily happen to lead to biological success *despite being false*. And third, there are representations that systematically confer some positive biological benefit *in virtue of being false*.

Let us take these in turn. For an example where falsity is the biological norm, the vervet monkeys will serve well enough. Let us suppose that the monkeys are designed to err on the side of caution and will alert the troop to an eagle threat on the slightest pretext, with the result that the vast majority of "eagle" calls are occasioned by clouds, airplanes, and so on rather than eagles. (I make no claims for the ethological accuracy of this supposition.)

We can see why the monkeys might have been set up in this way. The cost of a false-positive call – one prompted by a fast-moving cloud – is far less than that of a false-negative call – ignoring a real eagle. The former mistake only means a wasted upward glance, but the latter could well mean death. It is far better, in biological terms, to bear the cost of regular false alarms than to run the risk of being caught by an undetected eagle.

But it is simply a mistake to think that cases such as these are a problem for teleosemantics. Despite the frequency of false positives, falsity is no part of the biological *function* of the "eagle" signal, nor of the mechanisms that produce them. Biological functions are always advantageous effects, results that contribute to survival and reproductive success. There are no such advantageous effects occasioned by false alarms. Pausing and looking upward because of a cloud is pure wasted effort. The advantageous effects of the "eagle" signal accrue specifically in those cases where there really is an eagle around, and the signal enables the monkeys to avoid capture. *That* is the function of the signal, and it is fulfilled specifically in cases where the signal is true, just as the teleosemantic approach would have it.

The monkey example is just a special case of the point that the biological function of some trait need not be normally or even often achieved – provided the payoff when it does occur is big enough to outweigh the cost of failed attempts. Male sperm are the standard illustration. Nearly all sperm are fated to wither and die before fertilizing an egg. But that doesn't mean that fertilization is not the function of the sperm. We don't want to say that perishing without achieving fertilization is the *function* of sperm just because that fate is biologically overwhelmingly probable.

Let me now turn to the second kind of example. Sometimes the behavior prompted by a false representation can lead to biological success by luck. A thirsty monkey sets off in a certain direction, prompted by the belief that there is water in the stream. As it happens, the belief is false (the stream has dried up), but happily, the monkey comes across a pool of water halfway there. At first pass, this looks like a case of a belief serving its biological function even though it is false. After all, the belief is here consumed by the monkey's thirst-quenching mechanism, and it leads successfully to the satisfaction of that mechanism's biological end.

But this is not a problem for teleosemantics either. While the thirst-quenching mechanism has achieved its end, the belief itself has not served its own biological function. Its specific function is to coordinate behavior with the putative presence of water in the stream. It hasn't done that in this case because there was no water in the stream. It was just a matter of luck that water was found; it wasn't because the belief was working as it was supposed to.

In general, a biological trait can lead to some eventual biological success by luck in a particular case without serving the specific function for which it was selected. On some occasion the camouflage of an insect saves it from predation by delighting a child whose laughter scares away a bird. But this clearly wouldn't be a case of the camouflage serving its specific function of *hiding* the insect. Similarly, the fact that a belief can luckily engender success though false is no counter-example to the teleosemantic claim that its specific function is to coordinate its consumer's behavior with its truth condition.

Third, there are arguably some few cases where representations do genuinely serve a biological function in virtue of being false. Consider the phenomenon of "depressive realism": most psychologically healthy people have an inflated view of their own social standing, by compar-ison with objective measures; the only people with accurate beliefs about their status tend to be depressed. Let us suppose that these mistaken beliefs among nondepressed people have a biological pur-pose: the function of the widespread false beliefs is to stop people from retreating into their shells and to encourage them to be enterprising. (Again, I make no claims for the biological accuracy of this supposition.)

Now cases like these really do involve beliefs that serve a biological function *because* they are false. It is specifically when lower-status people think that they are higher status that they are encouraged to be enterprising. And this certainly seems in tension with the teleose-mantic idea that we can equate the truth-conditional content of a belief with that circumstance in which it serves its specific biological function.

Here the truth condition is that you are higher status, but the function is served when you are lower status.

To deal with this issue, teleosemantics needs to recognize that some representations can serve two different functions. This is a familiar enough biological idea. For example, large earlobes can facilitate both audition and thermal regulation and be selected for both these positive effects. Similarly, in our case, the belief arguably has the functions of both (1) guiding behavior aimed at satisfying currently active motivational states and (2) boosting enterprise by fostering self-esteem.

Once we recognize these two distinct functions, we can see that there is nothing here to undermine the teleosemantic approach to representation. The function that matters to the teleosemantics of belief is the former one: guiding behavior in a way that will serve whichever motivations are currently active. To fulfill this function, your beliefs still need to be true. (Suppose that you want to make money and enter a popularity contest in the belief that your high status will win you the $100 prize. You won't get what you want if your belief isn't true.) That some specific beliefs, such as this belief about your status, might also have some further function, such as bolstering self-esteem, that is fulfilled when the belief is false does not eliminate the former teleosemantic function any more than earlobes acquiring a thermoregulatory function eliminates their auditory function.[4]

Determinacy of Content

A standard objection to teleosemantics is that it is not able to explain the possession of fully determinate contents by representational states. Jerry Fodor (1990) has argued this point in connection with the states in frogs' brains that prompt frogs to snap their tongues in the direction of passing flying insects. Fodor challenges teleosemanticists to explain why these states should be regarded as representing flying insects rather than small, black, moving things. After all, Fodor maintains, we could as well take the frog's visual system to be biologically designed to respond to small, black, moving things as to respond to flying insects. A healthy frog will snap its tongue whenever it is presented with a small, black, moving thing, whether or not it is a flying insect.

---

[4] For further discussion of this last kind of case, see Papineau (1993, chap. 3).

An initial answer to Fodor is to respond that teleosemantics focuses on conditions that ensure biological success, and the alternative conditions he has in mind fail to satisfy this requirement. It is true that a healthy frog will respond to small, black, moving things even if they are not flying insects. But the conditions that teleosemantics is concerned with are not those that can be expected to cause the frog's state, but rather those that will ensure that resulting behavior causes success. The frog's state clearly has the function of helping the frog catch flying insects rather than small, black things: no selective advantage accrues to a frog that grabs some nonnutritious speck of passing dirt. And so, in line with this, defenders of teleosemantics can argue that the frog's brain state represents a flying insect in a given direction rather than small, black things. For it is precisely when its brain state is prompted by a flying insect, rather than any small black thing, that an advantageous effect will accrue (cf. Millikan 1993).

However, it is not clear that this response fully deals with Fodor's worry. I just said that the advantageous result of the frog's state is flying insects rather than small, black things. No reproductive advantage accrues when it catches a small, black thing that isn't an insect. But why stop there? The biological point of catching flying insects is to get them into the stomach. No reproductive advantage accrues if an insect is caught but isn't ingested. Again, the biological point of ingesting something into the stomach is to get nutrients into the bloodstream. No reproductive advantage accrues if an insect is ingested but yields no nutrients into the bloodstream. And so on. In the end, the ultimate point of all functional traits is survival and reproduction. No reproductive advantage ensues from any intermediate effects if they don't eventuate in survival and reproduction.

Given all this, it might not be obvious why we should interpret the frog's state as representing flying insects. Why not read it as representing stomach filler? Or as nutrient source? Or even as reproduction enhancer?

To resolve this issue, we need to go back to the idea of biological traits having specific functions of their own, in addition to those they share with other components of the whole organism. In the example used earlier, the specific function of the heart is to pump blood. It also has the functions of circulating oxygen and eventually of fostering survival and reproduction – but these functions it shares with other organs, such as the lungs, and so they are not specific to the heart as such.

We saw earlier how this concept of specific functions matters when we apply teleosemantics to organisms that have motivational states. It allows us to think of states like desires as themselves biological components with specific functions, namely, the production of the specific effects that will satisfy them. And then we can equate the truth conditions of representational states that inform the pursuit of desires with conditions that will ensure that those desires are satisfied.

So, if we could credit frogs with motivational states, then this would resolve our issue. If the frog's behavior is motivated by the desire for a flying insect as such, then the state that prompts tongue snapping in a certain direction would signify that there is a flying insect in that direction. Whereas if the frog's is motivated by a desire for a nutrient source, say, then the state would represent a nutrient source in that direction.

However, there seem to be no good grounds for attributing such motivational states to frogs. Modern physiologic research suggests that frogs lack any integrated decision-making system in which belief-like states serve desires. Rather, each of its behavioral systems is guided by its own proprietary information, which is unavailable to its other systems of behavioral control. One channel of sensory information guides its prey-catching behavior, another guides its obstacle-avoiding behavior, and yet another its ability to jump away from looming threats. Lesions of the frog's optical system can dissociate these different abilities (Milner and Goodale 1995, sect. 1.2.2).

Still, this lack of an integrated decision-making system does not mean that the idea of specific functions has no grip on the frogs at all. There is nothing to stop us from applying this idea directly to the prey-catching system as such. As in other cases, this system has a sequence of functions. It is designed to catch flying insects, thereby to have them swallowed, thereby to allow digestion to place nutrients in the blood-stream, and thereby ... to lead to reproductive success. But only the first of these is arguably the function peculiar to the prey-catching systems, considered as the visuomotor system that governs head turning and tongue snapping. It is not necessarily the fault of this system if a flying insect is caught but does not end up in the stomach (because the swallowing mechanism is not working) or if it is placed in the stomach but not digested (because the stomach is malfunctioning), or so on.

If we accept, on these grounds, that the prey-catching system has the peculiar function of catching flying insects, rather than any later effects, then we can view the sensory signals that prompt behavior in this system as indicating the circumstance under which that behavior will achieve the system's peculiar end – that is, as indicating the presence of a flying insect is such-and-such a direction.

There might seem a further issue here. I have been assuming that the relevant signal is part of the prey-catching system. And this by no means seems mandatory. After all, why not regard the frog's sensory signal as part of the prey-stomaching system or as part of the prey-digesting system or so on? The effects occasioned by the sensory signal don't normally stop with the flying insect being caught – when everything is working as it should, the flying insect will also promptly be swallowed, and digested, and . . ..

This then threatens to render the content of the sensory signal indeterminate once more. Each of the systems at issue – the prey-catching system, the prey-stomaching system, the prey-digesting system, and so on – will have a specific function of its own. But this won't give the sensory signal a determinate content if it is not determinate which of these larger systems is informed by the signal. It will leave it open that the signal tells the prey-catching system about insects, so to speak, but also the prey-stomaching system about stomach fillers, the digestive system about nutrient sources, and so on.

However, this point has in effect already been dealt with. The signal in question is properly seen as a component of the prey-catching system, not of the prey-stomaching or the prey-digesting system. When we first analyze the larger prey-stomaching system, say, into its component prey-catching and prey-swallowing systems, there is as yet no need to bring in the signal as such. The larger prey-stomaching system fulfills its specific function as long as its component prey-catching and prey-swallowing systems fulfill theirs, however they manage to do that. It is only when we analyze the prey-catching system itself that the signal comes into view, so to speak. The prey-catching system fulfills its specific function when its components fulfill theirs – which requires, *inter alia*, that the signal tracks the condition in which the resulting snapping behavior will secure a flying insect. The signal is thus specifically a component in the prey-catching system and its truth condition – the presence of a flying insect in a certain direction.[5]

Outputs over Inputs

According to teleosemantics, the truth condition of a representation depends on the output of the representation, on what behavior it prompts, and not on the input to it, on what circumstances cause it. The vervet monkey's state means "eagle" because it prompts the monkeys to behave in ways appropriate to eagles, even if most of the things that cause the state are not eagles.

Teleosemantics contrasts in this respect with causal theories of representational content, that is, theories that aim to explain truth conditions in terms of their characteristic causes. The obvious problem facing such theories is to distinguish causes that constitute truth conditions from other things that cause the representation. This problem is often termed "disjunctivism": what makes "eagle" the truth condition of the monkey's state rather than the disjunctive condition "eagle-*or*-low-flying cloud-*or*-airplane-*or*-anything else that causes the state"?

This problem is dealt with from the start by teleosemantics precisely because it understands truth conditions in terms of outputs rather than inputs. It doesn't start by looking at the causes of representational states and then seek somehow to narrow these down to the truth-conditional causes.[6] Rather, it simply asks what will ensure that the behavior resulting from the state will be successful.[7]

---

[5] In an earlier paper (Papineau 2003) I argued that the frog's state was indeed indeterminate, on the grounds that it could equally well be considered a component in all the nested prey-catching, prey-stomaching, and prey-digesting systems. It was only after writing that paper that I came to appreciate how Neander's notion of specific functions (1995) resolves this issue in favor of the first option. (Perhaps it is worth mentioning here that in her 1995 paper, Neander herself ends up arguing that the frog's state represents "small, black thing." I would say that she is driven to this conclusion by her mistaken general assumption that traits in healthy animals cannot malfunction simply because the environment is unhelpful, which then implies that a healthy frog cannot be misrepresenting simply because a black speck of mud shoots by.)

[6] Perhaps the best-known causal theory is Fodor's asymmetrical dependence theory (Fodor 1987).

[7] Sometimes teleosemantics is understood as equating truth conditions with circumstances that are biologically supposed to cause the relevant state in "epistemically ideal conditions" and then criticized because it has no noncircular way of defining "epistemically ideal." But this criticism presupposes that teleosemantics is

Not all commentators view this output orientation of teleosemantics as an advantage. If the monkey's representation is triggered as readily by clouds as by eagles, would it not be better to include clouds in its truth condition?

This reaction is bolstered by the following well-known thought experiment due to Paul Pietroski (1992). The kimu are simple creatures whose only enemies are the snorf. The snorf hunt the kimu at dawn. Then one day a biological mutation endows one of the kimu with an ability to register the presence of red things and an inclination then to approach them. This is an advantage to its possessors because it leads them to climb hills at dawn, the better to observe the red sunrise, with the result that they avoid the snorf, which are ill suited to climbing hills. As a result, the disposition spreads through the kimu population.

Now consider the state a kimu gets into when it is stimulated by something red. It is natural to credit this state with the content "red." But an output-based teleosemantics sees things differently. In general, nothing good happens to kimu when they approach red things. Most red-approaching behavior is a waste of time. It is only when it takes them away from the snorf that it yields a biological advantage. So an output-based teleosemantics will construe the state in question as representing "snorf-free" or "predator-free" or something like that. Pietroski argues that this is highly counterintuitive. After all, by hypothesis, the kimu's senses are tracking redness, not snorf.

But this argument is by no means conclusive. Defenders of teleosemantics can object that Pietroski's intuitions are reading more into the story than is warranted. As Pietroski initially tells it, the kimu evolve some state that is triggered by redness and that functions to keep them away from the snorf. But his subsequent discussion invites us to suppose that the kimu have some general-purpose visual *system* whose outputs might inform an open-ended range of behaviors directed at various possible ends (such as avoiding blood, or finding apples, or indeed wanting to see red things). However, this supposition adds significant extra structure to Pietroski's initial story and so makes room for teleosemanticists to argue that an organism with that extra structure would indeed be representing redness rather than snorf-freeness: if the kimu's visual system did inform a *range* of different

---

in the business of distinguishing good causes from bad ones, when in truth it doesn't care about causes but only conditions for success.

behaviors directed at different ends, then the content of its visual states would be conditions that ensured in the achievement of all those ends, and one such state might well come out as representing redness. By contrast, if we stick to a minimal understanding of the snorf, in line with Pietroski's initial story, as having only a special-purpose visual sensitivity that brings no advantage except snorf avoidance, then it's not so clear that there is anything wrong with reading their states as representing "snorf-freeness": after all, if these states never do anything except trigger simple avoidance behavior, it seems natural enough to read them as representing the danger they are designed to avoid.

Doing Without History

Teleosemantics is not the only theory of representation that explains content in terms of outputs rather than inputs. "Success semantics," the origins of which can be found in Ramsey (1927), focuses specifically on belief-desire systems and in that context agrees with teleosemantics in equating the truth conditions of beliefs with circumstances in which resulting actions will satisfy desires. And, more generally, various species of convention-based signaling theory agree with the structure of producer-consumer teleosemantics in equating the truth conditions of a signal with those circumstances in which the behavior performed by the recipient of the signal will satisfy the recipient's ends (Lewis 1969; Skyrms 1996, 2010).

Where these theories differ from teleosemantics is in not viewing these structures as necessarily involving *biological functions*. For teleosemantics, the satisfaction conditions of desires, and more generally the ends of consumers of representations, are equated with the effects that these systems are *biologically* supposed to produce. And correspondingly, the producers of representations are taken to have the *biological* function of producing representations in circumstances where resulting behavior will fulfill the functions of their consumers.

Success semantics and signaling theory avoid these biological commitments. They see no reason to bring biology into the understanding of representation. In their view, we can understand what it is for a desire to be aimed at some outcome or, more generally, what it is for consumers to have ends, independently of any appeal to biological function: these are perfectly good everyday notions, and it is not clear that they demand any further analysis. And even if they do, they can arguably be understood in terms of other nonbiological everyday notions, such as contributing to psychological or bodily equilibrium.

True, understanding representation in teleosemantic terms will automatically carry with it an explanation of *why* the world contains the representational systems it does. Teleosemanticists work with the standard etiological understanding of biological function: a trait T has the function F if it was designed by natural selection to produce F, so to speak – or, less metaphorically, if T is now present because ancestral versions of T were selected because they produced effect F (Wright 1973; Millikan 1989; Neander 1991). On this etiological account of function, to ascribe a function F to a trait T will therewith explain the presence of T in terms of its selectional past. And in the case where we are dealing with representational systems, as with teleosemantics, an ascription of biological functions will carry with it an evolutionary explanation of those systems.

But those who favor nonbiological alternatives to teleosemantics can retort that it is one thing to explain the existence of the representational systems and another to invoke such systems in giving explanations of action. We don't need an evolutionary (or any other) explanation of why representation exists, just because we invoke representation in explaining further things.

As a preliminary to addressing this challenge, it will be useful to clarify exactly how representational notions do help us to predict and explain further things. A first thought might be that they help us understand how internal cerebral states such as perceptions, beliefs, and motivations interact in generating bodily movements. However, this kind of "narrow" psychological explanation makes no real use of representational notions that relate internal cerebral states to features of the environment. After all, if our focus of explanatory interest were solely in predicting and explaining bodily movements, we wouldn't need to think of cerebral states as related to things outside the head at all – we could just think of them as internal components in a structure of causal pushes and pulls (Papineau 1993, chap. 3).

The real significance of representational notions is that they allow us to predict and explain *success*, that is, the achievement of distal results. My belief that there are lobsters in that bay leads me to place my traps there – and then, if that belief is true, it further leads to my catching lobsters. It is the explanation of this eventual result for which representation is crucial – hinging, as it does, on the way my belief is supposed to track the whereabouts of lobsters.

So the crucial pattern is this: behavior B in pursuit of end E is informed by representation R with truth condition C, and when C obtains (when R is true), not only is B performed, but in addition, E is achieved. In short, truth explains success. By ascribing representational contents, we are thus able to discern systematic patterns governing the achievement of distal ends.

Still, nothing in this, the opponents of teleosemantics can continue to object, depends on biological functionality. Teleosemanticists may pick out the relevant ends E as ones that fulfill biological purposes. But why does that matter? The representation-invoking explanatory pattern requires only that there be some C that systematically ensures that the behavior B prompted by R yields some E. And if this is so for the E's and C's identified by the teleosemanticists, it will remain so whether or not those E's fulfill biological functions. What matters are the present-day patterns relating the interlocking parts of representational systems, not the evolutionary history of those systems.

It is worth distinguishing two different ways of running this anti-evolutionary line. So far I have been assuming that the output-based alternatives will reject the teleosemantic appeal to biological functions altogether. But an alternative is to view representation as a matter of biological functions all right, but to understand function in some nonetiological way. There are various nonetiological approaches to biological functions available, united by the general thought that the biological function of a trait involves a contribution to *current* or *future* survival, reproduction, or other beneficial effect rather than to the *past* effects that *causally explain* the presence of the trait (Cummins 1975; Boorse 1976; Bigelow and Pargetter 1987). These nonetiological "forward-looking" accounts of function mean that there is room for output-based theories that agree with teleosemantics that representation is a matter of guiding behavior in a biologically functional way but which identify the relevant biological functions nonetiologically.[8]

---

[8] Some writers extend the term "teleosemantics" to theories that in this way explain representation in terms of nonetiological "biological functions" (Abrams 2005; Nanay 2014). There is nothing wrong with this, but for present purposes, it will be convenient to continue to restrict "teleosemantics" to etiology-based theories.

One issue here is the right way to understand claims about biological functions. This is a much-debated matter. One relatively uncontentious (if insufficiently remarked) point is that only etiological functions are suitable for *explaining* the presence of the traits that have them and so for taking such claims as *the function of hearts is to pump blood* at explanatorily successful, as intuitively they certainly are. Still, this does not mean that claims about biological functions might not also sometimes be properly understood in a nonetiological forward-looking way. Such forward-looking functional claims would not *explain* the presence of the items with functions, but they might for all that provide a useful way of categorizing some of their biological effects.

Fortunately, in the present context, we can bypass this issue of the proper understanding of claims about biological functions. Perhaps there are important philosophical issues that hang on which notion of *biological function* is most current in scientific and other contexts (though I rather doubt it). But in any case, and however that debate comes out, there is nothing to stop someone from *defining* a notion of biological function in some nonetiological way as involving contributions to current or future beneficial effects, or some such, and then maintaining that *representation* is best understood in terms of a contribution to biological ends so-defined – perhaps backing up this claim by arguing that this understanding best captures the present-day patterns we appeal to when we invoke representation to predict and explain success in achieving distal results.

The issue of biological functions thus turns out to be something of a red herring in the context of output-based theories of representation. Teleosemanticists want to explain representation in terms of etiological functions, effects that mattered for the selectional history of representational systems. Alternative output-based approaches deny that selectional histories matter, as opposed to identification of ends the achievement of which can systematically be tracked representationally. Whether these alternative approaches think of the relevant ends as biological functions is a subsidiary issue. Either way, we have the same challenge to teleosemantics: how can it matter to present-day predictive and explanatory patterns what roles the relevant ends played in the past?

Swampman

This challenge is highlighted by the well-known "swampman" thought experiment (Davidson 1987). Suppose that lighting strikes a steamy swamp in the tropical jungle and by miraculous coincidence a perfect molecule-for-molecule replica of a human being assembles itself from organic materials in the swamp. By hypothesis, this "swampman" will lack a history of natural selection and so, according to teleosemantics, will be incapable of representing anything. Yet, intuitively, it seems that swampman will be capable of at least some forms of mental representation. After all, it will be physically just like a normal human, so it will be able to visually register its surroundings and make appropriate behavioral responses. Given this, there would seem every reason to credit its states with truth conditions and use this to track when it will succeed in achieving its ends. So it looks as if teleosemantics has gone wrong somewhere if it denies that swampman has any representational capacities.

The standard teleosemantic response to this difficulty is to bite the bullet and conclude that swampman will indeed be incapable of representation. Maybe everyday intuition argues that swampman can represent, allow the bullet-biters, but a good theoretical account should be allowed to overturn a few everyday intuitions. Just as our modern concept of fish excludes whales, despite any naive intuitions to the contrary, so should a developed concept of representation exclude swampman. According to this line of thought, then, we should replace our naive concept of representation by the theoretically more powerful selection-based notion, even at the cost of overturning intuitions about swampman (cf. Millikan 1996; Neander 1996; Papineau 1996).

However, there is room for an alternative and more nuanced defense of teleosemantics against swampman worries. Rather than seeking to replace the everyday *concept* of representation with one that excludes swampman as a representer, teleosemanticists can leave that concept as it is and instead appeal to the status of teleosemantics as an *a posteriori reduction* of representational facts. On this natural way of under standing of teleosemantics, it is not offered as an analysis of our everyday concept of representation (after all, it was always implausible that this everyday concept should refer to selectional histories) but rather as a theoretical reduction, which appeals to scientific theory to uncover the important underlying features that bind different instances of representation together, just as scientific theory uncovers the underlying nature of water and other familiar chemical substances.

From this perspective, it is no argument against teleosemantics that a representationally competent swampman is consistent with our everyday concept of representation. You might as well argue against modern chemistry on the grounds that XYZ-composed water is consistent with our everyday concept of water. The fact that a swampman with representations can be *imagined* does nothing to undermine the central teleosemantic claim that in the *actual* world representational facts are *constituted* by selectional facts.

Of course, if a swampman were to exist in the actual world, then things would be different. Such a being certainly would display an important explanatory pattern covering the representationally guided achievement of its ends, in line with the everyday notion of representation. However, if a swampman were to exist, then teleosemantics would simply be false. Selectional histories wouldn't be an *a posteriori* theoretically important part of what representing agents have in common because it wouldn't be something they had in common at all. Still, none of this is relevant to how things are in the actual world, where swampman does not exist, and all representers do turn out *a posteriori* to share a selectional past. Given this, teleosemanticists can insist that an imaginary swampman is no more relevant to teleosemantics than imaginary molecular makeups are to chemistry (Papineau 2001).

Still, does this teleosemantic answer to the swampman challenge really scratch the itch occasioned by nonetiological output-based theories of representation? Maybe the general run of representers in the actual world do have selectional histories. But why think of this as the "important underlying feature" that *constitutes* representation rather than as an incidental past circumstance that happens explain why the actual world contains representational systems? After all, isn't the important thing that representational systems display present-day patterns of behavioral success, not what caused them to be like that? As long as they do display such patterns, they can fruitfully be viewed as representational, independently of any selectional history they may have. We need only think of swampman once more to see the point.

In response to this continued challenge, teleosemanticists can first make the initial point that it is no accident that all representational systems found in the actual world have a selectional history. Representational systems, even the simplest, are complex structures involving information-gathering producers and flexibly behaving consumers that are well suited to the needs of the organisms that possess them. Along with all other instances of apparent design in the natural world, the existence of such structures demands explanation. It beggars belief that such helpful complex mechanisms could in reality ever arise by chance, swampman-style. The only serious possibility is that they are results of past selection processes that preserved and refined structures that were biologically helpful to their possessors.

Teleosemanticists then can follow this with the point that anybody who wants to *know* about representational systems will inevitably also be interested in their selectional histories. This is a heuristic rather than a principled point. There is no absolute barrier to an investigator fully understanding some representational system directly, figuring out the operations of its various interlocking parts entirely from firsthand observation. But this is not how it works in practice. As with all cases of "reverse engineering" – figuring out the inner workings of some mechanism from observing its operations – it is practically essential to consider how the system was designed, to think about the way in which its parts are *supposed* to work together. Without considering the designed purposes of the parts, it wouldn't be feasible to distinguish important effects from incidental features. And these points apply to systems designed by natural selection just as much as they do to humanly designed systems. Throughout biology, investigators appeal to possible evolutionary histories to help them understand the current workings of biological systems.

In the case of representational systems, this point applies particularly to the causal transitions between representational states made within the system – the "syntactic" moves whereby some internal states lead to others and eventually to the selection of behavior. It is normally assumed without thinking that these syntactic moves will tend to respect the semantic values of the relevant states – that is, that they will be made in a way that conduces to true states generally leading to further true states and thence to the selection of behavior appropriate to goals. Without this assumption, it would be impossible to draw the information-processing flowcharts that are crucial to many theories in cognitive science and neuroscience. But this "syntax respects semantics" assumption is itself a design assumption. It rests on the idea that representational systems have been set up so as to ensure that the selection of behavior will be appropriately geared to circumstances.

The opponents of teleosemantics might wish to object that none of this really addresses their central point, which is that the current workings of representational systems are metaphysically independent of their causal histories, and that it is only the former that matter for predicting and explaining current behavioral success. However, by this stage, this point is wearing thin. If all actual representational systems have a design history – and anybody who is interested in understanding a representational system must also be interested in its design history – then why continue to insist that real representation only involves present-day patterns?

Consider watches. It is natural to think of watches etiologically. Watches are portable items that make the time of day visible and, what is more, have been designed for this purpose – not by natural selection, of course, but by the conscious watchmakers who constructed their complex workings specifically to ensure that they display the correct time.

However, one can imagine an antietiologist about watches. "What matters is only that watches correlate well with the time. It is these present-day patterns that we use to tell the time. True, all normal watches have probably been produced by conscious designers for this purpose. But that is a different matter. It might explain the existence of watches but is incidental to their predictive and explanatory significance."

Well, one could insist in this way that a design etiology is not a requirement for being a watch. But what would be the point? After all, every actual portable item that displays the time has been designed for that purpose, and moreover, if you want to understand the workings of a watch, you have no option but to try to figure out what the designer intended the components to do. Given this, nothing would seem to be gained by dropping the etiological requirement for being a watch apart from unnaturalness and an unnecessary purification of categories.

The analogy should be clear. We could, in principle, opt to think of representation as a nonetiological kind. But nothing would be gained. The same range of instances would still come out as representational, and we would simply have cleansed representation of something that is practically essential to understanding its workings.

Varieties of Selection

Let me conclude by addressing an issue that might have been worrying some readers for a while. As the last two sections will have made clear, teleosemantics is committed to regarding all original representation as deriving its functionality from past histories of natural selection.[9] But is this at all plausible? The most familiar kind of natural selection is the intergenerational selection of genes. However, it is highly unlikely that all original representations can be explained in terms of such *genetic* selection. For a start, most human mental representations are products of ontogeny rather than phylogeny. No genes have been selected specifically to foster my desire for a new iPhone or my belief that the Mets will win the World Series.

Fortunately for the teleosemantic project, the possession of etiological functions by biological traits does not always depend on the selection of genes that give rise to those traits. I shall distinguish three ways in we can have etiological functionality in nongenetic traits. The first, emphasized by Ruth Millikan, appeals to a many layered account of functions. The second involves nongenetic selection in learning. The third depends on the intergenerational inheritance of nongenetic items. Together these three processes greatly expand the range of items that can possess etiological-selectional functions.

Let us take multilayered functions first. Millikan notes that some biological items have a "relational" function, which is a function to do something when bearing a certain relation to something else. The chameleon's camouflage system has the relational function of matching the chameleon's skin color to its environmental background, whatever that may be. Given a specific background to adapt to, this mechanism then generates traits with *derived* functions. When the chameleon is crouching on a brown branch, its brown color has the derived function of matching it to the branch. The camouflage

---

[9] I say "original" because, of course, many *derived* representational systems – codes, computer languages, and artificial languages generally – have been constructed to serve their representational purposes not by natural selection but by conscious designers. But *original* representation needs natural selection to serve as a "blind" designer.

mechanism might never have produced that shade of brown before, but even so the skin color will have this derived function, courtesy of the fact that the overall camouflage mechanism has been selected to produce whatever color will match the background.

Such multilayered functions are relevant to the many representational systems that are *compositional*, in the sense that they construct complex representations out of simpler components. Such compositional systems often generate entirely novel representations, items that have a meaning even though they have no historical precedents.

Consider the dance whereby bees can "tell" other bees where to go to find nectar, with the direction of the dance indicating the direction of the nectar and the duration indicating its distance. Any particular dance will be adapted to the current location of nectar and so will have a derived function of guiding behavior in a way appropriate to that location. However, the dance that indicates this specific distance and direction might never have occurred before in bee history. If so, it would owe its functionality not to any beneficial effects of previous versions of that specific dance but rather to the beneficial effects of the overall representational *system* for indicating directions and distances.[10]

This general model can be applied to many representational systems, including human cognition. I doubt that anybody has ever thought that "old tables make good toothpicks" until I wrote that sentence down. But this won't stop this mental representation deriving its biological function, and hence its truth condition, from the functionality of the overall system that generates complex thoughts out of simpler concepts.

Still, what about the *elements* in this compositional system, like the simple concepts "table" and "toothpick"? Won't their functionality have to derive from their past contribution to the selection of genes, even if not in the specific combination "old tables make good toothpicks"? But it is scarcely credible that concepts like these can have any kind of gene-based function.

Here teleosemantics can appeal to the other two ideas mentioned earlier. The first was selection-based "learning." This doesn't involve the differential reproduction of organisms over generations but the

---

[10] Both these chameleon and honeybee examples are discussed in further detail in Millikan (1984).

differential "reproduction" of cognitive or behavioral items during the development of a given individual. Such ontogenetic selection takes place, for example, when cognitive responses are molded by experience during learning. In such cases, we can think of the items selected as having the function of producing those effects in virtue of which they were favored by the learning mechanism. It is arguable that concepts such as "table" and "toothpick" might gain representational functions via this route.

An alternative form of nongenetic selection relevant to teleosemantics is nongenetic "intergenerational selection." Many traits are passed from parents to children by nongenetic channels outside the sexual "bottleneck": these traits include the possession of parasites, the products of imprinting mechanisms, and the many cognitive and behavioral traits acquired from parents via social learning. A number of biological theorists have argued that such non–genetically inherited traits can be naturally selected through the normal Darwinian process of differential reproduction of organisms (e.g. Jablonka and Lamb 1999; Mameli 2004). Non–genetically inherited traits that become prevalent in this way will have functions, namely, the effects that favored their possessors. Again, it seems possible that functions of this kind could help to explain the contents of mental representations. After all, it seems a natural enough thought that certain non–genetically inherited ways of thinking are an advantage to their possessors because they make them sensitive to certain features of their environment.

Conclusion

Teleosemantics offers a powerful framework that promises to explain the nature of representation in nonrepresentational terms. At first sight, it might seem unlikely that a simple appeal to biological functionality can account for the significance of truth, or accommodate the determinacy of representational content, or deal with aspects of representation that have no genetic basis. However, once we appreciate the full extent of the resources to which teleosemantics can appeal, we see that it has the flexibility to deal with these and other objections.

6 | *The Methodological Argument for Informational Teleosemantics*

KAREN NEANDER

The Bare-Bones Version

Dennis Stampe (1977) and Fred Dretske (1986) proposed that mental reference to content supervenes on information-carrying functions.[1] Their proposal endorsed two main theses: (1) that mental reference to content is grounded in the normal-proper functions of components of cognitive systems (teleosemantics) and (2) that mental reference to content is grounded in the natural information processed by these systems (informational semantics). My aim here is to make explicit a methodological argument in support of this dual thesis ("informational teleosemantics"). The argument is methodological in the sense that it relies on certain claims concerning explanatory concepts and practices in the mind and brain sciences.

This first section gives the bare-bones version of the methodological argument. Later sections discuss each of the premises in turn, and then they discuss the kind and degree of support that the argument provides for the conclusion. Without further ado, here is the bare-bones version:

P1: A notion of normal-proper function is central to the multilevel componential analyses (aka "functional analyses") of the operation of bodies and brains that are currently provided by physiologists and neurophysiologists.

P2: The brain's normal-proper functions include cognitive functions.

P3: The same notion of function (mentioned in P1) is central to the functional analyses of cognition that cognitive scientists provide.

---

[1] For other informational versions of teleosemantics, see Neander (1995, 2012), Jacob (1997), Shea (2007), and Schulte (2012). Stampe (1977) did not see himself as offering an informational semantic theory, but causal and nearby relations have since been regarded as the basis of natural information relations.

P4: An assumption in the mainstream branches of the cognitive sciences is that cognition involves information processing.

P5: The (relevant) notion of information involved in talk of information processing in cognitive science is a notion of natural, factive information.

P6: Cognitive science posits "normative aboutness," with the norms derived from the normal-proper functions and the aboutness from the natural, factive information.

C: Some version of informational teleosemantics (broadly conceived) is supported by the explanations of cognition that the mind and brain sciences currently provide.

Though probably implicit in the suggestions made by Stampe and Dretske, this argument has not been fully articulated before. I try to cast some light on the reasons for this in the discussion that follows.

Premise 1

The first premise says that a notion of normal-proper function is central to the multilevel componential analyses (aka "functional analyses") of the operation of bodies and brains that are currently provided by physiologists and neurophysiologists. This is a descriptive claim about the kinds of functional analyses of living systems that biologists presently give. It is not a prescriptive claim about the functional analyses that they should give. It still might seem controversial, but it deserves to be considered uncontroversial for reasons to be explained.

The first thing to say in relation to this is that the relevant notion of a normal-proper function is here identified *ostensively*, as the one that is central to biological talk of normal and abnormal function(ing) – of systems functioning properly, of dysfunction, malfunction, impaired functioning, and functional deficits. So P1 does not commit us to any particular philosophical analysis of the notion. It is controversial among philosophers of biology how ascriptions of normal-proper function are best understood, but P1 completely sidesteps this controversy.[2]

A few points about the notion are (rightly) generally accepted. In this sense of "function," the function of a component in a system is not a mere effect of it. The heart has the function to pump blood but not to

---

[2] Neander (2012) provides an introductory survey of views on biological function.

make whooshing noises, although hearts do both, and whooshing noises are useful diagnostic aids. Nor need a (token) component perform its function. The appropriate occasion might not arise (an antelope born in a zoo might never need to use its long legs to escape from predators) or the environment might not cooperate (a diver's lungs will not absorb oxygen if she is deep-sea diving when her tank runs out). Also, a token component can malfunction. If it malfunctions, it lacks the ability to perform its function, or it lacks the ability to perform it to a normal degree of efficiency (a person's pancreas can have the function to produce insulin even if it is unable to produce insulin or to produce the right amounts at the right times). And there is no incoherence in the idea that functional impairment could become typical in a population for a time, in a pandemic, or due to an environmental disaster, for instance. Given that dysfunction could become typical for a time, the function-dysfunction distinction is not simply the typical-atypical distinction (even if statistical facts are somehow involved, as some claim).

Most self-avowed supporters of teleosemantics have supported the idea that such functions are (roughly) "Wright-style" functions. Larry Wright's (1973) seminal idea was that the function of an entity is what it does that explains *why* it is there or *why* it has the form that it has (e.g. why there are eye saccades or why we have pineal glands). A little more precisely, most self-avowed supporters of teleosemantics employ a more recent etiological theory (such as the one developed by the present author[3]), which differs from Wright's in significant ways. Usually, such a theory explicitly ties the functions of components to past selection. The core idea is that the function of an item is to do what it was selected to do or what items of the type were selected to do (depending on what type of selection is involved). On this view, normal-proper functions are "selected effects" or "selected functions." And everyone will agree that selected functions are, at least at first blush, plausible candidates for identification with normal-proper functions. The heart does not have the selected function to make whooshing noises. Tokens need not perform their selected functions because the appropriate occasion might not arise or the present environment might not cooperate. And there is also the possibility of both token and typical malfunction because the selected functions of tokens depend

---

[3] For details, see Neander (1991, 2016) and Neander and Rosenberg (2012).

not on their current dispositions but on the past selection that resulted from past dispositions.

Some dispute whether normal-proper functions are selected functions and offer alternative analyses, but the present point is that they need not dispute P1, which makes no commitment to an etiological theory of normal-proper function.

On behalf of P1, the next thing to say is that physiologists and neurophysiologists (and not just evolutionists) certainly use *a* notion of normal-proper function. This is not sensibly disputed. For example, a quick word count reveals that one paper published in the *American Journal of Physiology* uses the term "dysfunction" twenty-two times.[4] (Of course, the logic of the notion requires more work to ascertain, but I am assuming that its aforementioned generally accepted features are not in dispute here.) What can be sensibly and relevantly disputed is the theoretical purpose, if any, that is served by the use of the notion of normal-proper function. P1 says that physiologists and neurophysiologists advert to the normal-proper functions of components *in explaining how bodies and brains operate*. And, in saying this, P1 goes beyond what can be established by a word count. That physiologists and neurophysiologists speak of normal and abnormal functioning is clear, but why do they? To assess P1, a metascientific analysis of the general nature of the explanations that physiologists and neurophysiologists give when explaining how bodies and brains operate or function is needed. There will not be space for a full discussion of this here, but a few of the main points can be made.

To discuss this, we also need to take note of a second notion of function, variously called a "systemic function" or "Cummins function" (or sometimes a "causal-role function"). This is defined by a systemic capacity theory, the classic version of which is Robert Cummins' (1975).[5] On Cummins' original analysis, a function of a component in a system is its contribution to whatever complexly achieved (Z^*) capacity of the system happens to be under analysis. Cummins tells us that a component X of a system S "functions as a Z in S (or: the function of X in S is to Z) relative to an analytical account A of S's capacity to Z^* just in case X is capable of Z-ing in S and A appropriately and adequately accounts for S's capacity to Z^* by, in part, appealing to the capacity of

---

[4] Dayal et al. (2008).   [5] See also Craver (2001).

X to Z in S" (Cummins 1975, p. 762).[6] Let's call these functions, which are defined by Cummins' original analysis, "Cummins functions."

According to Cummins, it is *these* Cummins functions, and not Wright-style selected functions, to which contemporary biologists advert. And numerous philosophers of biology have agreed with Cummins about this, at least with respect to the biologists' answers to "how" questions (such as "how does the human visual system enable vision?" or "how are our diurnal cycles controlled?").[7] Yet everyone will also agree that, at least as Cummins originally defines them, such functions are not normal-proper functions. They do not, for instance, respect the usual function versus dysfunction distinction. Cummins (1975, p. 757) comments that "[i]f the function of something in a system S is to pump, then it must be capable of pumping in S" (and this is indeed entailed by his analysis). This precludes the possibility of token malfunction (let alone typical malfunction). In order for malfunction to be possible, it must be possible for something to have a normal-proper function that it lacks the capacity to perform, even in the right circumstances. Furthermore, components will sometimes have Cummins functions to malfunction. Because Z^* is determined by researcher interests, and physiologists are interested in explaining complex pathologic processes, such as the growth of tumors, components will have Cummins functions to contribute to pathologic processes (if they contribute to them) when pathologic processes are under analysis. Are we to conclude that physiologists and neurophysiologists advert to normal-proper functions for some reason or other yet to be determined, but not in explaining *how* complex organic systems operate?

The etiological and systemic capacity theories of function are sometimes presented as having quite different meta-analytic aims. For instance, Paul Sheldon Davies says, "The theories appear to have distinct explanatory aims. While selected functions explain the persistence and proliferation of a trait in a population, systemic functions explain how a system exercises some capacity" (Davies, 2001, p. 28). (Davies demurs from this division of explanatory labor to dispute whether selected functions are needed at all.) Along the same lines, Phillipe Huneman (2013, p. 2) says that supporters of the etiological

---

[6] Cummins' symbols are changed for convenience.
[7] See Godfrey-Smith (1993) and Millikan (1989) in contrast to Millikan (2002).

and systemic capacity theory "[b]oth acknowledge that 'function' is a concept used in some explanations, but they diverge from the first step because the etiological account thinks that the function of X being Y explains the presence of X whereas, for the causal role theorist, the function of X being Y explains or contributes to an explanation of the general proper activity of a system which includes X."

Of course, there is no obvious inconsistency in holding (1) that physiologists and neurophysiologists advert to normal-proper functions, but (2) they do not advert to them in explaining how complex organic systems operate. However, a careful look at the science reveals that biologists often do advert to normal-proper functions in this very context. Huneman's reference to the "general proper" activity of a system hints at why they do. If physiologists explain the "general proper" activity of complex organic systems, *some* notion of "proper" activity is involved.

Here is what I think: Cummins is right that, in explaining the operation of a complex organic system, biologists conceptually decompose the system into its components, at multiple levels of analysis, and describe how the system's complexly achieved capacities are produced by the diverse contributions that each of the diverse components make to them. He is also right that much of the work of the physiologists and neurophysiologists is aimed at explaining the *actual* functioning of living systems. Much of the work is experimental and devoted to understanding the actual activities of one or a few components in one or a few individuals. However, physiologists and neurophysiologists are also, as a collective, thereby contributing to the development of descriptions of so-called normal systems, which among other things involves descriptions of their functioning when they are functioning properly. For example, physiologists try to explain the functioning of the normal human immune system, and neurophysiologists try to explain the functioning of the normal human visual system. This collective enterprise involves the notion of normal-proper function(ing). (Note that I do not say that it *only* involves this notion. Of course, many other notions are used too, including, perhaps, other notions of function.) Research reports to physiology and neurophysiology journals are routinely framed in terms of reports about what occurs in normal or abnormal systems.

Some nonbiological sciences give multilevel componential analyses of recurring processes. Cosmologists, for instance, give multilevel componential analyses of the formation of planetary systems. They also may use statistics to generalize, describing the general causal roles of components as small as subatomic particles or as large as stars. However, adding the use of statistics to a multi-level componential analysis is not the same as adding the use of a notion of normal-proper function to it. Cosmologists do not ascribe malfunction-permitting functions to the components that typically contribute to planetary formation. Why do physiologists and neurophysiologists ascribe malfunction-permitting functions?

Scientific idealization is sometimes used as a means of simplification, in response to the complexity of what is being described. And describing the operation of a single cell or somatic system, let alone a single human brain, is a daunting challenge, owing to the sheer complexity of what is to be described. However, the idealization involved in describing the normal-proper function(ing) of a living system does not seem to serve the purpose of simplification, since the normal system can be at least as complex as an abnormal one. It is hard to describe the purpose of the idealization much more fully without taking sides on the issue of how normal-proper function is to be understood. But its purpose, as I see it, is in part to *generalize* in the face of the vast deal of variation in actual functioning within a species, and in part to provide a description of *organized* complexity, organized complexity being the kind that results from the coadaptation of components to each other as well as to the external environment.[8]

If physiologists and neurophysiologists use a notion of selected function for this purpose, the normal system that they try to describe is (among other things) one in which each component that was selected to do something is able to do (is disposed to do) what it was selected to do. But alternative philosophical accounts of normal-proper function can also try to capture the idea that the description of normal-proper functioning plays a useful idealizing role in explaining *how* complex organic systems operate.

Neither physiology nor neurophysiology is a one-concept science or a science with a single goal to which the relevant scientists are all single-mindedly and solely dedicated. So, in claiming that

---

[8] I describe this idealization in more detail in Neander (2015). For the point that normal-proper function serves this theoretical purpose, see also Boorse (1977), Neander (1991), Millikan (2002), Garson (2013), and Brandon (2013).

physiologists and neurophysiologists seek to describe how systems function when they function normally or properly, I am not making an exclusionary claim. Physiologists and neurophysiologists also want to explain pathologic processes (diabetes and Alzheimer's, for example). Plus, as indicated earlier, what is "general" and "proper" can pull apart. Some general activity that is not "proper" (at least on an etiological theory) is still described in creating the composite portrait of a species. For instance, it is useful to describe adaptive and maladaptive consequences of features that are hard to change owing to architectural or developmental constraints. And what is "proper" is not always general (at least on an etiological theory). Selection does tend to drive a great many adaptive traits to fixation, but there are also different ways to be normal (there are different sexes, developmental stages, local adaptations, and alternatives with frequency-dependent fitnesses and so on). Plus there are components that have the normal-proper function to adapt individuals within their lifetime to their peculiar circumstances (some involving ontogenetic selection processes, such as the antibody selection involved in immunity and the neural selection involved in learning). So even normal-proper functioning can in some respects be idiosyncratic, but describing ontogenetic adaptive processes (immune processes, learning processes, memory processes, and so on) is consistent with the aim of generalization.

To reiterate, P1 claims that physiologists and neurophysiologists use *a* notion of normal-proper function *in explaining how bodies and brains operate*. It leaves it open how the notion of normal-proper function is best understood. Someone could agree with P1 and support the etiological account of normal-proper function. Or someone could agree with P1 and instead support the idea that normal-proper functions are *modified* Cummins functions (perhaps along the lines of the friendly amendment suggested by Philip Kitcher [1993]). Or that someone could instead support some other account of normal-proper function, such as Christopher Boorse's (1997, 2002) biostatistical and cybernetic account of normal-proper function.[9] P1 does take a stand on whether normal-proper function ascriptions play a role in the explanatory

---

[9] Nanay (2010) offers a radically different alternative, but see Neander and Rosenberg's (2012) response.

descriptions physiologists and neurophysiologists give when explaining how bodies and brains operate, though it leaves it open precisely why they do. In my view, while there is some room for sensible debate over the details, P1 is true, if not obviously true.

Premises 2 and 3

P2 says that the brain's functions include cognitive functions. It does not say that every function of the brain is cognitive, that only brains perform cognitive functions, or that everything that a brain does is a function of it. One must hold a very nonstandard view of cognition to deny P2, at least with respect to the human brain. The "methodological argument" argues that informational teleosemantics is implicitly supported by the mainstream branches of the mind and brain sciences as they are now. It makes no claim about the support that the ideal, completed mind and brain sciences would provide. So any support that it provides for informational teleosemantics is anyway conditional on a very radical view of cognition (say, a Cartesian dualist, ontological behaviorist, or ultraradical version of an embodied cognition theory) not being true.

Paradigmatically, decision making, memory, learning, and deliberative reasoning are counted as cognitive processes, but perception and motor control are here counted as cognition, too. This is cognition in the liberal sense of "cognition." At least some aspects of motivation (and, arguably, of emotion) also count. But the present argument does not turn on precisely which processes count as cognitive. If someone wanted to object to this liberal use of "cognition," the argument would need to be reworded if it were to accommodate them, but this would only be a terminological alteration.

P3 says that the same notion of normal-proper function as was mentioned in P1 is central to the functional analyses of cognition that cognitive scientists currently provide. I believe that this is true. The way in which cognitive neuropsychologists, for instance, speak of neurologic impairment seems the same as the way in which they speak of the cognitive impairments that result. That is, they seem to use "functional impairment" in the same sense in both cases. It is hard to argue for this other than by an extensive linguistic analysis of the scientific literature (which I won't take on here), but I know of no good argument to the contrary.

Here is one argument in its favor: while cognitive science is not
neuroscience, the difference in the main aims of the two scientific fields
is one of emphasis rather than one of a sharp departure. The central aim
of cognitive scientists is to understand the normal information processing that is supported by the neural substrate. The central aim of the
neuroscientists is to try to understand the normal neural substrate that
supports the information processing. (By "neural substrate," I mean,
more precisely, the neural-plus substrate.[10] I do not mean to suggest
that the relevant substrate consists exclusively of neurons.) If the difference is, as just described, one of emphasis, the cognitive scientists
and the neuroscientists are working on the same overall functional
analysis of our cognitive capacities and their substrates, even though
they are working on different levels or aspects of this analysis or at least
with different emphases. If it is the same overall functional analysis and
seen by the practitioners as being the same, then the notion of normal-
proper function that is central to the analysis of the normal cognitive
system presumably will be the same as the one that is central in the
analysis of the brain (or the situated, embodied, evolved brain, if
speaking of the brain alone is too narrow).

This seems quite right to me. But I suppose that someone might think
that the autonomy of psychology or, alternatively, the uniqueness of
each adult human cognitive system could undermine this argument.
If so, they might also think that it could undermine P3.

In a useful review of the literature on the autonomy of cognitive
science, Victoria McGeer (2007) classifies some people as ultra pro-
autonomy and some people as ultra-ultra pro-autonomy, and it might
help to consider the views of the "ultra" and "ultra ultra" people with
this issue in mind.

McGeer classifies cognitive scientist Alfonso Caramuzza as "ultra
pro-autonomy." Caramuzza defends the investigative methods of his
field, cognitive neuropsychology, as allowing investigation into cognitive systems to proceed independently of neuroscience.[11] He argues
that inferences to the structure of normal cognitive systems can be
made from single-subject studies of functional impairment.[12]
Caramuzza argues that an accumulation of such single-subject studies

[10] I borrow the useful term "neural-plus" from Felipe de Brigard.
[11] See especially Caramuzza (1986).
[12] See also Caramuzza and Coltheart (2006). And, for an interesting example of
such a single-subject study, see McCloskey (2009).

can provide strong evidence concerning how cognitive capacities dissociate and so strong evidence concerning how cognitive capacities normally combine and interact, even in the absence of an understanding of the substrate that implements them. So, he claims, cognitive neuropsychology can proceed without any help from neuroscience. But Caramuzza (1992, p. 85) also voices the sensible opinion that there will be a "co-evolution of cognitive science and neuroscience, moved forward by multiple cross-adjustments at the level of results and theory." Note that a notion of normal system and a distinction between normal functioning and functional impairment are central to Caramuzza's research strategy.[13] He also allows that the findings of neuroscience may inform and constrain cognitive science and vice versa. His view is perfectly consistent with the view that the functional analyses to which the two sciences are working are ultimately to be integrated.

Max Coltheart, whom McGeer classifies as "ultra-ultra pro-autonomy," expresses a more extreme view in a passage (in which he describes himself as "ultra"). He says:

No amount of knowledge about the hardware of a computer will tell you anything serious about the nature of the software that computer runs. In the same way, no facts about the activity of the brain could be used to confirm or refute some information-processing model of cognition. This is why the ultra-cognitive-neuropsychologist's answer to the question, "Should there be any 'neuro' in cognitive-neuropsychology?" is "Certainly not; what would be the point?" (Coltheart 2004, p. 22)

This echoes the claims made by the "machine functionalists" of the 1960s and 1970s in ways that I would want to reject.[14] But, in any case, Coltheart invokes the usual notion of a "normal" system. For instance,

---

[13] A word count reveals that the word "normal" is used thirty-nine times in Caramuzza (1986). Although I did not try to count how many times it is used in the special sense of normal versus abnormal functioning, in the same work, Caramuzza also speaks of lesions, impaired performance, deviant performance, brain-damaged patients, and so on.

[14] Coltheart's comments invite us to picture the mind and its substrate as a two-tiered system, with the mind as the software and the substrate as the hardware. The idea of two distinct tiers seems hard to sustain given that learning and memory can involve more or less persistent changes in anatomic structures in the brain. See Squire and Kandel (2003). It also seems to require more plasticity and less functional specialization than the evidence seems to suggest.

when he describes the respective rationales for cognitive psychology and cognitive neuropsychology, Coltheart (2004, p. 21) says:

The aim of cognitive psychology is to learn more about the mental information-processing systems that people use to carry out various cognitive activities. Some cognitive psychologists do that by studying the performance of people whose cognitive processing systems are normal. Others do it by studying people in whom some cognitive processing system is abnormal: Such investigators are the [cognitive] neuropsychologists.

Nothing that Coltheart says (as far as I can see) contradicts the claim that he uses the same notion of normal and abnormal systems, whether he is discussing neurologic or cognitive normal systems or abnormalities.

I do not know what proportion of cognitive scientists today would support the autonomy of psychology. But those who do seem, as McGeer notes, to be supporting only *procedural* autonomy (complete procedural autonomy, in the case of ultra-ultra pro-autonomy people and partial in the case of the ultra pro-autonomy people). Procedural autonomy is perfectly consistent with the neuroscientists and the cognitive scientists each working – more or less independently – on different aspects of the same overall functional analysis of the same system.

If Coltheart"s ultra-ultra pro-autonomy stance were right, cognitive science would have no need of the "neuro" in "neuropsychology." His stance is too extreme, in my view. But even if it were right, neuroscience still would need the "psychology" in "neuropsychology." To try to completely explain the functioning of the human brain and nervous system without explaining cognition would be like trying to completely explain the functioning of the human immune system without explaining how it defends against disease.[15] It simply cannot be done given that the brain's functions include cognitive functions. So, for this and the previously mentioned reasons, I conclude that P3 is true. The same notion of normal-proper function as is used in physiology and neurophysiology, in explaining how bodies and brains operate, is also used in explaining the functioning of cognitive systems.

Turning now to the question of whether adult human cognition is so unique to each individual that P3 is undermined, it must be allowed that the description of the "general proper" activity of the brain, with

---

[15]   Figdor (2010) argues that neither multiple realization nor multiple realizability would entail the autonomy of psychology, and in my view, this is right.

respect to cognitive functions, might encounter special difficulties. The brain is a complex of mechanisms cobbled together and coadapted by phylogenetic selection processes, at least as a first approximation, but it is also a system that has been adapted by these processes to further adapt itself to the creature's individual environment over the shorter term in the case of memory and learning and over the even shorter term in the case of perception and the control of behavioral responses. Someone might think that at least *adult human* cognition is, as a result, so unique to each individual that talk of the normal-proper functioning of an adult human cognitive system is not useful. Clearly, there is much truth to the claim that some degree of plasticity, much acquired culture, and much individual learning *are* the norm, and there is much truth to the claim that each adult human mind is unique. But each adult human body is probably unique too, as a result of different genomes and different environments and a lifetime of different interactions between them. What is more relevant, with respect to P3, is whether there is normal-proper functioning of cognitive systems, in our case endowing us with a special degree of plasticity and a special capacity for culture and learning. Or what is more relevant, more precisely, is whether cognitive scientists assume there is.

Cognitive scientists tend to assume, as a defeasible starting position, that human cognitive systems share a more or less universal species design. They speak of the normal human visual system and of certain normal processes in human memory, for example. However, the possibility of progress in cognitive science does not depend on there being a single uniform species design with respect to cognitive capacities. There is no single uniform species design in the somatic case either. It could conceivably turn out that the universality assumption is so great an oversimplification in the case of cognitive systems, or at least human cognitive systems, that not much real progress in cognitive science is possible because there are so few useful true generalizations to be made. Optimism concerning future progress in cognitive science is premised on this not turning out to be true. But future progress in cognitive science does not require a universal species design.

Besides which, P3 does not commit us to thinking that progress in cognitive science is possible. Rather, it speaks of explanations that cognitive science currently provides and it makes no predictions about the future.

Premises 4 and 5

This brings us to P4, which says that a key assumption in the main-stream branches of the cognitive sciences is that cognition involves information processing. Again, this is a descriptive and not a prescriptive claim, and as a descriptive claim, it should not be controversial. Cognitive scientists standardly posit the transduction of information at the sensory receptors, the carrying of information by subsequent signals, the processing of information, the storing and retrieving of information, the use of information to enable adaptive actions, and so on and so forth.

According to P5, there is a natural, factive notion of information involved in this. There are (at least) two senses of "information." When we say that Lizzie misinformed the police about the location of the diamonds or that a government ran a misinformation campaign, the informational content of which we speak is representational content. In this sense of "information," misinformation is misrepresentation.[16] Cognitive scientists sometimes use "information" in this sense. However, they are also thought to use, and I believe that they generally see themselves as using, "information" in a second sense, too. In this second sense, information is akin to Gricean "natural meaning," which is attributed to natural signs.[17] It is *factive*. If the dark clouds on the horizon carry the information that a storm is on its way, then a storm must in fact be on its way. If Johnny's spots carry the information that he has the measles, then he must in fact have the measles.[18]

---

[16] Dretske (2008, p. 2) also denies that even the intentional notion of information permits misinformation in the following passage, when he remarks that "information, unlike meaning, has to be true. If nothing you are told about the trains is true, you haven't been given information about the trains. At best, you have been given misinformation, and misinformation is not a kind of information anymore than decoy ducks are a kind of duck." Dretske is denying that "mis" characterizes the information as false. Instead, he claims that the prefix is used to say, in effect, that the misinformation is not really information. If Dretske were right, this would not undermine the main argument here. It would only show that there is no sense of "informational content" that is synonymous with "representational content."

[17] See Grice (1989).

[18] The claim that cognitive scientists use a fully factive notion of information is vulnerable to challenge, given that the relevant concept of information has no agreed-on analysis. This complicates the present argument. For example, suppose that x is thought to carry the information that P just in case an occurrence of x makes P more probable. Then x could carry the information that P consistent

Because it is factive, there is no possibility of misinformation in this second sense of "information." If it turns out that no storm was on its way, then the dark clouds did not in fact carry the information that a storm was on its way. If Johnny does not in fact have the measles, then his spots did not carry the information that he had.

Most analyses of factive information try to analyze it in terms of causation, correlation, or conditional probability or the like. This contrasts with attempts to ground the meaning of words or traffic signs or sequences in Morse code, for example, in tacit social or legal or explicitly agreed on communication conventions. Factive information is in virtue of this said to be *natural* as well as factive.[19]

Unfortunately, there is no agreed-on analysis of natural, factive information. Like the notion of representation, the notion of natural, factive information is used as an explanatory primitive. It is used in explaining cognition, but its analysis is a metascientific project. While it is said to be "information theoretic," it is also well known that information theory as such does not provide an analysis of the relation in virtue of which one event or state of affairs (the sign) carries information about another (the signified). I am inclined to think that a simple causal analysis best suits the purposes of cognitive science and informational versions of teleosemantics. But I anyway wholeheartedly agree with Andrea Scarantino (2013, p. 64) when he says that "information is a mongrel concept comprising a variety of different phenomena under the same heading." As Scarantino adds, the best that one can do with a single analysis is to try to capture some interesting phenomenon invoked in some portion of its use.

Perhaps part of the point of using the term "information" in cognitive science is that it is an agreed-on term for a natural, factive relation between the inner and outer worlds, which components of the cognitive system are supposed to bring about, but whose precise nature is controversial. But the main point is to speak of how cognitive systems create inner natural signs – enduring or transitory – of variable features

---

with x occurring and yet not P. Further, this might seem to provide for the possibility of misinformation. So it might be argued that natural information is already nonfactive and misinformation permitting. I believe that this probabilistic analysis is too weak. However, I will not take the time or space needed to develop the argument against it here.

[19] In the tradition of Shannon and Weaver (1998).

of the environment and use them to modify the creature's responses in adaptive ways.

Despite the lack of an explicit and generally agreed-on analysis of what flows when information flows during the course of information processing, the information-processing paradigm remains dominant in cognitive science. Again, it is important to remember that P4 and P5 are intended to be descriptive, not prescriptive. Someone could agree with P4 and P5 and yet think that the information-processing paradigm should be replaced.

Premise 6

Finally, P6 says that cognitive science posits "normative aboutness," with the norms derived from normal-proper functions and the aboutness from the information involved.

Natural information has "aboutness" despite being factive. Dark clouds carry information *about* the coming of a storm and Johnny's spots carry information *about* his having measles. Along the same lines, activity in the visual cortex could carry information *about* the shapes, colors, textures, motions, and locations of visual targets; so-called traces could carry information about the past that allows memories to be reconstructed; and so on. Signs that carry natural, factive information can be said to have "informational content." By itself, this is not intentional or representational content.

Representational content is said to permit the possibility of misrepresentation, while mere natural, factive informational content does not by itself permit the possibility of misrepresentation. Not all mental representations can misrepresent, or not in every context of use. If I think, "Stop!" while I'm guzzling ice cream or listening to a speaker drone on, that injunction cannot misrepresent, although it can fail to be implemented. However, all mental representations contribute to satisfaction conditions. That is, representational mental states have truth conditions, correctness conditions, accuracy conditions, fulfillment conditions, implementation conditions, and so on, and the representational contents of the mental representations that are involved in these states contribute to these.

Representational content is said to be "normative" because it is (to adopt the usual shorthand way of putting it) misrepresentation permitting. Normal-proper function is also said to be "normative" because it is malfunction permitting. However, there is no intended implication that either the relevant semantic norms or the relevant functional norms are prescriptive as opposed to descriptive. What is meant is that there is correct representation and misrepresentation and there is proper functioning and malfunctioning. This is not true of natural, factive information or actual functioning as such.

However, wedded to the aboutness of factive information, in the cognitive scientists' explanations of cognitive capacities is the "normativity" associated with ascriptions of normal-proper function. Cognitive mechanisms are said to *have the (normal-proper) function to* do various things with information. They are said to have the function to transduce it, send it, carry it, process it, store it, retrieve it, and use it in various ways. And they can have the function to do so and yet can fail to do so properly, in the way that they are (so to speak) "supposed to."

It has quite frequently been alleged by supporters of teleosemantics who have rejected informational teleosemantics that normal-proper functions and natural, factive information are like oil and water and cannot be wed in this way. The short version of their claim is that the normal-proper functions of things concern their effects (their selected effects), while any information they carry concerns their causes, and so there cannot be functions to be caused in a certain way to do something. But this is simply a misunderstanding concerning the notion of function. Normal-proper functions necessarily involve effects, but they can involve causes too. There can be response functions, which are functions to respond to certain causes (to specific stimuli) by changing state in certain ways and thereby producing different effects in different situations.[20]

Thus, a kind of normative aboutness is born as a theoretical posit in the mind and brain sciences. This is the simple, elegant insight of Stampe and Dretske, although they delivered it without the long preamble. The insight is that normative aboutness – nature's way of making a mistake, as Dretske put it, is born of this marriage between the normal-proper functions ascribed to components of cognitive systems and the natural, factive information that some components are said to have the function to transduce, send, carry, process, store, retrieve, use, and so on.

[20] For extended discussion, see Neander (2013).

From Methodology to Metaphysics

The conclusion is that some version of informational teleosemantics, broadly conceived, is supported by the explanations of cognition that the mind and brain sciences currently provide. Broadly conceived, teleosemantics is the thesis that that mental reference to content is ontologically grounded in the normal-proper functions of components of cognitive systems. Broadly conceived, informational semantics is the thesis that mental reference to content is grounded in the natural, factive information that is processed by these systems. Informational teleosemantics, so conceived, puts these two claims together but makes no specific claims about how normal-proper functions or natural, factive information is to be analyzed. Their analysis is a metascientific project, not a scientific project (although scientists can, of course, participate in it).

However, informational teleosemantics constrains the analyses of normal-proper function and natural, factive information. It won't do, for example, to claim that functions are ontologically grounded in the explanatory aims of researchers (as Cummins does) and to then turn around and explain intentional mental phenomena (such as the explanatory aims of researchers) as grounded in such functions. It won't do, at any rate, while there is still hope for a naturalistic explanation of intentional mental phenomena, a hope that is not to be abandoned lightly.

This reminds me of the saying that one person's *modus ponens* is another's *modus tollens*. If we have grounds for thinking that informational teleosemantics is true (or is anyway supported by the current science) and expect intentionality is to be explained in nonintentional terms, the relevant notions of normal-proper function and natural, factive information will be expected to have nonintentional analyses. However, someone else might argue from the opposite direction, that we have independent grounds for thinking that one or both of the relevant notions are intentionally laden (e.g. must be analyzed in terms of a researcher's explanatory aims) and therefore that we have reason to believe that informational teleosemantics is not true or that intentionality cannot be analyzed in nonintentional terms. Depending on where different philosophers begin, they tend to take different paths. But perhaps the methodological argument will help to tip the scales for those who have not yet chosen either path.

Short of a fairly radical revision of the way that physiologists, neurophysiologists, and cognitive scientists give idealized descriptions of how complex systems operate or of the information-processing approach to explaining cognitive capacities, some sort of informational teleosemantics enjoys the implicit support of the sciences most nearly concerned. The sciences most nearly concerned with explaining cognition posit (natural, factive) information-processing (normal-proper) functions. In doing so, they posit normative aboutness, where the aboutness comes from the information and the normativity from the functions. At a minimum, the argument shows that informational teleosemantics, broadly conceived, is a highly conservative thesis. In invoking the notion of natural, factive information and the notion of normal-proper function, and in bringing them together to explain an aspect of cognition, informational teleosemantics is only invoking what is already invoked and is only bringing together what is already wed by the sciences devoted to explaining cognitive capacities.

Of course, this does not prove that informational teleosemantics is true. The support provided for informational teleosemantics is conditional and defeasible. It is conditional because teleosemantics is a theory about the real nature of mental reference to content, and the mainstream branches of neurophysiology and cognitive science might be on the wrong track in relevant respects. Perhaps they will be revised or are already being revised in relevant respects. I've not tried to predict the future of the mind and brain sciences or tried to arbitrate in already existing disputes. Nor have I tried to clarify the role of the notion of normal-proper function or the notion of natural, factive information in the less mainstream approaches to explaining cognition that already have some support today. But the methodological argument does not rest on either, especially unstable or noncentral aspects of the mind and brain sciences, relatively speaking. Both are central and well established.

The support for informational teleosemantics is also defeasible. If one looks at how the relevant sciences explain cognition, an informational version of teleosemantics seems to hang from its branches, ripe for the picking. I think that in some vague sense of "we," we philosophers should not let it go to waste. Those who want to understand how meaning emerges in a fundamentally meaningless world ought to take note. *Considered as a theoretical posit,* this is how normative aboutness arises, from the perspective of the mainstream branches of the sciences most concerned with explaining cognitive capacities. This places us under some obligation to take a good, long, hard look at informational teleosemantics. It would be stupid if, as a profession, we did not seriously consider the possibility. But, of course, being philosophers, we also should check the fruit for hidden worms. And I fully appreciate that many think that they have already found worms that make the fruit unpalatable. The methodological argument does not show that informational teleosemantics is problem-free. It is even consistent with it being hopelessly problematic. But the methodological argument supports a certain amount of stubborn and optimistic perseverance in seeking any solutions that it might have.

And even if informational teleosemantics were true, its scope would still be up for debate. Perhaps it most plausibly applies to what some like to think of as unconscious neural representations as opposed to conscious mental representations, or to nonconceptual representations as opposed to conceptual representations, or to the kinds of mental representations that infants and nonlinguistic creatures can possess but not, except as a kind of more primitive underpinning, to the mental representations that are special to linguistically and culturally endowed adult humans.

Concluding Remarks

I hope that this attempt to make the methodological argument for informational teleosemantics explicit casts light on the motivation behind attempts to develop the best possible version of it. Different philosophers will favor different scientific approaches to understanding the mind. Certainly, some will reject the scientific approach here described as "mainstream." Some will think that it is already "old hat" or will prefer approaches that I think have already been rejected for good reason (such as overly behaviorist or Gibsonian approaches). But, whichever approach is accepted or rejected, the science and the philosophy cannot be completely decoupled if the content ascriptions that a philosophical theory of content generates are to be relevant to scientific explanations of cognition. If the more mainstream branches of the relevant sciences ascribe normal-proper functions to cognitive mechanisms to carry, store, and use natural, factive information, then it makes perfect sense to try to understand how far these already posited information-related functions will take us in understanding mental reference to content.

7 | Nature's Purposes and Mine

RONALD DE SOUSA

Whether or not we find what we are seeking
  Is idle, biologically speaking.

> – *Edna St. Vincent Millay*

No thank you. I don't think nature intended us
  to drink while flying.

> – *Passenger refusing a drink in Gardner Rea cartoon*

"What is by nature proper to each thing," wrote Aristotle, "will be at once the best and the most pleasant for it" (1984b, pp. 6–7). This chapter may be described as a meditation on the question of what can be made of Aristotle's sunny optimism in a post-Darwinian age.

Aristotle's maxim immediately raises four questions. First, given that philosophers have long attempted to elucidate ways in which humans *transcend* nature, what does it mean to say that anything is "by nature proper" to us? Second, talk of bitter medicine and mottoes such as "no pain, no gain" suggests that Aristotle is here at odds with common sense. Why should what is best be expected to be also most pleasant? Third, best and most pleasant for whom? Because we are social beings, as Aristotle himself famously stressed, should the maxim not be supplemented by a reminder that what is best and more pleasant for you might be neither for others? Even if strictly correct, the maxim would be of limited use to one who wishes to be a good citizen as well as a happy child of nature. And fourth, just what is the relevant sense of "a thing" in the maxim as we might now understand it? We are composites of living parts, and controversy has raged over the question of what "thing" we should be talking about when we discuss what is "best": species, populations, groups, individuals, cells, genes, even mitochondria, and "all of the above" have been candidates for the role of beneficiaries or "units" (whether these be equivalent or not – also matter of dispute) of natural selection.

I am grateful to this volume's editor and to Christopher Clarke for excellent comments on previous versions of this chapter.

141

All these questions are pertinent to what follows, if only implicitly, but I will be more narrowly concerned with the questions of how biological knowledge can have a bearing on our philosophical conception of ourselves as human beings. This question can be regarded from a metaphysical point of view and from the point of view of ethics broadly conceived. I care mostly about the latter, for although a philosophy of the human person can't avoid being metaphysical, I am not interested in the sort of metaphysics that has no conceivable relevance to how we should live. Barring speculation about the possibility that quantum effects in microtubules might enable and explain free will (Penrose 1994), for example, it is unlikely that facts about the inner constitution of the atoms of which we are made will have philosophically interesting consequences. There is a slightly higher likelihood of philosophical payoff in the fact that we are made of cells, the ancestors of which lived solitary lives for one or two billion years before teaming up to form multicellular organisms. More clearly pertinent is the scientific refutation of the popular belief that individual consciousness will survive the annihilation of the brain, though what we should infer from this about how to live is less obvious. The philosophical tradition contains a wide variety of possible attitudes to mortality ranging from a cheerful endorsement of Epicurean *carpe diem* to nihilism about value. At the end of this chapter I shall ask some concrete questions and hazard controversial answers concerning what biology and psychology might teach us about our traditional ideologies of love and sex. But I begin with an old controversy concerning the very idea of inferring anything about value from natural facts.

Nature and the Naturalistic Fallacy

Many of us were brought up to think that there is something called a "fact-value" or "is-ought gap" and that any attempt to bridge this gap commits the "naturalistic fallacy." While it would be tedious to go over the debates that have swirled about this claim, it is worth noting that the existence of such a fallacy would entail that no justification of ethics is possible. To see why, consider the ambiguity of the word "nature" as neatly encapsulated by J. S. Mill:

In the first meaning, Nature is a collective name for everything which is. In the second, it is a name for everything that is of itself, without human intervention ... while human action cannot help conforming to nature in one meaning of the term, the very aim and object of action is to alter and improve nature in the other meaning. (Mill 1874, p. 12)

In the sense in which "we cannot help conforming to nature," the reference is simply to the totality of facts about the actual world. That (call it N_1) includes everything that humans bring about. Mill's second sense (call it N_2) is the status quo, which it is the aim of any action to modify. The difference between N_1 and N_2 is the sum of everything that we actually do. Call it A for "action." Some members of A are things we ought to do. Other members of A are things we should have refrained from doing, and still others are deontologically and axiologically indifferent. Because values can conflict, a single action, event, or situation might be positioned differently on different evaluative dimensions. Recall E. M. Forster's famous remark, "If I had to choose between betraying my country and betraying my friend, I hope I should have the guts to betray my country" (Forster 1951, p. 68). Caring for one's friends and caring for one's country represent different values, differently correlated with other scales of value, depending on the priority accorded to individuals or community. But what justifies such judgments? Where should we seek *reasons* for claims about values?

By hypothesis, N_1 refers to all actual facts. N_2, by contrast, lists only facts that existed before we acted, as well as counterfactual possibilities that would have been actual had we not acted. If no normative statement can be derived from any statement of fact, then no normative proposition can figure in either N_1 or N_2.[1] So what, if not a fact (for all facts are contained in the union of N_2 and N_1), can constitute a *reason* to justify a normative claim? If that reason cannot consist of any *facts*, must it consist of some nonfact?

Unless the question is rhetorical, we presumably have in mind something other than mere falsehoods (although many moral precepts may well rest on nonfacts in precisely that sense – nonfacts about God and his commands, for example). What other nonfacts could there be?

A vigorous philosophical tradition, going back to Hume's distinction between matters of facts and relations of ideas (Hume 1975),

---

[1] For the purposes of this chapter, I make no distinction between norms, "oughts," and values. "Normative" is used generically for anything that faces facts across the alleged gap.

distinguishes empirical facts, discoverable only by experience, from *a priori* truths, which owe their status to logic or meaning alone. But logical or analytic truths, even though they are not always transparent to intuition, are unlikely to entail claims about the desirability of some ways of life over others or about the rightness of some actions and the wrongness of others.

In short, no normative statements can be justified at all unless we relax the constraints on the range of statements admissible in their support. One way to relax those constraints is to select a normative major premise so mild that it might command universal assent. In past ages when all could take religious belief for granted, the precept that one ought to follow God's command might serve, though securing agreement on the content of such commands was another matter. In a postreligious age such as we might optimistically assume ourselves to have reached, by contrast, basic facts about us, such as biology might disclose, might constitute the privileged class of facts apt to provide guidance about how to live.

But among the myriad facts of biology, which are we to select for inclusion in that privileged class? What biology teaches about the sorts of beasts we are can be viewed in either a minimalist or in an expansive mode. A minimalist interpretation would collect only facts about what is possible. A person might run a mile in four minutes, but no one can leap unaided over tall buildings. Morality can neither require nor forbid the impossible, and if we are to get guidance from natural facts about what is possible, these will have to be characterized more expansively as not only possible but also more or less conducive to a worthwhile life. If we all agreed on what counts as a worthwhile life, we might hope to find novel and relevant knowledge in evolutionary theory, psychology, and brain science.

There are many working illustrations of how useful such knowledge could be if one could only persuade politicians to take it into account. Recent books by Patricia Churchland (2012) and Sam Harris (2011) have attempted to do just that. Both have been accused of attempting to leap across the fact-value gap, oblivious to the philosophers standing guard to stop them. But if we grant a broad consensus on certain basic values, such as autonomy, happiness, and the development of capabilities conducive to the realization of these values, biological and social sciences offer much information to improve the lot of human beings (Nussbaum 2000). There is increasingly compelling evidence, for example, that poverty is bad for your health, and extreme inequality is correlated with a slew of other social ills (Wilkinson and Pickett 2010; Atkinson 2015). From a philosophical point of view, however, any argument premised on facts such as these remains an enthymeme, and its silent evaluative major premise is of just the sort that antinaturalists reject, namely, that human thriving and happiness are inherently good and that pain and unjustified coercion are inherently bad. We must either renounce the enterprise of justifying any statement of value or else relax the strictures imposed by the nonnaturalist principle that bans any inference from fact to norm or value.

Relaxing the Prohibition Against Naturalism

How would such a relaxation work? I see two ways, based on different principles for selecting a privileged class of facts that straddle the fact-value distinction. The first treats values as response-dependent properties and looks for the privileged facts among human emotional responses. The second, which has a much longer pedigree, privileges certain facts about nature as representing not merely what happens but what is *supposed* to happen.

On the first option, the values of existing things in the world are something like "Cambridge properties," not inhering in the world but derived or projected from properties inhering in something else, namely, human responses. These are, of course, facts about human beings, but on this view they do not presuppose the independent objective reality of value. Hence they can count as reasons for judgments of value. The appeal to emotional responses illustrates the subjectivist response to the question raised in Plato's *Euthyphro*, whether we prefer things because they are inherently good or whether good is so-called because it is preferred. Variants of this proposal have come to be known as "sentimentalism" (Kauppinen 2014). There are two things to note about it. First, while the privilege accorded to actual emotional responses is a form of relativism, it is not incompatible with the objectivity of the value properties in question. This is so for two reasons. First, although the responses that constitute the privileged class of facts are subjective states, the fact that they occur is an objective fact about observers. Their occurrence can be assessed from an axiological point of view. Second, on the model of Locke's view of the relationship between secondary qualities and the primary qualities that underlie them, we can postulate some inherent properties of the world that normally give rise to the responses in question. Those too are objective, but they are not inherently value laden. To be yellow is not to have a determinate property defined in terms of any specific light frequency but to have the capacity to produce, in normal viewers under normal circumstances, the impression of yellow. Yellowness supervenes on objective properties that are not identical to it. Because circumstances can be abnormal, this allows for mistakes and illusions. Similarly, the view that ethical properties are response dependent allows us to regard them as both relative and objective.

But relative to what? The question leads to a second way of relaxing the nonnaturalists' strictures. This is the principle of "natural law," which goes back to Aristotle and Aquinas and still forms the basis of most of the edicts that come out of the Vatican. It is also advocated, among contemporary philosophers, by "virtue theorists" (Hursthouse 1998). Virtue theory posits a substantive equation between the good, the pleasant, and the thriving in the spirit of Aristotle's observation with which I began. Although is not clear whether virtue theory requires us to believe in objective, human-independent moral truths, it does seem committed to the existence of a universal human nature.

But how can we discover what human nature, in the relevant sense, actually consists of? The answer to this question, which constitutes the key move of natural law theory, in effect promotes statistical norms to a normative status, on the basis of Aristotle's criterion that what happens "always or for the most part" is what nature intends (Aristotle 1984a, Met. 1027a20). The strategy is bait and switch, playing with the ambiguities in both the words "nature" and "law." It relies for its normative force on making sense of the idea that not everything in the set of facts N_1 is good: certain things that actually occur in nature are deemed *unnatural*, aberrations of nature rather than what nature "intended." The "bait" is the promise that nature itself will somehow reveal what it "intends," allowing us to uncover its laws in the sense in which that term is understood in science. The "switch" occurs when encountering exceptions to the alleged law: instead of regarding these as falsifications of a hypothesis, the natural law theorist condemns them as normatively unacceptable on the basis of their incompatibility with that "law" – thus begging the question by switching from the scientific to the legislative use of the word.

Despite its theological component, Aquinas's modification of Aristotle's scheme is more congenial to the modern mind that Aristotle's own. Aristotle thought that teleology was inherent in nature, without any need for an intelligent intention to explain it. Whether this applies to nature as a whole is controversial (Broadie 2007), but it certainly applies at the level of individual organisms, regarded as members of a species with a fixed nature. If teleology is inherent in nature itself, then we should be able to derive at least some normative statements from those natural teleological facts. But despite recent attempts by Thomas Nagel to resuscitate the notion of a natural teleology without intelligent design (Nagel 2012), that idea now strikes most of us as unintelligible. So the assimilation of the designs of nature to the purposes of God makes it easier to accept that nature actually has intentions. Its drawback is that it requires those not privileged to read God's mind to decipher those intentions from the empirical facts around us.

As Ruth Millikan (1984, 1993) and others have shown, the concept of objective teleology – independent of human interests and purposes – does not require intelligent design after all. Natural functions can be identified with those effects of an organ's activity that resulted in its being selected for and hence explain its present existence. Though refinements and objections have not been lacking (Allen, Bekoff, and Lauder 1998), I venture to think that the etiological explication of natural function marks one of the few genuine advances in philosophy in the past hundred years. But it does not answer the crucial question of what natural functions we should endorse as valuable and which ones we should regard – in the words of Katherine Hepburn's character in the movie *African Queen* – as "what we are placed in the world to rise above." Or rather, it does answer the question for natural law theorists. But it does so quite arbitrarily, in much the way that self-proclaimed biblical literalists interpret some pronouncements as the word of God, such as the prohibition of homosexuality, while dismissing others – such as the permissibility of selling your daughter into slavery – as reflecting mere accidents of history.

We may come to think better of natural law theory on the day that the Vatican reverses its ban on homosexuality after noticing the existence of gay penguins, but even that policy, taken to its logical conclusion, would lead only to the equally rebarbative endorsement of the doctrines of the Marquis de Sade. For the "divine Marquis" proved himself to be the only consistent naturalist philosopher by scrupulously following every natural inclination (Sade 1810). Just as we cannot avoid making choices among different elements of N_1, so we cannot evade specifying the criteria on which such choices are made.

The fact that some process serves an objective function does not imply that we should value it. Conversely, the fact that a capacity lacks a natural function is no reason not to prize it. Consider male and female orgasms. While the male orgasm serves reproduction by the ejaculation of sperm, the female orgasm does not appear to have any clear function. Like male nipples, it arises primarily as a side effect of the homology between the penis and the clitoris (Lloyd 2005). One indication of this is that among some of our relatives, such as the rhesus monkey *Macaca mulatta*, female orgasm has been shown to be possible but extremely unlikely ever to occur in the wild (Burton 1971). This implies that it cannot have been visible to natural selection and hence cannot have been selected for. Unlike male nipples, however, female orgasm has value. Some of the best things in life are spandrels.

One last reason for dismissing natural law theory is that Aristotle's criterion makes sense only if species remain unchanged. Applied to evolving species, it entails that millions of our ancestors must be condemned as perverts. For among our ancestors, all those that came a step closer to being human necessarily were exceptions to whatever was true "always or for the most part" of their peers. Human beings are descended from millions of freaks. If all our ancestors had been normal, we would be unicellular organisms.

Evolutionary Ethics

Natural law theory is based on an expansive interpretation of the lessons of biology at the level of behavior. Expansive interpretations of biology on the scale of evolution have not been lacking. Although philosophers and biologists generally regard the attribution of purpose to the universe as absurd, most laypeople regard evolution as a teleological process of ever-greater refinement and improvement, by which organisms got closer and closer to the ideal represented by the human species (or what the human species is *destined* to become). A few serious thinkers have also adopted this view and attempted to extract from the idea of evolution itself some sort of suggestive pattern that we could then use as a guide to life. Some, like the Jesuit paleontologist Teilhard de Chardin (1961), have regarded the process of evolution as implementing a long-term design tending to ever greater complexity, destined to achieve ultimate perfection in an "omega point" featuring some sort of higher collective consciousness. Surprisingly, some, more recent thinkers have construed this as a prophetic foreshadowing of the Internet (Kreisberg 1995). Biologist Julian Huxley wrote an introduction to Teilhard de Chardin's book endorsing the general idea that evolution is bound to yield ever-greater complexity. (T. H. Huxley, "Darwin's bulldog," was more tough minded than his grandson Julian. He maintained that "the ethical progress of society depends, not on imitating the cosmic process, still less in running away from it, but in combating it" [Huxley and Huxley 1947].) Again, some have speculated that optimal body plans and human-like intelligence were destined to result from natural selection (Conway Morris 2003).

It is true that a random walk that begins at the lowest possible degree of complexity has only one direction in which to move – namely, away from the wall of zero complexity. But some have worried that the human genome has reached a stage where any further increase in complexity would incur "mutational meltdown": barring an increase in the already remarkable fidelity of DNA copying, disruptive mutations would claw back any further increase in complexity (Ridley 2000). There are also reasons to think that an unlimited increase in complexity may eventually issue in a formless chaos of maximum entropy; with all interesting patterns that include those that implement living things lying somewhere between the stillness equivalent to absolute zero and the "edge of chaos" (Langton 1992). Nevertheless, a number of people have quite recently continued to try to make good on the promise of grounding ethics in evolutionary theory in one way or another. This is attested by the contributions to a volume on the subject edited by Paul Thompson (1995). Thompson himself has proposed that we can define "evil" in evolutionary terms. According to Thomson, we can give the word a biological sense:

Evil ... is the attempt to enhance one's own individual fitness at the expense of the short- or long-term perpetuation of the population to which the individual belongs. That expense ultimately reduces one's own fitness since population collapse thwarts the perpetuation of that individual's lineage along with everyone else ... A framework of behaviors that is evolutionarily stable constitutes a viable, implicit social contract. The basis for this social contract arises from the essential feature of neo-Darwinian fitness – a propensity for self-preservation ... In cognitive agents this – in part – manifests itself as rational self-interest. The term "evil" simply designates behaviors that break the rules of the social contract, that is, that work against the maintenance of an evolutionary stable system ... behavior ... that, were it generalized, would reduce the long-term fitness of all members of the group (even the perpetrator of the evil). (Thompson 2002, p. 246)

It might be complained that his definition doesn't completely capture what the word means in ordinary language. But a good precedent exists for giving a common term a slight technical twist: the biological treatment of the word "altruism" is compatible with egoism in the common psychological sense (Sober and Wilson 1998). The fatal objections to Thompson's proposal are of a different sort.

Both pro-social behavior and long-term fitness – in the guise of individual human beings' interest in having progeny – are conveniently things that we generally tend to approve of. But both are, for interestingly different reasons, contestable.

First, while it is certain that we have innate dispositions compatible with the development of psychological altruism, our dispositions to antisocial behavior are no less natural. What commends pro-social behavior is not the fact that it has been favored by natural selection. Rather, it is the very fact that it is pro-social. A preference for nice people over nasty ones has no need of support from evolutionary theory. To suppose otherwise is to violate a sound methodological principle that enjoins us to avoid wheeling in dubious propositions such as "Evolution favors pro-social behavior" in support of perfectly obvious ones like "Nice is better than nasty."

As for the desirability of progeny, the fact that most people find it obvious does not insulate it from the charge of question begging. David Benatar, for one, has argued that having progeny is always immoral, on the grounds that never being born at all is better absolutely than even what we would, when alive, regard as a good life (Benatar 2006).

The verdict on evolutionary ethics, in short, is that its various versions are all unconvincing. Neither the frequency of a given behavior nor the detection of any trend or pattern in evolution would be sufficient reason to think it good. On the contrary, as T. H. Huxley suggested, we might have reason to "combat the cosmic process," however quixotically, in the name of some more important value. But where would such more important values come from, if not from our nature as humans?

The Multiplication of Possibilities

To answer the question in earnest, we should look beyond the limited range that has occupied me so far – the question of what might be inferred from natural facts about what it is to be human – and focus on the *possibilities* that are afforded us by the one capacity we do not share with other animals. I refer, of course, to our capacity for speech, which provides us with a virtually unlimited potential for generating new values. Once one goes beyond the minimalist view that is content with identifying impossibilities, one can begin to make distinctions between different kinds of possibility.

In the abstract, it may seem as if kinds of possibility are related like Russian dolls, of which each is contained by the last and contains the next. What is logically possible is compatible with the laws of logic. What is mathematically possible is logically possible but constrained by the laws of mathematics. The physically possible is constrained by all those but also by the laws of physics. Chemical possibility further constrains what is logically, mathematically, and physically possible. And, prima facie, we might think that biological possibility similarly constrains chemical possibility.

Unfortunately, the neatness is only apparent. There may not be such a thing as a biological law. Suggested examples, such as the Hardy-Weinberg law and Mendel's laws of inheritance, are either mathematical principles that happen to be applicable to some biological phenomena, or they are just not true. Or both. Biological possibility, I want to suggest, is constrained not by a further set of laws but by specific circumstances of the sort exemplified in Szathmáry and Maynard Smith's "major transitions" of evolution (Maynard Smith and Szathmáry 1999).

Examples of major transitions include the "invention" of prokaryotic and, later, of eukaryotic cells. Another key transition is from asexual to sexual reproduction. Here the reliability and stability of cloning is traded for a risky but potentially much more diverse exploration of radically new forms because sexual reproduction is really not "*re-production*" at all but radically new production in which every individual is novel. Again, with the coming together of unicellular organisms into cooperative systems, first in homogeneous temporary bodies such as cellular slime mold and then into stable metazoans, individual cells lose their autonomy, becoming confined to specific roles. They must submit to the drastic process of apoptosis for the sake of the collective organism. In exchange for the loss of cellular autonomy, the resulting organisms acquire a rich new range of possible forms, behaviors, and potential niches. Later still, something similar happened when individual organisms formed societies, whether on the model of eusocial insects or that of hypersocial humans.

Several of these transitions involve a tradeoff between novel constraints and an enlargement of the range of concrete possibilities. In the latest of Maynard Smith and Szathmáry's key transitions, from elementary signaling to language, the new possibilities come not so much with new constraints as with new dangers. Language exposes one to manipulation and triggers an arms race between deception and detection. But what is most remarkable is the explosion of possibilities that it affords. Through discussion, debate, and inference, language makes possible the creation and transformation of values. In this process of proliferation, some values come in conflict with nature's basic imperative of replication, such as when an individual sacrifices herself and her chance of progeny for the sake of some idea. The whole process has a good claim to be regarded as the specific human differentia, which the existence of language merely enables.

The proliferation of values, which relies essentially on our capacity to talk, to debate, to make correct or fallacious inferences, involves a process that leads us to respond emotionally to new possibilities. Our beliefs, our desires, and the very nature of our interpersonal relationships are no longer simply determined by the emotional predispositions that we have inherited from our mammalian ancestors. We transcend biology, but as Daniel Dennett has pointed out, "This fact does make us different, but it is itself a biological fact" (Dennett 2006, p. 4).

In short, we might conclude that the main philosophical implication of biology is that we should be existentialists. Insofar as it is a biological fact that we have crossed that threshold beyond which we are faced with an indefinitely large set of possibilities, there is a sense in which our existence precedes our essence both as a species and as individuals.

Why Natural Selection Is Not Providence

While it may indeed be a biological fact that we transcend biology, this doesn't mean that we are not subject to deplorable atavisms. Many of our emotional dispositions prepare us in often astonishingly subtle ways to respond efficiently to life's challenges; at the same time, however, they can constrain and hamper our choices. The more optimistic perspective tends to dominate in the popular mind, where evolution is often credited with having assumed the role of Providence. Although sadly negligent in some particulars, Providence was nevertheless trusted to have done most things for the best, and both science and philosophy have flirted with that Panglossian perspective. While emotions used to be regarded as inimical to reason, much interdisciplinary work now stresses their functionality. The rehabilitation of shame, for example, is well under way (Deonna, Rodogno, and Teroni 2011), and even what looks to be an unequivocally nasty emotion, spite – the desire to harm another at high cost to oneself – has recently been commended for having an important part to play in the evolution of fairness (Forber and Smead 2014). Spite is also closely related to altruistic punishment: a willingness to incur some cost on behalf of a social group in order to punish an offense against the group, even if the offense has not directly affected the punisher (Boyd and Richerson 2005). These are just different aspects of the much-studied disposition to altruism, the exact explanation of which is still highly controversial (Nowak, Tarnita, and Wilson 2010; Wilson 2015). Whatever emerges as the resolution of these controversies, they illustrate a number of ways in which the workings of natural selection, parti-cularly on our emotions, result in dispositions that we might have reason to deplore – and which undermine Aristotelian optimism. Here are three more examples.

McDonald's Emotions. McDonald's food is relished on first acquaintance by any child, from any culture, who might otherwise resist unfamiliar foods. Clearly, it is the food God or Nature intended for humans. The cravings that it satisfies originate in its provision of four nutrients that natural selection programmed us to seek when they were scarce: fat, sugar, salt, and protein. Our native emotional equip-ment has much the same problem: some of it, including perhaps dis-positions to rape and violence, probably spread genes for their own perpetuation. (It has been claimed that one man in 200 is descended from Genghis Khan.[2]) What was once adaptive may not be valued under now changed conditions.

Individuals Are Expendable. More generally, we have no reason to believe that evolution has any "interest," however metaphorically understood, in individual organisms, including humans. Individuals are just one way that replicators use to replicate, and the type of sexually reproducing individuals we are constitutes only a very small proportion of the biosphere (de Sousa 2005; Clarke 2012). Whatever we might think about the relative importance of genetic, epigenetic, or extragenetic inheritance, individuals are never, as such, beneficiaries of natural selection. They are expendable. If evolution is based on the survival of the fittest, those fittest can't be individuals, for no individual survives. What survives is information, carried by whatever replicators there turn out to be. Given that one of the values that we have instituted, in recent liberal Western societies, is the supposedly priceless value of the individual, the system of values that we claim to live by provides us with a strong reason not to take evolution's choice of beneficiaries too seriously as a guide to what we should regard as important.

Frequency-Dependent Fitness. A third reason to mistrust the gifts of natural selection that can be illustrated in terms of a further problem for Paul Thompson's proposal about biological evil. As we saw earlier, Thompson appealed to evolution's supposed fostering of what was good for society. The implicit assumption was that if a trait is, from some applicable point of view, a "good thing," then natural selection will bring it to fixation; if it is a "bad thing," it will eventually be purged from the population. For many, if not most, traits, however, the fitness of the trait or gene depends in part on its frequency. When fitness is *frequency dependent*, alternative traits are in equilibrium, in the manner memorably modeled by Maynard Smith's fable of hawks and doves. When hawks dominate, doves have the advantage; when doves dominate, hawks have the advantage (Maynard Smith 1984). This sort of equilibrium is known to be at the root of the rather wasteful proportion of males to females in sexually reproducing populations. It may be the sort of equilibrium that also sustains the existence of psychopaths

---

[2] See http://blogs.discovermagazine.com/gnxp/2010/08/1-in-200-men-direct
     -descendants-of-genghis-khan/.

among us. Being a psychopath is probably a good strategy for an individual living among people capable of altruism and empathy. Conversely, in a society of psychopaths, mutants capable of empathy might well have an advantage similar to that of rare doves in a virtually all-hawk environment in Maynard Smith's thought experiment.

Global Reflective Equilibrium

The moral of these last reminders is that we cannot assume that what natural selection has made possible is also desirable. Conceptions of morality or, more broadly, of the best ways to live, whether they are modeled on what appears natural to humans or merely inspired by what is possible, will remain essentially contested. One point does emerge, however, from recent work on the biological origins of morality. That is that the responses that count for or against certain moral stances are emotional ones (Haidt and Bjorklund 2008). Because our emotions are far from forming any coherent unity, anyone who is committed to finding the best answers to questions about how to live is condemned to allow mutual confrontation among all the members of a chaotic set of emotions and dispositions. It seems highly unlikely that we can discover anything like a simple vectorial sum of all our emotional responses. This is bad news for anyone who aspires to find a rational justification for ethical principles, even in my loose sense of justification, in some set of natural facts. Yet there is no serious alternative to bringing into mutual confrontation our conflicting intuitions about general principles, specific cases, valuable activities, legitimate responses, and beneficial behavior. That process will not yield to scruples about the "naturalistic fallacy." The question is not whether any logically valid reasoning processes can carry you from a set of facts to one or more evaluative judgments. Rather, it is about our *emotional inclinations* to prize certain things and despise others in response to the contemplation of facts. This by no means excludes rational deliberation and logical reasoning. On the contrary: reasoning is essential to the process and itself subject to its own set of epistemic feelings, such as the despair and the feeling of recognition described in Plato's *Meno* or the "clarity and distinctness" promoted as criterial in the Cartesian project of grounding knowledge (de Sousa 2008).

In that perspective, Mill's assertion that "the sole evidence it is possible to produce that anything is desirable is that people do actually desire it" (Mill 1991) looks entirely reasonable. We don't need biologists to confirm the biological fact that we desire pleasure. Mill's claim has been criticized for fallaciously exploiting an ambiguity in the suffix "-able" or "-ible" that indicates worthiness in "desirable" but signals mere possibility in "visible." Semantically, the criticism is valid, but it is also beside the point. What counts is that we are strongly inclined to take desire as a reason for judging something to be desirable. If no inference is any *better* than that, then Mill's inference seems to be reasonable, even though it is sanctioned neither by logic nor by semantics.

If this seems dissatisfying, we should recall Hume's demonstration that inductive inference cannot be provided with any noncircular justification. Inductive inference is just what we do in virtue of the way our minds are constructed. Actually, the same holds for deductive inference: in a mode of reasoning that looks "flagrantly circular," as Nelson Goodman pointed out, "[a] rule is amended if it yields an inference we are unwilling to accept. An inference is rejected if it violates a rule we are unwilling to amend" (Goodman 1983, p. 64). This is essentially similar to the quest for "reflective equilibrium" recommended by John Rawls (1977) as the test for ethical practice and principle. In the absence of a consensus on foundations, nothing else is going to be either required or possible in ethical reasoning than the pragmatic endorsement of reflective equilibrium. The lineage of this idea goes back, before Goodman and Rawls, to Nietzsche and Hume. Rawls's appeal to reflective equilibrium is of a piece with Goodman's characterization of the predicates we commonly use as "entrenched" in existing projective practice; in turn, it reflects Nietzsche's (1967) contention that instead of vainly attempting to justify ethics, we should attend instead to its genealogy. It is also clearly in harmony with Hume's reduction of our inductive knowledge of cause and effect to "custom and habit" (Hume 1975, sect. V, pt. 1).

Our search for a philosophical reflective equilibrium that takes account of biology must be grounded in our emotional responses not only to the facts of biology but also to the models these provide. These can be valuable even when they are merely metaphors. They may even derive from facts (or ways of thinking about facts) about species other than our own. Recently, for example, a movie about penguins was taken up as a model for human behavior by certain fundamentalist groups, who enthused that it "passionately affirms traditional norms like monogamy, sacrifice and child rearing" (Miller 2005). Rather than inferring that the two species exemplify two very different types of sociality, this approach bizarrely derives norms for humans from facts about a species with which we have no common ancestor for millions of years. As it happens, the penguin example was (like many of the "facts" that Aquinas thought to notice in nature) an invention bearing little relation to reality.

For an example to which one might feel more sympathetic, consider the idea of *individuality*. As mammals, people, unlike some other forms of life, are individuals in a sense that can be made quite specific and differs from the mode of life of other life forms, including most plants but also some parthenogenetically reproducing metazoans. In addition to being unique at the genetic level (with the exception of monozygotic twins), the sort of individuals we are enjoy an extraordinarily large potential for becoming even more different from one another in the course of development and learning. This is a fact of biology, but our attitude to it and what we make of it are obviously not determined by that fact. One might, for example, insist that in order to compensate for that unfortunate diversity, we need a strong dictatorial power that will bend us all to the same mold. But we may also be inspired to think that we *should*, in some nonmoralistic sense of "should," take advantage of the opportunity this affords us to make of ourselves, in a phrase once used by the French writer André Gide, "Ah, the most irreplaceable of beings" (Gide 1942, p. 186).

With individuality, we can reflect, comes diversity. Diversity in forms of life is attractive from both an ecological and an individual perspective. On the one hand, when plant species disappear, we may lose potential cures for diseases yet unheard of. But, on the other hand, we also value diversity for its own sake. The living world's astounding range of forms of life is awe inspiring. Analogously, the multiplicity of possible experiences appears as a gift bestowed on us by nature herself, which it would be churlish to reject.

Monogamy

If human diversity is deemed of intrinsic worth, why should diversity in relationships not seem equally desirable? And yet, in practice, we pigeonhole everyone and every relationship into one or two of a small number of categories: straight, gay, or bi and single, married, engaged, or "just friends." Why should this be? I conclude with some very brief remarks on this somewhat controversial question, intended to illustrate how we might actually take seriously certain findings of biology about possibility. A reasonable application of the strategy of reflective equilibrium, I want to suggest, might lead us to a fresh conception of often unquestioned assumptions about the role of the erotic in our lives.

The dominant ideology governing our normative conceptions of love, sex, and marriage is grounded in the ideal of monogamy and guarded by the social endorsement of the emotional sanction of jealousy. In Western society, the ideal of sexually exclusive monogamy, though honored more in the breach than the observance, is recognized officially in the institution of marriage and unofficially in the hypocrisy of shocked responses to celebrity scandals. The characterization of human beings as by nature a monogamous or "mildly polygynous species" (Barash and Lipton 2001, p. 41) is frequently brought out to explain or excuse a sexual double standard and is conveniently supported by a standard story told by evolutionary psychologists. That story starts from the discrepancy in gamete size between males and females and by a suspiciously swift chain of inference deduces that men and women should differ in many ways, supposedly traceable to different strategies of reproduction, resembling r-reproducing and k-reproducing species, respectively. Sexual jealousy should be more intense in men because of their uncertainty about paternity; women, however, are supposed to be more likely to experience emotional jealousy on account of their need for continuing support in the upbringing of offspring. Unfortunately for the standard story, that difference, although it seemed borne out in the United States (Buss 1994), tends to vanish altogether in countries with more gender equality (Harris 2004). Much the same is true for other alleged sex differences: on further examination, most turn out to be effects of the very stereotypes that they supposedly justify (Tavris 1992; Fine 2011). If anything, it now seems likely that the biological facts about female sexuality are closer to the traditional view of women as sexually insatiable than the nineteenth-century view, still prevalent in some circles, of coy females requiring to lie back and think of the Empire in order to serve the needs of procreation (Baker and Bellis 1995; Ryan and Jethá 2010).

Helen Fisher (1998, 2004) has shown that what we call love tends to conflate three very different syndromes, each of which has its own characteristic phenomenology, neurochemical correlates, and duration. These three are lust, obsessive romantic love or "limerence" (Tennov 1979), and long-term attachment. By conflating these, the monogamist ideology comes very close to requiring what even a minimalist biological perspective might judge to be simply impossible. George Bernard Shaw put it thus:

When two people are under the influence of the most violent, most insane, most delusive, and most transient of passions, they are required to swear that they will remain in that excited, abnormal, and exhausting condition continuously until death do them part. (Shaw 1986, pp. 34–5)

Multi-million-dollar industries of couple counseling, prostitution, and pornography bear witness to the fact that the resulting norms are unenforceable and exact a severe toll from individuals attempting to conform to them. Among the multifarious possibilities brought on by the human capacity for language, one might therefore infer that an alternative ideology might be more likely to fulfill a modestly expansive view of how nature might best be recruited to promote thriving relationships. For those who might still see this as "incompatible with human nature," proofs of possibility can be found both in anthropology and in experiments conducted by minority explorers in ordinary liberal society. Anthropology affords examples of societies, such as the Mosuo, where marriage is unknown. Mosuo women choose their lovers as they please, and men's interest in their progeny is dealt with not by jealous sequestering of the mothers of their children but by raising their sisters' offspring (Yang and Mathieu 2007). Increasingly visible polyamorous communities bear witness to the fact that to recognize the factual separability of attachment and sexual attraction enables many people to reject the bizarre conception of loyalty or "fidelity" in terms of sexual exclusion (Easton and Hardy 2009).

It will not be easy to adjust social norms, even in the light of the undeniable fact of diversity in individual temperaments and preferences. But the considerations just alluded to suggest that different ideologies of sex and love are possible. Racism, slavery, sexism, and the fanatical opposition to gay marriage offer instructive precedents. All were supported by an abundance of allegedly scientific evidence about "human nature," now plainly seen to be worthless (Gould 1981). In only a couple of hundred years – or a surprisingly short fifty in the case of gay marriage – the stronger arguments have won out with the majority of the society as a whole. Perhaps a similar regestalting might, in another fifty years, result in the current normative ideal of sexually exclusive monogamy being seen as resting on objectively false dogmas about human nature. The ideology of monogamism, just like racism, slavery, sexism, and heterosexism, might then come to seem almost unintelligible.

That would be one way of implementing the deepest philosophical lesson of biology. That lesson, which should come as an ironic rebuke to the army of fulminating biophobes who think they are defending humanism by attacking a supposed "biological determinism," is that we should all be existentialists.

8 | Biology and the Theory of Rationality

SAMIR OKASHA

Introduction

Philosophers since antiquity have been interested in the nature of rationality. A central concern in epistemology is to assess the rationality of our beliefs, while a central concern in practical philosophy is to assess the rationality of our actions. These topics are interesting partly because it is not clear what the relevant standards are for assessing the rationality of beliefs and actions. For example, it is often said that rational beliefs are ones that are "apportioned to the evidence," but what exactly does that mean? Does it imply that two individuals with the same evidence must be in identical credal states on pain of one of them being irrational? Similarly, it is often said that rational actions should reflect an agent's beliefs about how best to bring about the consequences she most desires, but what exactly does this mean? What if the agent does not know the likely consequences of the different courses of action open to her? What if the agent desires things that are harmful for her? Though we all have an intuitive grasp of what rational belief and action consist of, producing substantive analyses of these concepts has not proved easy.

Some progress on these issues comes from the theory of rational choice, the mainstay of modern economics. Rational-choice theory offers a precisely defined, albeit rather "thin," notion of rationality. A rational agent's beliefs, on the standard picture, can be modeled by a subjective probability function over some set of alternatives ("states of the world"); when the agent gets new evidence, he update his probabilities by Bayesian conditionalization. As regards action, a rational agent chooses between alternative actions using expected utility maximization, that is, by assigning utilities to the possible consequences of each action and picking an action that maximizes expected utility with respect to his probabilistic beliefs. This picture

This work was supported by the European Research Council Seventh Framework Program (FP7/2007–2013), ERC Grant Agreement No. 295449.

of rationality involves a healthy dose of idealization because real-life agents rarely have explicit probabilistic beliefs and almost never consciously compute expected utilities. However, an ingenious argument, due originally to Ramsey (1931) and Savage (1954), shows that an agent whose binary choices satisfy certain fairly intuitive conditions necessarily behaves *as if* he had explicit probabilistic beliefs and an explicit utility function and was aiming to maximize his expected utility.

Many philosophers define "rationality" in a richer sense than this – to mean that an agent has good *reasons* for his beliefs and actions and that these reasons have been instrumental in causing the beliefs and actions. (Some would go further and require that a rational agent be aware of these reasons.) Understood this way, rationality requires fairly sophisticated cognitive abilities and so is presumably the preserve of a few species, perhaps only *Homo sapiens*. By contrast, conforming to the consistency requirements of rational-choice theory could in principle be achieved by an organism that lacked "reasons" altogether but was capable of making behavioral choices. In a useful discussion, Kacelink (2006) refers to rationality in the sense of acting or believing on the basis of reasons as "PP-rationality" (standing for "philosophers and psychologists") and contrasts it with the "E-rationality" of "economists," by which he means satisfying the standard principles of rational choice, such as expected utility maximization.

Can a biological perspective shed light on the nature of rationality? Scholars from a number of disciplines have suggested that it can. In philosophy, naturalistically inclined thinkers at least since Quine (1969) have suggested that human rationality is the result of Darwinian selection; thus, for example, Dennett (1987) claims that "natural selection guarantees that most of an organism's beliefs will be true, most of its strategies rational" (p. 7). More recently, Sterelny (2003) argues that belief/desire psychology, which arguably underpins our capacity for rational thought and action, can be considered an adaptation to a "hostile environment" and has sketched an account of how it might have evolved; Godfrey-Smith (1996) argues similarly. In a different vein, authors such as Skyrms (1996) and Binmore (2005) have argued that evolutionary considerations can illuminate a variety of phenomena that traditional rational-choice theory struggles to explain, such as the human sense of fairness and our capacity for altruism. A useful survey of philosophical work on the evolution/rationality connection is Danielson (2004).

In psychology, a number of authors have advocated a Darwinian approach to human cognition and decision making by focusing on the question of adaptive function. Thus Gigerenzer and colleagues argue that many aspects of human cognition that appear defective by traditional rationality criteria may generate adaptive behavior in particular environments and so are thus "ecologically rational." A recent paper in this vein by Hammerstein and Stevens (2014a), entitled, "Six Reasons for Invoking Evolution in Decision Theory," argues that instead of the traditional axiomatic approach to rational decision making, we should study decision making using an evolutionary approach. They suggest that considerations about what is adaptive rather than what is "rational" according to some idealized theory will shed more light on how humans actually make decisions. A related argument is made by the evolutionary psychologists Cosmides and Tooby (1994), who argue that the mind comprises evolved "modules" equipped for specific tasks, which enable "better than rational" behavior. Useful surveys of this area include Gigerenzer and Selten (2001) and Hammerstein and Stevens (2014b).

In behavioral ecology, the branch of evolutionary biology that studies animal behavior from a Darwinian basis, rationality concepts play an interesting role. Though this field focuses mainly on nonhuman animals, it has often borrowed models and concepts from rational-choice theory and given them a biological twist. Typically, this involves reinterpreting the utility function as a biological fitness function and allowing natural selection rather than a rational agent to do the optimizing. Thus, for example, models of optimal foraging often assume that animals foraging for food behave like rational Bayesian agents, updating their "beliefs" on receipt of new information and choosing fitness-maximizing strategies (Houston and McNamara 1999). Similarly, Maynard Smith (1982) famously used concepts from classical game theory to shed light on social interactions among animals, giving rise to the field of biological game theory (see below). It is striking that rational-choice models, which have often been criticized for assuming "superhuman" reasoning abilities, should prove so useful for understanding the behavior of animals with only limited cognitive powers.

In cognitive and comparative psychology, there is considerable discussion of whether, and in what sense, the behavior of nonhuman animals qualifies as rational. Researchers in this area often give intentional or "belief/desire" explanations of the behavior of animals, including mammals and birds. For example, Clayton, Emery, and Dickinson (2006) argue persuasively that the food caching and recovery behavior of Western scrub jays is most naturally explained by attributing to them beliefs and desires, that alternative nonintentional explanations fail, and that the jays' behavior is therefore rational. Against this, it might be argued that the birds do not have beliefs in the full sense (perhaps because they lack language) or that, even if they do, their behavior is not rational because it is not reason based in the requisite way. This issue turns in part on the correct interpretation of the empirical evidence and in part on the precise concept of rationality that's in play. A useful collection of papers in this area is Nudds and Hurley (2006); see also Andrews (2014, sect. 2.3).

In economics, there is a growing literature on the biological foundations of preferences. While most economic theorizing takes an agent's preferences (e.g. over consumption bundles) as a given, this literature asks what sort of preferences we should expect to evolve by Darwinian selection. The underlying assumption is that human preferences stem from our evolved psychology and so should admit of a Darwinian explanation. Thus, for example, Robson (1996) studies the evolution of attitudes toward risk, producing the striking finding that in certain circumstances, agents whose preferences violate the axioms of expected utility theory should enjoy a selective advantage (see below). In principle, this type of argument could help to explain why the actual behavior of humans seems to systematically depart from the predictions of rational-choice theory. A useful overview of work in this field is Robson and Samuleson (2011).

A proper survey of the diverse lines of investigation described earlier would be beyond the scope of a single chapter (and, most probably, author). Here my focus is on overarching philosophical and conceptual issues. In the next section I examine the idea that biology supplies an alternative evaluative yardstick for assessing beliefs and actions, distinct from the yardstick employed in traditional discussions of rationality. In the section that follows I look briefly at the concept of ecological rationality and its implications for the study of the human mind. Then I examine the link between evolution and rational choice, focusing on the idea that Darwinian fitness can supply some "meat" to the abstract utility function of rational-choice theory. In the final section I examine the idea that evolution and rationality can "part ways," that is, that evolutionarily successful behavior may fail to coincide with rational behavior.

Biology and the "Yardstick" of Rationality

Rationality is a normative notion. Rational beliefs and actions are ones that conform to the norms of belief formation, belief change, and choice of action, whatever exactly they are. (This is so whether we are talking about "PP-rationality" or "E-rationality" in Kacelnik's terms.) Thus, to call a belief or action rational is not simply to describe it but also to evaluate it. Indicative of this normativity is the fact that it makes sense to ask what beliefs a person *should* have, given her evidence, and what action the person *should* choose, given her aims (or, perhaps, given the aims that we think she should have). The source of this normativity is a deep philosophical issue according to some authors, but fortunately, we can leave this matter aside. For the moment, the point is simply that inherent in the idea of rationality is the idea that an agent should believe or act in a certain way and thus the possibility that the agent's actual belief or action will fail to be as it should, hence irrational.

One way to see the relevance of biology to rationality theory is to note that evolutionary biology suggests its own normative standard by which to assess actions (and, indirectly, beliefs). Consider a male organism in a sexual species who is trying to attract a mate. A number of possible mating strategies exist, for example, performing a showy display, engaging in male-male combat, or trying to take control of another male's harem, each of which will have different consequences for the organism's reproductive success (or "fitness"). This suggests a natural way of normatively evaluating the organism's choice of strategy. As well as asking which mating strategy the organism *does* actually adopt, we can also ask which strategy it *should* adopt, that is, which strategy will be fitness maximizing in the relevant environment, or evolutionarily optimal. If the organism chooses a suboptimal strategy, it makes sense to say that the organism has failed to do what it should have done or has failed to achieve the "goal" of maximizing its reproductive success.

The fact that evolutionary biology supplies its own yardstick of normative evaluation, based on the calculus of Darwinian fitness, yields a notion of "biological rationality" (cf. Kacelnik 2006) that is logically distinct from the rationality notions used in other disciplines but may nonetheless bear interesting relations to them. Because biological rationality is all about enhancing one's fitness, the notion applies in the first instance to behaviors or choices. As such, the notion is applicable to any organism capable of behavioral plasticity. A bacterium that swims toward a chemical gradient has made a "choice" about which direction to swim in, and it makes sense to ask whether its choice is the "correct," that is, fitness-maximizing, one. The notion also can be applied to beliefs and desires as long as the organisms in question are capable of having them, for these mental states give rise to behavior. Thus, in principle, various aspects of human cognition and decision making can be evaluated by the yardstick of biological rationality (see below).

Biological rationality appears logically independent, in both directions, of rationality in the sense of having reasons for one's beliefs and actions. An agent's beliefs and actions might be suitably reason based and yet not enhance his biological fitness; conversely, an agent's beliefs and actions might be fitness enhancing and yet not based on good reasons, perhaps because the agent lacks the capacity to have reasons at all. What about rationality in the sense of conformity to the norms of rational-choice theory? It seems obvious that this need not imply biological rationality: an agent may have consistent preferences that are detrimental to his biological fitness, as many modern humans arguably do. But the converse inference, from biological rationality to conformity to rational-choice norms, has often been defended (e.g. Gintis 2009, p. 7; Kacelnik 2006; Chater 2012). This inference seems reasonable: if an organism displays adaptive behavior and so chooses actions that maximizes its fitness, then presumably it is behaving like a utility-maximizing agent whose utility function is simply its fitness function? In fact, matters are not quite so simple, for reasons discussed in the last two sections of this chapter.

Some philosophers might dispute whether biological rationality counts as a genuine species of rationality on the grounds that it is really just another name for adaptiveness or fitness maximization. According to this objection, the sense of "should" in which an animal should perform a biologically rational action carries no real normative force and is not interestingly similar to the sense of "should" in which humans should base their beliefs on the evidence or conform to the dictates of rational-choice theory, for example. After all, wherever the notion of adaptive function applies, then it makes sense to talk about malfunctioning or not doing the "correct" thing, as proponents of teleosemantics have long stressed, but malfunctioning is not usefully equated with irrationality. So biological rationality does not deserve its name, the objection goes.

In response, it must be granted that the notion of adaptive function applies in contexts where talk of rationality or irrationality would be inappropriate. If an organism's digestive system malfunctions, for example, it makes good sense to say that the system has not done what it should do, but this is not a *rational* shortcoming. The operation of the digestive system is too automatic for such a characterization to be useful. However, matters are different when we are dealing with behaviors or actions, particularly if the organism in question displays considerable behavioral plasticity or is capable of learning about its environment and modifying its behavior to suit the circumstances, as many birds and mammals can. Animal behavior of this sort is objectively similar (at a suitable grain of description) to human behavior and in some cases is homologous with it, despite the desire of some philosophers to see a chasm between humans and nonhumans. Where such behavior is concerned, the normativity that derives from the notion of adaptive function is plausibly regarded as a type of rationality, or proto-rationality.

This point can be bolstered by recalling two facets of the traditional rationality concept discussed by philosophers. First, rational action is *goal-directed* action, in which an agent is trying to achieve an end. An action qualifies as rational to the extent that it serves the agent's end (or is believed by the agent to do so). Much animal behavior appears unambiguously goal directed – think of a bird collecting sticks in order to build a nest, or a primate sharpening a tool in order to crack a nut, or a honey bee performing a waggle dance in order to communicate the location of a nectar source. It is difficult to make sense of such behavior without the assumption that it is goal directed (certainly in "as if" sense and arguably in a stronger sense). In recognition of this, behavioral ecologists frequently use an intentional idiom (e.g. "wants," "tries," "knows," "communicates") to describe and explain animal behavior; this idiom is typically regarded as neither metaphorical nor dispensable. Calls to banish the intentional idiom from the study of animal behavior (e.g. Kennedy 1992) have been noticeably unsuccessful. From this perspective, the evaluation of behavior in terms of its biological rationality looks like a bona fide species of rational evaluation.

Second, as McDowell (1994) argues, following Davidson (1984), when we give a folk psychological explanation of an agent's action or belief, our explanation makes the belief or action intelligible by rationalizing it; this is quite different from an explanation in physics, in which a phenomenon is made intelligible by showing that it had to happen as a matter of natural law. I suggest that this lends support to the view that biological rationality is a genuine type of rationality,[1] for when we explain an organism's behavior in terms of its biological rationality, for example, a chimpanzee fashioning a tool from a twig in order to catch termites, this yields exactly the sort of intelligibility that McDowell treats as definitive of rationalizing explanations. We are able to see how the behavior makes sense, or is appropriate, in terms of the organism's goal (acquiring food), which itself subserves the ultimate goal of maximizing fitness. The type of understanding that we get of the organism's behavior is more akin to the type of understanding we get from intentional explanation than from physical explanation.

One distinctive feature of the biological rationality concept is that it is externalist. A behavior counts as biologically rational if it is fitness maximizing or adaptive, which depends on the environment. Craving high-calorie foods was adaptive in the Pleistocene environment of our hominid ancestors but for humans in today's environment is not. By contrast, rationality in the sense of having reasons for one's beliefs and actions, or in the sense of conforming to the consistency conditions of rational-choice theory, are internalist matters. Whether an agent is rational in either of these senses depends on how things are "in its head," not in the external environment, so a suitably intelligent agent should be able to achieve rationality simply by a process of self-

---

[1] This is somewhat ironic, given that both McDowell and Davidson treat rationality as the preserve of human beings.

reflection and amelioration. For biological rationality by contrast, the world must cooperate, too.

Though biological rationality may be logically independent of rationality, in the other senses discussed earlier, it is tempting to think that empirically it must be related somehow to them. Creatures that act and believe for good reasons or whose choices conform to rational choice norms will generally enjoy a selective advantage over ones that do not, the suggestion goes; thus rationality in these senses and the cognitive equipment necessary for them are themselves Darwinian adaptations. Therefore, for the most part, beliefs and actions that are rational in the philosophical or economic senses are also likely to be biologically rational – or else natural selection would never have led to them. Dennett (1987) gives voice to this sentiment in the preceding quotation when he asserts that natural selection ensures that most of our beliefs will be true and most of our strategies rational (cf. Stephens 2001).

This conjecture may be correct, but it is an empirical issue, and potential counterexamples abound. One interesting counterexample comes from D. S. Wilson's work on the evolution of religion. Wilson (2002) opposes the modern liberal idea that religious belief is simply a rational pathology or the result of our usually accurate belief-forming processes going awry. Instead of assessing religious beliefs against the yardstick of factual truth or epistemic rationality – by which they inevitably fall short – he argues that we should instead use an adaptationist yardstick. Wilson claims that religious believers are motivated to engage in pro-social actions to fellow group members, resulting in group-level benefits. Thus a process of between-group selection would have favored religious over nonreligious groups, he argues. If Wilson's (controversial) theory is true, then it renders religious beliefs and practices intelligible by showing that they "make sense" when judged by the criterion of fitness maximization despite violating the usual norms of rational belief formation.

To summarize so far, evolutionary biology suggests a way of normatively evaluating actions and beliefs by how well they promote an organism's fitness in its environment that is distinct from the type of normative evaluation traditionally invoked in philosophy and in rational-choice theory. Though *sui generis*, biological rationality is still a bona fide type of rationality because it enables us to "make sense" of the beliefs and actions of both humans and nonhumans by showing how they help to fulfill their evolutionary goal.

Humans and Ecological Rationality

Proponents of "ecological rationality," notably Gerd Gigerenzer and Peter Todd and colleagues, focus primarily on human psychology and cognition (Gigerenzer 2010; Todd et al. 2012). Their theory incorporates aspects of biological rationality, in that it emphasizes successful performance in particular environments, but it has a distinct focus. Gigerenzer's point of departure is Herbert Simon's concept of "bounded rationality," which stresses that humans do not have unlimited computational abilities and so cannot implement sophisticated optimization algorithms. Thus we rely on heuristics, or rules-of-thumb, to make decisions and solve problems. These heuristics are special purpose and are tailored to specific environments, allowing them to exploit environmental regularities. (For example, the "recognition heuristic" says that if choosing between two objects, one familiar and the other unfamiliar, choose the familiar one. In an environment full of dangerous objects, this heuristic makes sense.) These "fast and frugal" heuristics are computationally cheap but get the job done.

Ecological rationality theorists emphasize the domain-specific nature of the heuristics that guide human decision making. A heuristic helps us with a particular task, for example, determining whether a social partner is honest. Different tasks call for different heuristics, so the human mind is an "adaptive toolbox," Gigerenzer and Selten (2001) argue. By contrast, traditional rational-choice theory is a domain-general approach: the "maximize expected utility" rule can be applied to any choice problem, and the rules of probability can guide uncertain reasoning about any subject matter. This emphasis on domain specificity is also a theme in the work of evolutionary psychologists Cosmides and Tooby; they argue that on general Darwinian grounds we should expect the mind to be composed of specialized modules because this allows more efficient problem solving than applying an all-purpose "general intelligence" (Cosmides and Tooby 1994). This inference – from a Darwinian premise to a conclusion about the structure of the mind – seems plausible, but ultimately the issue must be settled by direct psychological and neurobiological evidence.

Ecological rationality theorists paint an optimistic picture of human psychology. This contrasts with the emphasis on "cognitive biases" by theorists such as Kahneman and Tversky, who document systematic departures from the norms of rational-choice and probability theory (Kahneman 2011; Kahneman and Tversky 2000; Kahneman, Slovic, and Tversky 1982). According to these theorists, humans commit basic probabilistic errors, exhibit time inconsistency in their intertemporal choices, commit the base-rate fallacy, display "loss aversion" and "uncertainty aversion," and are prone to an alarming variety of "framing effects." These results, which are well confirmed experimentally, are often interpreted as showing that humans are "just not irrational." From a biological perspective, this is somewhat puzzling because it is hard to see why evolution would favor creatures prone to such biases. However, ecological rationality theorists offer a different picture. Relying on simple heuristics, rather than attempting to implement optimization, is an efficient way of solving problems. In our evolutionary past, there was a premium on making quick decisions and choices, so using a simple heuristic (rather than searching through all the options looking for the "best," for example) was an adaptive strategy given our limited computational powers. In their natural settings, such heuristics work well, but applied out of context, they can make us look irrational.

To the extent that this line of argument is successful, it makes it more intelligible, in broad biological terms, why human reasoning and decision making exhibit some of the features they do. However, this is different from showing that the *specific* violations of rational-choice precepts found by Kahneman, Tversky, and others were to be expected. It is one thing to be able to explain, as Gigerenzer and colleagues arguably can, why humans do not make choices by explicitly trying to compute expected utilities, relying instead on simple shortcuts, but this does not explain the specific violations of expected utility maximization that have been found, such as displaying the Ellsberg preferences in choice under uncertainty or using hyperbolic discounting in intertemporal choice, for example. It is conceivable that these and related phenomena could be accounted for in terms of ecological rationality, but to date, they have not been.

Proponents of ecological rationality are often rather disparaging of probability theory and rational-choice theory. They regard the latter as *a priori* philosophical and mathematical exercises, which do not help the quest to understand real-life decision making and cognition. Gigerenzer and colleagues argue that an agent who relies on ecologically rational heuristics for making choices will often outperform an agent who tries to conform to the decision-theoretic ideal, at least in the particular environments for which the heuristics were tailored. So not only is rational-choice theory unhelpful for scientists seeking to understand human psychology, but it is also unhelpful for agents themselves. Cosmides and Tooby (1994) argue similarly.

At times ecological rationality theorists go further and argue that probability theory and rational-choice theory are incorrect even as normative ideals, not merely that they are poor descriptions of how actual human cognition works. Thus Gigerenzer and Selten (2001) say that their theory "provides an alternative to current norms, not an account that accepts current norms and studies when humans deviate from these norms ... bounded rationality means rethinking the norms as well as studying the actual behavior of minds and institutions" (p. 6). The suggestion, in short, is that the norms of traditional rational-choice theory constitute an inappropriate standard by which to judge creatures that have evolved to be ecologically rational.

This negative attitude toward rational-choice theory is by no means mandatory for those persuaded that adaptive considerations can illuminate the study of rationality. Indeed, it is perfectly possible to hold, with the philosophical mainstream, that deductive logic and probability theory yield correct norms of rational belief and that decision theory correct norms of rational action while at the same time holding that biological or ecological rationality constitutes a different standard by which our beliefs and actions can be normatively evaluated. I suggest that this attitude – permitting a plurality of valid rationality concepts – is more reasonable. We should allow that rationality in the sense of having good reasons for one's beliefs and actions and rationality in the sense of conformity to rational-choice precepts are both valid forms of normative assessment, while also allowing that our beliefs and actions can be assessed in terms of ecological/biological rationality.

I suspect that the hostility of some ecological rationality theorists toward rational-choice theory stems from the tendency, particularly among economists, to use the assumption of ideal rationality to build what are meant to be descriptively accurate models of human behavior. This is certainly a questionable way to proceed, given that experimental work shows clearly that humans systematically violate the rational-choice norms in at least some contexts (e.g. Ariely 2008). Given this fact, the idea of basing a science of human decision making on Darwinian principles rather than on the abstract axioms of decision theory is undeniably attractive. However, it does not follow that we should jettison decision theory and probability theory as normative ideals, but only that we should not assume without evidence that they are descriptively valid.

Utility and Fitness

Rational-choice theory is sometimes criticized for relying on a purely abstract utility concept. To say that rational agents maximize their utility is not to say much, the criticism goes, because "utility" is in effect defined as whatever an agent wants. One version of this criticism goes further and alleges that utility maximization is both empirically empty and normatively silent because virtually anything that an agent does can be reconciled with it. This criticism is arguably overstated, particularly for choice under uncertainty, because the axiomatic conditions that an agent's choice behavior must obey for the agent to be describable as an expected utility maximizer are not trivial, but it is nonetheless true that the doctrine of utility maximization provides little insight into why agents act as they do or the reasons behind their choices. This is partly why traditional philosophical work on "practical reason" makes little use of utility theory.

If we are persuaded by the idea of a Darwinian approach to rationality, then a natural hope is that Darwinian fitness may put some "meat" on the abstract utility function of the rational-choice theorists. To see why, consider a typical case of goal-directed animal behavior: a foraging bird moving from one food patch to another as its rate of food intake declines. Moving patch incurs significant costs and risks but may still be the best thing to do if food becomes too scarce in the current patch. So the bird needs to settle on a strategy for when to move from one patch to another. Suppose that the bird's foraging behavior has been honed by natural selection and so is biologically rational: it implements the strategy that will maximize its expected reproductive success, given the information it has. Armed with this knowledge, a scientist-observer can make precise sense of the bird's behavior, which might otherwise appear inexplicable or random.

Table 1 *A Game with Two Pure-Strategy Nash Equilibria*

Left		Right
Top	(2, 2)	(0, 0)
Bottom	(0, 0)	(1, 1)

The key point is this: the bird's behavior becomes explicable once we posit a *specific* goal, namely, maximizing reproductive success (or some proxy for it); we know from evolutionary theory that behavior directed toward this goal is a likely (though not inevitable) outcome of Darwinian selection. Thus our foraging bird is behaving like a utility maximizer of the sort described by rational-choice theory, *but whose utility function is of a very specific sort*. The bird behaves "as if" it cares about maximizing its expected reproductive success. Merely hypothesizing that the bird's behavior maximizes expected utility, *modulo* some utility function or other, so satisfies the canons of rational choice on its own explains rather little. It is the additional hypothesis that the bird's utility function is its fitness function that enables us to explain and predict its behavior. This is the sense in which a biological perspective can put flesh on the bones of the utility function.

This observation tallies with the way that game-theoretical models, in particular, have been deployed in a biological context. Consider a simple two-player simultaneous game as depicted in Table 1, in which each player has two (pure) strategies at his or her disposal. The entries in each cell denote the payoffs to (player 1, player 2). In traditional game theory, these payoffs are assumed to be utilities; the assumption is that each player wants to maximize his or her (expected) utility. This game has two pure-strategy Nash equilibria, (Top, Left) and (Bottom, Right), yielding payoffs of (2, 2) and (1, 1), respectively. Classical game theory offers these as the "solutions" of the game and predicts that one of them will be observed; at such an equilibrium, each player is choosing the strategy that maximizes his or her payoff conditional on his or her opponent's strategy and so has no unilateral incentive to deviate.

Beginning with Maynard Smith (1974, 1982), biologists have taken models of this sort and given them a biological twist by interpreting the payoffs as fitnesses rather than utilities. So interpreted, the model describes a social interaction between two organisms, the outcome of which augments each organism's fitness by the relevant amount, so an organism's overall fitness depends both on its own strategy and that of its social partner. On the simplest assumption, each organism's strategy is genetically hardwired and faithfully transmitted to its offspring. (Alternatively, the organisms may exhibit behavioral plasticity and be capable of choosing a strategy depending on an environmental cue.) Biologists typically imagine a large population of organisms, evolving by natural selection, and ask which strategy will come to dominate the population. Under reasonable assumptions, it can be shown that the population will usually reach an evolutionary equilibrium corresponding to a Nash equilibrium of the original game.[2] Unlike in classical game theory, where an equilibrium is meant to result from a process of rational deliberation by intelligent agents, in biological game theory an equilibrium is reached as a result of a dynamical process, namely, the differential proliferation of the fittest strategies.

This illustrates the fact that utility and fitness play isomorphic roles in rational and biological game theory, respectively. The former is the quantity that determines which strategy a rational agent will choose; the latter is the quantity that determines which strategy Darwinian evolution will program organisms to choose. When rational agents choose utility-maximizing strategies, this leads to an equilibrium in rational deliberation; when organisms choose fitness-maximizing strategies, this leads to an equilibrium of an evolutionary process. This consideration, and, more generally, the close analogy between the fitness-maximizing paradigm of evolutionary biology and the utility-maximizing paradigm of economics, has led many authors to see a deep connection between evolution and rationality theory (cf. Maynard Smith 1974; Stearns 2000; Grafen 2006a; Orr 2007; Okasha 2011).

The suggestion that utility and fitness are isomorphic in this way is appealing but needs qualifying for three reasons. First, it is not always clear what the analogue of the rational agent actually is in a biological context. Usually it is individual organisms that engage in goal-directed behavior and whose choices thus may be evaluated in terms of

---

[2] See, for example, Weibull (1995) for a careful account of these assumptions.

biological rationality, but in other cases it is groups of organisms (or "superorganisms") that are the locus of goal-directed action, for example, the coordinated behaviors of certain social insect colonies (cf. Seeley 1996, 2010). In other cases still, involving conflicts of interest between the genes within an organism, the entity that has a "strategy" and is thus akin to a rational agent is the gene itself (cf. Haig 2012).[3] The question of which biological unit should be treated as agent-like (and why) is closely related to the discussion of "levels of selection" in evolutionary biology (Okasha 2006; Gardner and Grafen 2009).

Second, utility and fitness are measurable on different scale types. In rational-choice theory, utility is generally taken to be either ordinal or cardinal depending on the problem at hand; in biology, fitness is generally treated as a ratio-scaled quantity, for the zero point of fitness is meaningful, so it makes sense to say that one strategy (or genotype) is twice as fit as another (cf. Grafen 2007). One *might* think that there is a further disanalogy in that utility is usually taken not to be interpersonally comparable, while the whole point of the fitness concept is to compare the fitness of different individuals. On the most natural way of formulating the utility/fitness connection, though, there is a single fitness function for all individuals in the population, mapping strategies (or profiles of strategies in the game-theoretic case) onto fitness. Different organisms play different strategies and hence receive different fitness payoffs, but this is simply the analogue of a rational agent receiving a different utility payoff from different outcomes, which involves only *intra*personal comparison.

Third and most important, the appropriate definition of "fitness" is a subtle issue in biology and depends on modeling assumptions. In the simplest evolutionary scenarios, expected lifetime reproductive success is the right fitness measure; natural selection favors organisms whose behavior maximizes this quantity. Many phenotypic traits can be understood in terms of their contribution to maximizing expected reproductive success. However, in more complicated scenarios, matters are different. For example, if organisms engage in social interactions, then it is necessary to take account of the effect of an organism's actions on its genetic relatives, so Hamilton's "inclusive fitness" becomes the relevant measure (cf. Hamilton 1964; Grafen 2006a). If there is class

---

[3] This is known as "intragenomic conflict" and arises because the genes in a sexually reproducing organism are not transmitted en masse to its offspring.

structure in a population, for example, individuals belonging to different age cohorts, then the appropriate fitness measure is different again, for it is necessary to weight offspring by their "reproductive value" (Charlesworth 1994; Grafen 2006b). Thus we cannot assume *a priori* that we know which quantity (if any) of evolved organisms will behave as if they are trying to maximize (cf. Mylius and Diekmann 1995).

This consideration complicates the fitness/utility analogy but does not invalidate it altogether, for the basic Darwinian idea that natural selection will often give rise to adaptive behavior is a mainstay of evolutionary biology and enjoys broad empirical support. Many organismic traits, including behaviors, are manifestly there because they enhance the organism's "fit" to its environment. The fact that the appropriate quantitative measure of "fit" depends on the details of our evolutionary model shows that natural selection is a more complicated process than was once thought but does not undermine the idea that adaptation to the environment, which results from selection, is a pervasive feature of the living world. To the extent that such adaptation occurs, it is legitimate to regard adapted organisms as akin to utility-maximizing rational agents trying to maximize their fitness, with the caveat that "fitness" must be defined appropriately for this idea to work and that different definitions may be needed in different cases.

Can Evolution and Rationality "Part Ways"?

I noted earlier that philosophers such as Quine and Dennett have argued that rational beliefs and behavior are the likely outcome of natural selection. However, against this, a number of authors have argued that considerations of rationality may sometimes "part ways" from considerations of fitness maximization, to use an expression from Skyrms (1996). This is a striking suggestion, raising the prospect of an evolutionary explanation for why humans sometimes depart from traditional canons of rationality.

Skyrms illustrates this "parting of ways" with a simple "prisoner's dilemma" game, as in Table 2. In a rational-choice setting, in which the payoffs denote utilities, it is widely agreed that in the one-shot game the rational agent should play *D* (defect) because it strongly dominates *C* (cooperate). Thus the expected utility of playing *D* must exceed that of *C*. This is so even if the agent believes that its opponent is likely to

Table 2 *Prisoner's Dilemma*

		C	D
Player 2	C	(6, 6)	(0, 10)
Player 1	D	(10, 0)	(2, 2)

play the same strategy as itself, presuming the truth of "causal decision theory" (Lewis 1981) because the two players are causally isolated.

Suppose that we now transpose to an evolutionary setting and consider a large population of organisms engaged in a one-shot pairwise interaction; the payoffs now represent increments of (personal) fitness. Which type has the higher fitness? As Skyrms observes, this depends on the pairing assumption that we make. Under random pairing, in which the probability of having a C partner is the same for both types, it is obvious that type D must be fitter. The expressions for the fitnesses of the two types are then

$$W_C = 6.P(C) + 0.P(D)$$

$$W_D = 10.P(C) + 2.P(D)$$

where $P(C)$ and $P(D)$ denote the probabilities of being paired with a cooperator and a defector, respectively; these probabilities are given by the overall frequency of each type in the population. As Skyrms notes, these expressions for expected fitness are identical to the corresponding expressions for the expected utility in the rational-choice context, calculated using standard (Savage-style) decision theory. Under random pairing, the type with the highest expected fitness chooses the action that confers the highest expected utility, so evolutionarily optimal behavior is identical to rational behavior.

Skyrms observes that matters are different if there is correlated pairing. We must then calculate the expected fitness of each type using the conditional probabilities of having a partner of a given type, which may differ for cooperators and defectors. The resulting expressions are

$$W_C = 6.P(C/C) + 0.P(D/C)$$

$$W_D = 10.P(C/D) + 2.P(D/D)$$

where $P(X/Y)$ denotes the probability of having a partner of type X, given that one is of type Y oneself. It is easy to see that if the correlation is strong enough, that is, the conditional probability of having a C partner is sufficiently greater for C types than for D types, then the C type may be fitter overall and so spread by natural selection.[4] Skyrms concludes that with correlated pairing, "rational choice theory completely parts ways with evolutionary theory. Strategies that are ruled out by every theory of rational choice can flourish under favorable conditions of correlation" (1996, p. 106).

Sober (1998) develops the same point slightly differently in the context of discussing what he calls the "heuristic of personification" in evolutionary biology. This heuristic is the idea that "if natural selection controls which of traits T, A_1, ..., A_n evolves in a given population, then T will evolve, rather than the alternatives, if and only if a rational agent who wanted to maximize fitness would choose T over A_1, ..., A_n" (p. 409). Sober maintains that this heuristic is usually unproblematic but fails in certain contexts, one of which is the one-shot prisoner's dilemma. The rational agent will never play cooperate because it is strictly dominated, Sober reasons; however, it is possible that natural selection will favor cooperate over defect if the requisite correlation exists. Thus the heuristic of personification fails: the rational strategy and the evolutionarily optimal strategy do not coincide.

These arguments are intriguing, but there is an obvious response, developed in detail by Martens (forthcoming). In the Skyrms/Sober model, there is no particular reason to equate the rational agent's utility function with its personal fitness function. Indeed, evolutionary biology teaches us that in social settings, the relevant fitness measure is not personal fitness but *inclusive* fitness, as noted earlier. To calculate an organism's inclusive fitness, we need to take account of the effect of the organism's action on other members of the population, weighted by the "coefficient of relatedness" (denoted r) between them. This coefficient is a measure of the genetic (and thus strategic) correlation between them; in the current context, the natural measure of r is $[P(C/C) - P(D/C)]$.[5] It is straightforward to show that if a rational agent's utility

---

[4] This is an instance of the statistical phenomenon known as "Simpson's paradox."
[5] This is a special case of one standard definition of r in evolutionary theory, namely, the linear regression of recipient genotype on actor genotype.

function depends suitably on its inclusive fitness, then the Skyrms/Sober "parting of ways" disappears.

This particular "parting of ways" argument therefore does not succeed. Skyrms and Sober's model does not show that irrationality will evolve but rather that "other-regarding" preferences will evolve – organisms will appear to care about the biological fitness of others as well as themselves. However, there are other suggestions in the literature for how irrational behavior may evolve. For example, Robson (1996), in an intriguing analysis, shows that organisms whose choice behavior violates the axioms of expected utility theory will often enjoy a selective advantage and so will evolve in a population. This remarkable result arises from the existence of "aggregate risk," which refers to risks that are correlated across members of a biological population, for example, bad weather. From a rational-choice perspective, it should not make any difference to an agent whether a given risk is aggregate or not, but from an evolutionary perspective, it does, given that what matters in evolution is reproductive success relative to the rest of the population. This is why Robson's model appears to yield the evolution of irrationality.

As with the Skyrms/Sober argument, however, it has proven possible to restore the connection between evolution and rationality in Robson's model by judicious choice of the utility function (though the necessary "fix" in this case is far from obvious). Grafen (1999) and Curry (2001) both show that if an organism's utility, in each state of nature, is defined as its fitness divided by the average population fitness in that state, that is, its relative fitness, then evolution will in fact favor maximization of expected utility after all because expected relative fitness is the appropriate criterion of evolutionary success in the presence of aggregate risk. The key point is that with aggregate risk, behavior that fails to maximize an organism's expected absolute fitness may nonetheless maximize its expected relative fitness. So, in theory, Robson's "parting of ways" also can be eliminated, though, empirically, the idea that evolution could program an organism to care about its relative fitness is questionable, given that relative fitness depends on the actions of others and so is not within an individual organism's control (cf. Okasha 2011).

It is tempting to suggest that the moral of the two preceding cases generalizes; that is, that *any* putative "parting of ways" between evolution and rationality can in principle be avoided by suitable choice of utility function. However, there is no theoretical reason to think that this must be true. Moreover, a number of authors have successfully developed models in which clearly irrational behaviors, for example, intransitive choices, are favored by natural selection and in which there is no obvious way to "restore" rationality by suitable choice of utility function (Houston, McNamara, and Steer 2007). Thus it would be premature to conclude that a "parting of ways" argument cannot succeed, even given the latitude of defining an agent's utility function as we please. This issue needs to be judged on a case-by-case basis.

The "parting of ways" idea just discussed should be sharply distinguished from the quite different idea that humans derive positive utility from things that do not enhance their biological fitness (personal or inclusive). Empirically, this clearly seems to be so: modern humans often have preferences for things that are neutral or detrimental to their fitness, for example, skydiving, contraception, or reading philosophy books. This is an interesting phenomenon, but it need not involve any irrationality in the sense of a violation of rational-choice norms and so does not involve any "parting of ways" in the preceding sense. I conclude by briefly discussing the phenomenon.

From a biological perspective, is it possible to explain why humans derive utility from things that are detrimental to their fitness? Opinions on this issue differ. One response is that human preferences are heavily dependent on learning and culture, exhibiting extensive cross-cultural variation; thus preferences are not under tight genetic control and so are not susceptible to biological explanation. This may be partly correct, but it pushes the question one step further back. Why did evolution make humans susceptible to acquiring preferences by learning or cultural transmission, which would cause them to behave in ways that harm their biological fitness? Was it an unintended side effect of selection for the ability to learn, for example?

One interesting take on this issue comes from Sterelny (2012), who argues that at a certain point in hominin evolution, we changed from being "fitness maximizers" who desired things that are good for our genes to being "utility maximizers" who desired things that are nonadaptive or even maladaptive. Sterelny attributes this change to the shift from small-scale to mass society. In traditional small-scale societies, cultural transmission is primarily vertical, from parents to offspring, but as societies got larger, horizontal transmission becomes dominant. So individuals became susceptible to acquiring maladaptive beliefs and preferences by horizontal means. Moreover, in mass society, the power of cultural group selection declines, so the filtering mechanism by which socially disadvantageous traits would be selected out was weakened. The upshot, Sterelny claims, is that humans retained their powers of instrumental reasoning but came to have preferences for things that did not enhance genetic fitness.

A different take comes from work by Samuelson and Swinkels (2006) and Rayo and Robson (2013). They argue that the challenge is to explain why humans derive utility from *anything* other than biological reproduction itself. Food, sex, and shelter, for example, obviously causally promote our fitness, but our desire for these goods is not purely instrumental. We desire tasty food as an end in itself, not simply because we know that consuming food will enhance our survival and hence our fitness. From an evolutionary viewpoint, this seems odd. Given that biological fitness is what really matters, surely Mother Nature should have produced organisms who care noninstrumentally only about their fitness and whose desires for "intermediate goods" such as food and sex are purely instrumental? Yet modern humans are not like this. So, according to this view, the challenge is not so much to explain why humans derive utility from things that are detrimental to fitness, but rather to explain why we derive utility from anything other than fitness itself.

The answer, according to the preceding authors, depends crucially on lack of information. Organisms are not born knowing the causal structure of the world and can only learn some causal regularities by trial and error within their lifetime. Plausibly, the causal consequences for fitness of consuming different foodstuffs, having sex, and so on are not something that our ancestors could have learned. If these causal consequences could be learned, then Mother Nature could make each organism care only about fitness itself. After learning the relevant causal facts, organisms would then produce biologically optimal behavior. But, given that this is impossible, Mother Nature instead equips organisms with intrinsic (noninstrumental) desires for intermediate goods. Therefore, humans have the utility functions they do precisely to compensate for their bounded rationality, that is, the limitations on what can be learned. This intriguing theory puts the connection between evolution, learning, and rationality into a new perspective.

Conclusion

Traditionally, the topic of rationality has been discussed without the benefit of a biological perspective by philosophers, psychologists, and economists. This traditional approach has undoubtedly yielded much interesting work. However, as this brief survey shows, a biological and, in particular, a Darwinian perspective offers the potential for new insights into the nature of rationality, both human and nonhuman, and suggests interesting new questions to ask. This is so for three main reasons. First, Darwinian fitness suggest a new normative yardstick – biological rationality – by which to evaluate beliefs and actions. Second, the cognitive capacities underlying rational thought and action are presumably evolved, raising the specter of a Darwinian explanation of aspects of human rationality and of our rational shortcomings. Third, the science of evolutionary biology itself has drawn extensively on ideas from rational-choice theory, suggesting a deep isomorphism between the fitness-maximizing paradigm of the former and the utility-maximizing paradigm of the latter. Each of these three topics is an ongoing field of enquiry.

9 Evolution and Ethical Life

PHILIP KITCHER

In 1975, E. O. Wilson (1975) famously declared that "the time has come for ethics to be removed temporarily from the hands of the philosophers and biologicized" (p. 27). Wilson's own program for "biologicizing" ethics proposed to align fundamental maxims for conduct with the demands of natural selection. Although that particular venture has not attracted many adherents, Wilson reintroduced an important question: "How exactly does ethics relate to evolution?" During recent decades, this question has been addressed not only by evolutionary biologists, primatologists, and anthropologists but also by members of the community that was supposed, temporarily at least, to let it go.

My aim is to consider two traditions in approaching the relation between evolution and ethics. The first, prominent in recent analytic philosophy, debates whether various standard metaethical positions can be reconciled with the operation of natural selection. Ironically, the philosophers engaging in these debates, like Wilson before them, take the crucial evolutionary connection to be with Darwin's notion of natural selection. Unlike Wilson's discussion, however, their involvement with biology ends once that notion has been introduced. As I'll argue, this influential philosophical movement is even more vulnerable than human sociobiology to complaints about its overly narrow conception of evolutionary mechanisms.

The alternative approach, pursued by writers from a variety of disciplines, cleaves more closely to Darwin's own treatment of the evolution of ethics. In his *Descent of Man* (1871), Darwin was principally concerned to show how the "moral sense" might have emerged from capacities present in nonhuman animals. He aimed to make genealogical connections, and the putative operation of natural selection was left in the background. The traditions, then, can be characterized in terms of the two great contributions of the *Origin*: one, the "selectionist" emphasizes the mechanism(s) of evolution; the other, the "genealogical," approaches ethics in light of the interconnectedness of living organisms.

Because I am among those who have attempted to articulate the genealogical approach, it should be unsurprising that I shall argue for its superiority. My concern, however, is not so much with my particular version as with a cluster of related, sometimes mutually supporting lines of research: the following discussion is (mostly) intended to be ecumenical toward my fellow genealogists. Moreover, there may be little surprise in the need to study historical connections first – after all, it's hard to make serious use of appeals to natural selection until you have some clear conception of the traits that were supposedly selected for. When the philosophical embrace of evolution is divorced from biological and anthropological details, philosophers all too easily lose contact with the phenomena they are out to explain. For all the admirable precision they bring to some of the issues they address, no link with the history of life is ever forged, and, in consequence, no "[l]ight will be thrown on the origin" of our moral practices.[1]

To provide a genealogy for some aspect of human behavior is to specify a series of transitions out of which the current form of that aspect emerged. If the genealogy is to be Darwinian, it must be possible for Darwinian mechanisms to bring these transitions about. Hence, proposing a Darwinian genealogy will typically require the proponent to indicate how this constraint can be met. Almost inevitably that is done by offering "how possibly" explanations, hypothetical accounts that reveal how the transitions envisioned *might* have emerged, without any commitment to supposing that these explanations tell the *actual* story. Genealogists will maintain that there are many alternative hypotheses between which the evidence available cannot discriminate. It is important to realize, however, that the "how possibly" explanations are not part of the genealogy but rather ancillary material used to defend it against a reasonable concern.

By contrast, the selectionist tradition puts the evolutionary mechanisms front and center. Its guiding idea is that Darwinian mechanisms rule out the emergence of certain types of traits, traits to whose existence some target metaethical view is committed. A seminal

---

[1] The quoted phrase is part of Darwin's single sentence about human beings in the *Origin* (1859, p. 488).

contribution to the selectionist tradition is Sharon Street's widely discussed article, "A Darwinian Dilemma for Realist Theories of Value"
(Street 2005). Street starts from the thesis that "[e]volutionary forces
have played a tremendous role in shaping the content of human evaluative attitudes" (p. 109). She then develops her dilemma by claiming
that value realists have two options: they can either suppose our
evaluative attitudes to be uncorrelated with the pressures of natural
selection, or they must see natural selection as favoring a capacity for
grasping value-theoretic truths. The former horn leads to an untenable
form of skepticism, while the latter asserts an inferior scientific (evolutionary) hypothesis.

Street's initial thesis is vague in two important respects. What "evolutionary forces" does she have in mind – those merely of biological
evolution or those pertaining to cultural evolution as well? Second,
what exactly is the trait whose emergence is problematic? Although
Street makes some serious efforts to explore evolutionary options –
unlike most of the philosophers who have responded to her article, she
engages with some of the recent discussions of human evolution and of
primate behavioral biology – her picture of the workings of evolution
remains in the vicinity of a much-criticized oversimplification of evolutionary analysis. According to that picture, would-be theorists can pick
any trait that strikes their fancy and explain its presence by hypothesizing advantages it would have conferred in the struggle for existence.[2]
Or, to adapt the point to the selectionist program, they can argue that
a particular trait would have been somehow disadvantageous in reproductive competition and, without considering that there is more to
Darwinian evolution than individual natural selection and that traits
are often developmentally bound together in suites, conclude that it
could not have emerged. The obvious riposte for Street's target, the
value realist, is to seek an evolutionary explanation of the relation
between our evaluative attitudes and independent values by supposing
that selection shaped a collection of psychological dispositions that,
when jointly employed, could give rise, in social environments with
considerable cultural transmission, to a process of inquiry that could
disclose those independently existing values. Neither Street nor any of
the realists who have responded to her has articulated any such

---

[2] A classical source of the critique of this "adaptationist program" is Stephen Jay
Gould and Richard Lewontin (1979).

explanation. In the absence of a candidate genealogy, the question of its possibility is simply unsettled.

Subsequent philosophical discussion has brought some of the issues that Street raises into sharper focus.[3] Yet, despite the clarity achieved for parts of the debate, the traits taken to have evolved remain nebulous, and there is no attempt to use what is known about human evolution to explain how they might have emerged. Distinguished value realists have been driven to apparently desperate positions. Derek Parfit (2011) confesses his inability to articulate either the metaphysics or the epistemology of the realm of values; moved by Street's dilemma, Thomas Nagel (2012) concludes that something fundamental must be missing in the Darwinian picture of life.

In the *Descent*, Darwin expressed the hope that tracing the connections between traits in nonhuman animals and the human moral sense would "[throw] light on one of the highest psychical faculties of man" (1871, p. 95). In effect, he conjectured that constructing a genealogy might enable more exact specification of the capacities underlying ethical life. The genealogical tradition shares this hope. It is also motivated by a breathtakingly obvious point: it's folly to debate possibilities of evolutionary explanation until you have some relatively clear conception of what is to be explained – first, catch your *explanandum*.

Moreover, the genealogical program has obvious advantages in being far less vulnerable to the charge that it has replaced the rich framework of contemporary evolutionary theory with a simplistic caricature. Today's evolutionary analysts recognize that natural selection acts on traits that are often developmentally linked, that cultural selection is an important force that operates on a significant number of animal lineages, and that cultural selection can favor traits that would have been eliminated by natural selection. The failure to appreciate these points posed difficulties for human sociobiology, and the problems are even worse for the selectionist program. That program aims to show that certain – imprecisely characterized – traits (such as a "capacity for tracking moral truth") *couldn't* arise out of Darwinian mechanisms. Its tools for the supposed demonstration are analyses showing the disadvantages of the traits. But, to repeat, a trait reproductively disadvantageous for its bearer might either belong to

---

[3] Two excellent articles probing the metaphysical issues are Clarke-Doane (2012) and Shafer-Landau (2012).

a suite of developmentally linked traits that are overall advantageous or might be favored by cultural selection. Genealogists, by contrast, only invoke natural selection in the ancillary work of defending their sequence of transitions. They can shrug off the complaint that they are spinning "just so" stories about selection, for given that we cannot know the *actual* evolutionary causes, what is *required* is a story, compatible with the available evidence, demonstrating how a transition *could* have occurred.

Interestingly, selectionists seem to think ventures in genealogy are committed to storytelling. Shafer-Landau expresses his reluctance to venture into genealogy by referring to a "non-negligible amount of speculation" in reconstructing the emergence of ethical life (2012, p. 26).[4] He is correct to suspect that some important facts about the human past probably will never be known. However, there are other facets of the emergence of our individual psychological capacities and the forms of our social life about which we can be reasonably confident. The fact that we cannot know all we would like to know about the route from the Upper Paleolithic to the present should not incline philosophers to throw away the information and the clues that we do possess.

Darwin's own genealogical account elaborates a basic proposition he takes to be highly probable: "[A]ny animal whatever, endowed with well-marked social instincts, the parental and filial affections being here included, would inevitably acquire a moral sense or conscience, as soon as its intellectual powers had become as well, or nearly as well developed, as in man" (1871, p. 95). The story Darwin reconstructs supposes that many animals, including those with which we share recent common ancestry, associate in small groups and sometimes respond to one another's behavior in cooperative ways. Darwin recognizes that positive responses inside the band are far from ubiquitous and stop at the boundary of the local group. He hypothesizes that our ancestors already had the ability to recall past actions bringing significant positive or negative consequences for them. After the acquisition of language, those consequences would have included openly expressed

---

[4] The remark is made in connection with my own genealogical account (kindly characterized as "perhaps the best"). Even here, however, he blurs the difference between a genealogical reconstruction of ethical life and an attempt to show how it evolved (or, more precisely, how some transitions evolved and how others could have evolved).

judgments from fellow group members. Because of individuals' "sympathy" for those around them, they would have been especially sensitive to judgments of this sort (our ancestors had, he claims, a "regard for the approbation and disapprobation of [their] fellows"). So they acquired a "habit" of reinforcing or overriding their desires and impulses in ways that would improve the conformity of actions "to the wishes and judgments of the community" (1871, p. 96).

Darwin's outline genealogy ascribes traits to human precursors and to the communities to which they belong. Our preethical ancestors are assigned the following characteristics:

1. Living in enduring communities with others.
2. Sometimes, but not always, responding cooperatively to the needs of their fellows. This involves both cognitive abilities (to detect the needs) and sympathy (to modify behavior).
3. A capacity for restraining some behavioral tendencies and reinforcing others.

The route away from the preethical situation is seen as extending two and three: Darwin's story supposes that the original sympathies are enlarged. We can see this as involving

4. An ability to identify patterns of behavior eliciting positive or adverse reactions.
5. The extension of item 3 to some of the identified patterns, thus increasing the frequency of positive reactions and decreasing the frequency of negative ones.
6. The use of language to describe such patterns and to mark some of them as approved and others as repudiated.

Once items 4 to 6 are in play, the hominid/human lineage has arrived at a primitive form of ethical life. The individuals involved meet Darwin's criterion for possessing a "moral sense": "[a] moral being is one who is capable of reflecting on his past actions and their motives – of approving of some and disapproving of others" (1871, p. 605).[5] Nevertheless, the scope of the actions subject to such reflections is surely limited, and the *grounds* for engaging in them are crude. Ancestral humans are

---

[5] In fact, when items 4 to 6 are operative, individuals not only reflect on the past but use their reflections as guides to current action.

troubled by the adverse reactions to some of their behaviors – more simply, they dislike being punished.

Darwin sees that the list of extensions so far offered is incomplete. He adds

7. The expansion of powers of sympathy beyond the local group.
8. The recruitment of other forms of motivation to reinforce or override impulses and desires (going beyond the "low motives").[6]
9. Extension of approbation and disapprobation to conduct that does not directly affect others (self-regarding virtues).

With these further extensions, something like genuine ethical life has emerged. The individuals with the expanded psychological capacities are closer to people we know. They act by recognizing particular features of the options before them, responding to those features with sympathy for others, with a sense of solidarity with a group, and with respect for the patterns of conduct it has commended. Sometimes, when matters are complicated, they weigh up aspects of the situation or they discuss with others. Even though they continue to dislike disapproval and to fear punishment, those emotions are embedded in a broader cluster of potential springs of action. If philosophers are tempted to seek in this cluster some core constituting the "moral point of view," that *is* temptation. Darwin's well-known gradualism extends to his treatment of ethical life, decomposing capacities philosophy tends to elevate as single powers: that is, after all, the point of the relevant parts of the *Descent*.

The genealogical tradition has deepened and further articulated the evolutionary narrative by studying the capacities appearing on Darwin's list and showing how they vary across species and intraspecific populations. Patricia Churchland's work offers us a detailed account of some neural mechanisms contributing to sociality and a rich description of variation in social capacities. She provides a thorough analysis of item 1 (in her terminology: the "platform" on which ethical life is built).[7] In a series of books and articles, Frans de

---

[6] Darwin discusses the motives of fear and self-interest in *Descent* (pp. 112 and 127), supposing the cooperative behavior originally undertaken for its selfish benefits may strengthen the capacity for sympathy.
[7] *Braintrust* (Churchland 2011); also "The Neurobiological Platform for Moral Values," in Vol. 151 of *Behavior* (I shall henceforth cite this issue of *Behavior* as B).

Waal identifies in apes and monkeys many of the capacities on the Darwinian list: complex forms of sympathetic response to conspecifics, restraint in the presence of higher-ranking animals, abilities to recognize patterns of behavior and their consequences, and a capacity for using that recognition to adjust conduct.[8] Michael Tomasello's research into cooperation and shared intentions in chimpanzees and young children provides establishes continuity between nonhuman and human awareness of some behavioral patterns as normative (see items 4 to 6 on the Darwinian list) (de Waal 2009). Christopher Boehm (1999, 2012) presents a picture of human social evolution that synthesizes a wide range of anthropological findings; a particularly relevant feature of his account is the emphasis on normative discussion, undertaken by all adult members of hunter-gatherer groups on terms of rough equality.[9] Kim Sterelny explores the consequences of cultural selection for hominid and human social life.[10]

This body of recent work shows that a Darwinian genealogy of ethical life is not simply speculation but a plausible hypothesis about the human ethical past. Shafer-Landau is on firmer ground, however, with respect to some later developments in the envisioned sequence. It is uncontroversial that during the late Paleolithic and early Neolithic, human communities grew in size. Deposits at sites reveal that 20,000 years before the present (20KYBP) human bands of the standard size (thirty to seventy members) were coming together for short periods of time. By 15–10KYBP, there is evidence of larger communities (up to 200 members), and the first cities (containing a thousand inhabitants or

---

[8] For sympathy, his *Good Natured* (1996) offers a treasure trove of instances; for restraint, see *The Bonobo and the Atheist* (2013) – the example of turn taking at pp. 149–50 is especially cogent; for the exercise of cognitive abilities in social contexts, *Chimpanzee Politics* (1982) and *Peacemaking Among Primates* (1989) provide a rich collection of examples. de Waal's synthesis of these contributions to Darwinian genealogy, which he sees as exhibiting the "building blocks" of morality, is given in *Primates and Philosophers* (2006). A concise summary, with some refinements of his earlier discussions, is given in "Natural Normativity," B, pp. 185–204.

[9] Boehm presents his own concise version of a Darwinian genealogy (coupled with selectionist explanations) in "The Moral Consequences of Social Selection," B, pp. 167–83.

[10] *The Evolved Apprentice* (2012a); he discusses my own approach in "Morality's Dark Past" (2012b) – his genealogy and mine offer two ways of elaborating the Darwinian version (although I think Sterelny makes too much of the differences between them).

more) date from 8KYBP. For the more distant past, the clues are much scarcer, although archeologists have built a strong case for the existence of trading networks stretching over a few hundred kilometers active even before 20KYBP.[11] Nevertheless, we cannot reliably fill in the details of how groups came to extend their moral frameworks to embrace outsiders (at least in trading contexts) or how exactly sympathies were expanded beyond the local band.[12]

With respect to items 8 and 9 on the Darwinian list, direct evidence from the Paleolithic and early Neolithic is even harder to obtain. The first written documents (from Mesopotamian cities, 5KYBP) demonstrate that citizens were expected to exhibit some self-regarding virtues and to be motivated by social solidarity and respect for the law.[13] Hence, at some point between the beginnings of ethical life and the invention of writing, our ancestors developed conceptions of self-regarding virtues, of social solidarity, and of respect for the law. It is, nevertheless, quite impossible to pinpoint the changes that occurred or to make responsible estimates of when they happened.

Attention to documents that have survived from the first two millennia of writing inspires two further additions to Darwin's list. The preambles that frame the legal codes, as well as myths and stories, provide conceptions of human relationships and explore the conditions under which human lives go well. Thus we might extend the list:

10. The development of a conception of relationship, identifying some relationships as valuable.

[11] The hypothesis of Paleolithic trading networks was originally advanced by Colin Renfrew (see Renfrew and Shennan 1982), based on the discovery of obsidian tools used at sites quite remote from the nearest source. Contemporary excavations provide extensive documentation of trading routes in the late Paleolithic (15KYBP), and some archeologists argue for significantly earlier development of trade in Africa (McBrearty and Brooks 2000).

[12] Darwin's own account of this is uncharacteristically flat footed. He suggests that enlarged sympathies came about as the consequence of uniting two small groups to form a larger social unit (1871, p. 119). This puts the cart before the horse. Sympathies would have required enlargement, at least to the extent of providing the local moral protections more broadly, before any such coalescence could occur.

[13] These features pervade what we think of as ancient legal codes (the Lipit-Ishtar Code, the Code of Hammurabi, and so forth) but which are plainly addenda to vastly more extensive systems of agreed-on rules, transmitted orally for hundreds of generations. The preambles often provide insight into expected forms of moral motivation.

11. An emerging conception of the worthwhile human life including more than the satisfaction of physical needs.

Here, too, the exact character of the intermediate steps cannot be determined. We may "speculate" – as Darwin did – that engaging in systematic cooperation with another human being might intensify the capacity for sympathy so that from an initial stage at which human beings approve cooperative interactions as a means for realizing benefits (for both parties or for one), they may come to see adjustment of conduct to the behavior of another as desirable in itself and ultimately even to be valued in the absence of the first-order benefits.[14] Thus they may have expanded the range of goals to which they aspired, envisioning a worthwhile life as demanding more than physical satisfactions.

A Darwinian genealogy must meet a condition: the steps it posits must be ones the recognized forces of evolution could have generated. As already explained, the elements of narratives offered by the genealogical tradition fall into two classes: those lending themselves to relatively detailed specification and those for which the clues are too scanty. With respect to the former class, it needs to be shown how the envisioned transition could have occurred – we need a "how possibly" explanation (explaining how the change was actually caused would be even better, but showing that it is compatible with evolutionary mechanisms suffices). For the latter class, it is necessary to construct a series of steps through which the psychological trait (often an expanded capacity) might have emerged and to show that those steps are consistent with evolutionary forces. For stages at which we have reason to believe that cultural transmission and selection might have played a role (including, but not limited to, the transitions occurring after the full acquisition of language), the appropriate complex of forces includes both natural and cultural selection. Because that combination typically allows for more possibilities than natural selection acting alone, it suffices to provide a "how possibly" explanation appealing only to natural selection.

Providing such accounts is not difficult. Churchland's review of the physiological bases of sociality across many groups allows a relatively precise analysis of the emergence of the "moral platform." The other

---

[14] See *Descent* (1871, p. 106) on strengthening sympathy by habit. I discuss enhancement of sympathy in *The Ethical Project* (2011, pp. 135–7) and at greater length in "Varieties of Altruism" (2010, pp. 121–48, particularly 133–6).

capacities on the list – increased cognitive powers to recognize and predict the behavior of conspecifics, expansions of sympathy, improved memory, and an enhanced ability to override or reinforce impulse to action – come with obvious advantages for individual performance or for social stability and increased cooperation.[15] From Darwin on, genealogists have had an easy time turning back the challenge that the sequence of transitions they envision is not compatible with the mechanisms of evolution.

Is the task *too* easy? Are the ground rules *too* lax? The answer is straightforward. To the extent that we can identify changes that must have occurred in the development of ethical life – changes pertaining both to individual psychology and to the organization of human societies – we can effect that illumination of moral practice for which Darwin hoped. Even without detailed knowledge of the timing and causes of the transitions, simply picking out those transitions as crucial decomposes the nebulous "trait" about which many philosophers speculate. Items 1 through 11 form the data for a philosophical account of ethical life. In the next section I'll offer part of my own preferred version (with apologies to fellow genealogists who might want to diverge at various points). We can then end the disputes that are the inevitable stuff of selectionism.

The question of progress in evolution is much debated. Many influential evolutionary theorists have been skeptical about claims to progress in the history of life. Skepticism concerning the "advance" represented (say) in the emergence of multicellular organisms is, however, quite compatible with supposing more limited types of progress – there are no general difficulties in making sense of the idea of progress in the ability to fly or to digest particular types of plants. A genealogy of ethics might also allow a limited notion of progress, supposing some transitions to constitute genuine advances. Some of the changes history records – the abolition of slavery and the abandonment of public performances in which human beings and other animals were forced to fight to the death – make the idea of an asymmetry in the direction of change hard to resist. Giving up the practice and returning to it are not equivalent.

---

[15] As Darwin himself saw. Embryonic explanations of the possibility of various transitions are scattered through chapter 4 of the *Descent*.

Moral realists can offer an apparently simple account of moral progress. Societies make moral progress when they replace false moral beliefs with true ones.[16] Questions about progress should be separated from issues about rationality or justification: to count a transition as rational is to claim it to be generated by processes with a privileged status (perhaps processes with a reliable tendency to generate replacement of false beliefs with true ones). Once this point is appreciated, there can clearly be rational moral transitions that are not progressive (the tendency for replacement of falsehood with truth is not exemplified in the case), as well as progressive moral transitions that are not rational. In principle, moral realists might suppose all our ethical ancestors to have been sleepwalkers, stumbling toward moral truth without any sense of where they were going. Yet moral realism cannot escape attributing *some* capacity for "tracking the moral truth." At least at the end of the genealogy, enlightened philosophical analysts have some such basis for making their judgments of progressiveness.

I propose to retain the concept of moral progress without the realist metaphysics.[17] I am motivated by wanting to hew closely to the phenomena revealed in the Darwinian account of the preceding section. Progress is not viewed as *toward* some hypothetical independent moral truth but *away* from the predicaments in which consecutive groups of our ancestors found themselves.[18] People make progress by solving problems.

We often think of progress in teleological terms – when we travel, we measure our progress by the diminishing distance to our destination. Yet there are familiar examples of progress *from*. Doctors make progress by finding ways to cure, treat, or palliate the diseases afflicting

---

[16] This rough formulation will do for present purposes, even though there are other modes of progress (replacing a state where a question is unsettled with a belief in the true answer, for example). More important, what is believed (part of the official code) may not be what guides action. For some complexities, see my article, "On Progress" (2015).

[17] Making sense of moral progress is the principal task of chapters 5 and 6 of *The Ethical Project* (Kitcher 2011), from which the material of the next few paragraphs is largely drawn. The views of that book are refined (and aligned with those of some of my fellow genealogists) in "Is a Naturalized Ethics Possible?" (Kitcher 2014, pp. 245–60).

[18] A concept of progress of this sort is deeply embedded in the pragmatist tradition and is especially evident in Dewey, an early subscriber to the genealogical approach to morality.

their patients. No teleology is required to make sense of their accomplishments: we don't need "medical realists" to invoke an ideal of perfect health, successively approximated by advances in research.

Pragmatic progress, progress as problem solving, is an obvious way to introduce a notion of progress into the Darwinian genealogy, for it simply replicates the ways limited notions of progress apply in thinking about evolution. The descendants of *Eohippus* are not taking (ever faster) steps toward the ideal horse but overcoming problems posed by their environments. Similarly for our own species in its preethical state. At the root of our problems lie the psychological features Darwin identified. Human beings are social animals, drawn together (in ways Churchland illuminates), but imperfectly equipped for their social environment. Our ancestors had a limited capacity to respond to those around them, a capacity sometimes leading them to discern the plans and intentions of other group members and to respond in ways that furthered those plans (sometimes by forming new joint intentions, as Tomasello has shown). Because that capacity was *limited*, however, they were often unresponsive, even persisting with their own prior activities in ways that thwarted the plans of their sometime allies. The ur-problem that sparked a sequence of transitions introducing partial solutions to it was the problem of "limited responsiveness."

Darwin's account implicitly recognizes the limitations in pointing to human desires to avoid "disapprobation." His contemporary successors offer a rich body of evidence about the existence and limitations of responsiveness in our evolutionary cousins: thanks to de Waal, Tomasello, and Boehm, we now understand far more about how chimpanzee and bonobo societies both cohere and break down and recognize what are plausibly the similarities between those societies and the hominid and early human modes of community life. The capacity for responsiveness is developed enough to preserve a group for relatively long periods of time, but its limitations require constant work of mutual reassurance and peacemaking; they constrain the amount of cooperative activity and set bounds on the size of a functional group. Occasionally, even prolonged activities of peacemaking are not enough, and the society falls apart. Although the lives of our actual ancestors were poor and relatively short, they were not always nasty or brutish and certainly not solitary – but they were often plagued by social tensions.

The ethical project responded to the ur-problem by introducing patterns of approved conduct, effectively amplifying the capacity for responsiveness and thus facilitating cooperation and decreasing the frequency and intensity of social disruption. Even before the acquisition of language, some such patterns may have been introduced and yoked to the prior ability to restrain or reinforce impulses to action.[19] After the emergence of language, I hypothesize that troublesome patterns could more easily be identified and that the moral practices of the small human groups (which remained at the chimpanzee-bonobo size until 20KYBP) were codified as they are in surviving hunter gatherers: through discussions among all adults, on terms of equality, in the "cool hour." This hypothesis does not simply assume that today's hunter gatherers continue a social arrangement once universal among our ancestors (who all interacted in similar ways with their environment). The ur-problem stems from our limited capacities for responding to others, and the most straightforward way of addressing that problem, once band members could talk to one another, was to foster a discussion in which all voices were heard.

As is the way with medicine and with technology generally, partial solutions to a problem generate further problems to be solved. So, too, with the ethical project. The genealogy of the preceding section recognizes important changes, both in individual psychology and in the structure of human societies. The latter is evident not only in the increase in group size but also in the differentiation of roles and the emergence of institutions that address specific contexts in which failures of responsiveness might arise (marriage and private property are obvious examples). The social changes provide material for new desires and emotions – human beings have come to want new forms of recognition and to yearn for particular types of relationship. Our conception of the good life has expanded beyond the horizons of the first participants in the ethical project, providing a far richer field on which human aspirations may conflict.[20]

[19] In *The Ethical Project*, I overemphasize the importance of linguistic formulations of norms. Tomasello (*Why We Cooperate*) has shown how norms can be recognized in the absence of language; Sterelny ("Morality's Dark Past") is right to chide me on this point.

[20] As *The Ethical Project* argues, the emergence of new problems out of partial solutions to the ur-problem causes complications for the notion of ethical progress, as well as being the source of many practical ethical difficulties. I argue that the ur-problem remains fundamental.

When some problems are addressed at a cost to partial solutions obtained for others, it is not always possible to make determinate judgments of progressiveness. My concept of ethical progress is pragmatic (progress *from*), local (focusing on transitions from one social state to its successor), and locally incomplete (not all transitions are awarded a determinate status). With respect both to the Darwinian genealogy and to recorded history, my concept categorizes some episodes as progressive and others as regressive, abstaining from judging the rest.

Moral realists view progress as acquisition of moral truth. I invert the relationship. In accord with Peirce's well-known approach to truth, moral truths are viewed as emerging as we make ethical progress.[21] More exactly, a moral statement counts as true just in case it, or some counterpart prescription, is introduced into ethical practice in a progressive transition and would be preserved throughout any indefinitely proceeding subsequent sequence of progressive transitions. A corollary of my thesis that the notion of ethical progress is locally incomplete is that some moral statements are neither true nor false.[22]

I'll close my attempt to provide a philosophical interpretation of the genealogy with a brief response to an important objection.[23] My approach to ethical progress may seem irremediably subjective. For the notion of a problem, or of a situation's being problematic, appears to depend on someone's regarding it as something from which relief is needed: the problem arises because the person wants to escape it. A "Berkeleyan" view of problems – in which to be a problem is to be perceived as one – cannot be quite right, for we recognize problems of which subjects are totally unaware (and, by extrapolation, the potential existence for problems of which nobody is aware). The deeper reason for rejecting the charge of subjectivism lies in the fact that problems often have an objective face, leading us to move beyond the subject's wishes to features of the environment that generate them. Cystic fibrosis patients in crisis, struggling for breath, have a problem to which we cannot – or *should* not – react by declaring that it's only

---

[21] See Peirce (1934). William James (1978) expresses a similar thought in his suggestion that "truth *happens* to an idea" (p. 97).

[22] This is elaborated in section 38 of *The Ethical Project*, where I relate my treatment of ethical truth to Isaiah Berlin's pluralism.

[23] Originally posed to me by Sharon Street, to whose thoughtful critique I am indebted.

a problem because they perceive it as such. The desire to breathe is not idiosyncratic. Place any developmentally normal person in the predicament and that person would want relief. The objection rightly appreciates that problems vary in their objectivity, according to the inclusiveness of the class of subjects who would want relief. Because the difficulties generated by our limited responsiveness to others pervaded the social environment in which our ancestors lived, any developmentally normal human subject would desire to escape that situation. My ur-problem achieves the maximal degree of objectivity. That is all a defense of my analysis of moral progress could wish for.

In conclusion, let's return to the selectionist approach to metaethics and to the debate generated by Street's "Darwinian dilemma." I'll proceed in two stages. First, assuming that the Darwinian genealogy does not already reveal any capacity for "tracking moral truth," I'll ask whether some possible extension of it could have yielded the favored capacity. Second, I'll outline an argument related to Street's, suggesting that it provides challenges for moral realism.

Earlier, I offered a strategy for the moral realist: claim selection shaped psychological dispositions, whose joint employment could give rise, in social environments with cultural transmission, to a process of inquiry able to disclose independently existing values. Implementing this strategy is easy: we need only attend to a facet of the evolution of morality about which I've hitherto deliberately been silent. The linkage of morality to religion is a well-known feature of the contemporary world: it provides the most popular "folk theory" of morality. For more than a century, anthropologists have recognized that this link pervades almost all known societies (Westermarck 1926). Its selective value is readily recognized. Ask tribespeople why they follow moral rules when they are no longer visible to their fellows, and they explain that they are *always* under observation (by ancestors, spirits, local deities, or whatever); if they break the rules, disaster will descend upon them, their families, or even the whole tribe. If, as genealogists typically suppose, conformity to agreed-on rules increases cooperative benefits and thus (ultimately) reproductive success, devices fostering conformity are favored by natural selection. So, introducing a version of the folk theory, a kind of moral realism, is not at odds with Darwinian mechanisms.

Once the folk theory is accepted, the course of moral deliberation can be modified. When we sit around in the "cool hour," sharing our perspectives as equals, nobody can impose his or her favored ideas about the rules. But what if one of us devised a means of convincing the rest that he or she had a special power, an ability to fathom the will of the transcendent police officer? It's easy to understand how, under cultural selection, a special role might emerge, one held by people who *claimed* the capacity to fathom the moral order. As, in fact, it has.

Moral realists need only amend the story a bit. Members of an extremely sophisticated tribe come to suppose that the rules they have constructed in their deliberations conform to an independent moral order. Tribespeople add the less sophisticated belief that "the moral order strikes back": violate it, and there will be trouble. Sometime later, fortunate individuals actually acquire the capacity to detect the features of this order. They convince others (or maybe most of the group comes to have the ability?) and from this point on, the code is modified not by the old style of deliberation but by careful detection of what is objectively morally required. Some previous rules might be overridden – and perhaps there are protests from people whose perspectives suffer in consequence. Would this lead to a regression in cooperation and thus selective disadvantages? Not necessarily. For the environment is now adjusted, most probably by introducing punishment for breaches of the new code, and correctly convinced of the powers to detect the moral order, erstwhile protesters fall into line. As the emotion of respect for the moral order becomes more prevalent, the primitive belief in the vengefulness of that order is discarded. The descendants of this tribe correspond to the realist's image of the enlightened contemporary moral subject.

This story is easy to tell partly because a variant of it belongs to the actual genealogy (and the divergent features are simple to adjust) and partly because of a "folk theorem" of contemporary evolutionary studies. Cultural selection is by no means restricted to *Homo sapiens*, and it was surely a significant force in human evolution long before we acquired language. Thereafter, it has been a dominant influence on our psychological traits and on our social organization. Once this is understood, there should be no astonishment that Street's dilemma is toothless – or that it doesn't need complicated abstract principles (typically controversial) to reply to it.

Yet moral realism is not entirely off the hook. There is an argument in the vicinity of Street's dilemma posing a serious challenge to it. The argument starts from observing how vague accounts of the independent moral order and our access to it typically are. How exactly is moral realism connected with the genealogy of ethical life? Consider an obvious parallel. We cannot understand the growth of our understanding of other aspects of nature without some view of what those aspects are and how people have entered into different cognitive relations to them: explaining our increased grasp of heredity involves recognizing what genes are, how they relate to chromosomes, what the genetic material is, and how Mendel, Morgan, Watson, Crick, and many others interacted with these constituents of reality. The Darwinian genealogy and the philosophical interpretation of it in reveal that nothing similar is required in explaining ethical progress. Where does moral realism fit in?[24]

At this point, moral realists have three options. They might accept the genealogy and argue that it is better interpreted by viewing our ancestors as successively discovering more about the moral order. Alternatively, they might suppose the genealogy to be incomplete – somewhere along the way, people really did acquire capacities for tracking independent moral truth. This latter approach might be developed in either of two ways: by supposing that the capacity became available to all or that it was only fully developed in certain privileged people.

To pursue the first option would be to view the discussions through which past societies have worked out solutions to the ur-problem as embodying a collective method for fathoming moral truth. This is effectively to reintroduce an unnecessary teleology, to envision some ideal of collective human flourishing against which ethical progress is to be measured – an analog of my imagined "medical realist's" picture of perfect human health. Moral realists who aim to pursue this option must address two challenges: Can they either avoid this teleological conception or justify it? What, if anything, does it add to the (constructivist) account of ethical progress as problem solving?[25]

The second option, probably that favored by the majority of moral realists, hypothesizes an individual capacity for discovering moral

---

[24] Similar questions arise about moral changes in recorded history. I use these to make a similar point in chapters 4 and 5 of *The Ethical Project*.

[25] In a forthcoming essay, "Tracking the Moral Truth," Paul Bloomfield offers a version of moral realism that pursues this first option. His proposals deserve a more detailed treatment than I can offer here, and I shall content myself with issuing these two challenges.

truths that has been overlooked in the genealogy I have presented. Realists who favor this idea face the challenge of specifying the capacity and moral reality to which it allows access, identifying when it arose, explaining how progressive transitions made prior to its emergence were able to disclose facets of the moral order, and showing how it operated once it had been acquired. That is a tall order.

But the realist's predicament is worsened by the need to decide between the democratic and the elitist versions of the preferred story. If the capacity is supposed to be present in all of us, realists should explain the stubborn persistence of disagreement, especially at times of moral advance. Interestingly, the great ethical theorists who have emphasized democracy with respect to individual moral capacities have inclined toward a form of constructivism easily linked to the genealogical interpretation I have offered.[26]

The elitist version of individualistic moral realism is worse, for it introduces the idea of a moral judgment that could override the best collective attempt at resolving the conflicts besetting our lives together. Moral realists will surely have been unhappy with my way of resolving Street's dilemma on their behalf – the story told at the beginning of this section. Yet, if they insist on an individual "tracking" capacity that can trump any ideal collective deliberation, they are effectively acquiescing in that historical modification of ethical life in which authority passed from the group to the religious teacher. They are accepting a distortion of the ethical project (albeit in a more abstract form).[27]

The genealogical account I have presented portrays human communities as engaged in a collective project to solve the problems descending from the ur-problem. Deweyan pragmatism extends the Darwinian hope of illuminating ethical life by supposing the illumination can guide our modification of the practices we have inherited: we see what we have been up to, and that helps us to

---

[26] I have in mind the approaches taken by Adam Smith and Kant (especially in the third formulation of the categorical imperative). In both instances, the preferred methods might be seen as attempts to simulate the collective discussion in the mind of an individual subject.

[27] Ironically, distinguished moral realists sometimes appreciate the point: Ronald Dworkin's *Religion without God* (2013) sees the independent order of values as replacing traditional theism; similar considerations seem to underlie Thomas Nagel's *Mind and Cosmos* (2012).

see how to go on.[28] Because human interconnections are now so vast and complex, we can no longer pursue the ethical project in the way it was developed through most of its history. We cannot sit down, as our ancestors did, working for a solution acceptable to all. Although it is possible to envision a collective ideal, a conversation in which all perspectives are represented, factual errors corrected, and all parties concerned to address the claims of others, that ideal cannot be realized. At their best, the great ethical theorists of the past supply tools with which we might try to fashion acceptable approximations to it.[29] Darwinian genealogy helps us to reframe their efforts. Within that frame, I suggest, moral realism comes to seem either irrelevant or, worse, a continued distortion of the ethical project.

[28] Appreciating this point is the key to addressing the frequently voiced complaint that (Darwinian) naturalism must commit a fallacy. For a brief response, see "Is a Naturalized Ethics Possible?"

[29] In effect, Adam Smith's treatment of sympathy and his construction of the impartial observer, as well as Kant's conception of legislation in the kingdom of ends, are built into the conditions of the ideal ethical discussion. It should thus not be surprising that genealogists like Darwin and Dewey incorporate ideas from these traditions in their accounts. Genealogists also can learn from those who have explored ways in which historical developments have constrained human lives: Rousseau, Marx, Nietzsche, and Foucault are all pertinent.

10 | *Human Nature*

EDOUARD MACHERY

According to the traditional "essentialist notion of human nature," human beings have in common a set of properties that are separately necessary and jointly sufficient for being human. Following trenchant criticisms by biologists and philosophers of biology (Hull 1986; Ghiselin 1997; Kitcher 1999), a consensus emerged in philosophy that this essentialist notion is untenable. At the same time, many think that it is important to develop a notion of human nature that is consistent with advances in genetics and evolutionary biology.[1] This "successor notion" should meet two conditions of adequacy. First, it should not fall prey to the objections formulated against the essentialist notion (Machery 2008; Griffiths 2009, 2011; Stotz 2010; Samuels 2012; Ramsey 2013). Second, it should not be anemic: it should fulfill some or, if possible, many of the functions the traditional essentialist notion of human nature was supposed to fulfill (Samuels 2012). In previous work (Machery 2008; see also Machery and Barrett 2006), I have argued that the "nomological notion of human nature" is immune to Hull's and others' objections, but I did not show that this notion can fulfill at least some of the functions that the essentialist notion of human nature was supposed to fulfill. The goal of this chapter is to show that the nomological notion of human nature is a satisfying successor notion to the essentialist notion: in addition to being consistent with advances in genetics and evolutionary biology, it fulfills exactly the roles that should be fulfilled without fulfilling those roles that, I argue, should be left unfulfilled (for a very similar dialectical setting, see Samuels 2012).

Here is how I will proceed. In the first section of this chapter, I will introduce the nomological notion of human nature. In the second

Thanks to Liam Bright, Grant Ramsey, and David Livingstone Smith for comments.

[1] For review of recent debates, see Downes and Machery (2014), and Kronsfeldner, Roughley, and Toepfer (2014). Some remain skeptical of the notion of human nature, for example, Lewens (2012).

section, I describe five functions that the essentialist notion of human nature was supposed to fulfill. In the remainder of this chapter, I will examine whether the nomological notion fulfills the functions that are worth fulfilling. The third and fourth sections discuss whether the nomological notion of human nature can be useful to characterize human beings and to distinguish them from nonhuman animals. The fifth section examines whether, and in which sense, the nomological notion can make sense of explanations of human characteristics that appeal to human nature. And the final section examines whether, and in which sense, one can turn to human nature to circumscribe human flexibility.

Why Develop a Successor Notion?

I will not repeat here the well-known arguments against the essentialist notion of human nature, but it is worth asking why one would want to develop a successor notion. Why not simply eliminate the notion of human nature from science, as some have suggested (e.g. Hull 1986)? There are at least three reasons for developing a successor notion of human nature.

First, the notion of human nature is used in some influential and successful research programs in the behavioral sciences – including generative linguistics, which attempts to determine the universals (if any) that underlie the diversity among languages (e.g. Chomsky and Foucault 2006), the nativist research program in developmental psychology (e.g. Carey 2009), the work of many comparative psychologists with an evolutionary bent (e.g. Frans de Waal 2009), and much of the research in the evolutionary behavioral sciences (e.g. Tooby and Cosmides 1990; Richerson and Boyd 2005).[2] Let's consider first some quotations. Psychologist Paul Bloom argues that fairness is part of human nature:

What we do see at all ages ... is an overall bias toward equality. Children expect equality, prefer those who divide resources equally, and are strongly biased to divide resources equally themselves. This fits well with a certain picture of human nature, which is that we are born with some sort of fairness instinct: we are natural-born egalitarians. (2013, pp. 65–6)

---

[2] It is not uncontroversial whether these research programs are really successful (see, for example, Evans and Levinson (2009) on generative linguistics).

Similarly, evolutionary biologist and cognitive scientist Tecumseh Fitch notes that the search for linguistic universals in generative linguistics is connected to the development of a theory of human nature. "Understanding this broadly shared basis for language, whatever it might be, was seen as central to understanding human nature by many eighteenth-century philosophers" (2011, p. 378).

Evolutionary biologists and anthropologists Peter Richerson and Robert Boyd describe their overall project in the following terms:

In the case of ordinary learning, individuals must have some way of weighting the importance of the value of L [the trait acquired] that they acquired by imitation against the value that their experience indicates is the best. Do they rely on their experience or on imitation? In the case of biased transmission, individuals must have some criteria of success – do they imitate wealthy individuals? ... *Ultimately, these are questions about human nature.* The answers must be thought in the long-run processes that govern the interactions of cultural and genetic evolution in our species. (Richerson and Boyd 2005, pp. 392–3, emphasis added)

In an article published in *Science*, economist Herbert Gintis describes some recent modeling of cooperation in human beings as follows:

The standard view holds that human nature has a private side in which we interact morally with a small circle of intimates and a public side in which we behave as selfish maximizers. Herrmann et al. [2008] suggest that most individuals have a deep reservoir of behaviors and mores that can be exhibited in the most impersonal interactions with unrelated others. This reservoir of moral predispositions is based on an innate prosociality that is a product of our evolution as a species, as well as the uniquely human capacity to internalize norms of social behavior. Both forces predispose individuals to behave morally even when this conflicts with their material interests. (Gintis 2008, p. 1345)

The notion of human nature earns its keep from the successful nature of the theories in which it is embedded. Because it is unclear, however, how to understand it – scientists rarely pause to explain what they mean by this notion – philosophers of science ought to explicate it.

Naturally, the fact that some scientists characterize their own research as an inquiry into human nature does not mean that the notion of human nature actually plays a role in their theories and is thus an ineliminable part of the conceptual apparatus of their research. It is not uncommon in the history of science to find scientists appealing to some notions that, on closer examination, turn out to be eliminable. For instance, Kitcher (1993, pp. 148–9) shows that in the nineteenth century the notion of ether was often used in thermodynamics but in fact played no role, which explains why thermodynamics was able to provide genuine explanations and predictions despite employing an empty notion. As Kitcher put it

The ether is a prime example of a presuppositional posit, rarely employed in explanation or prediction, never subjected to empirical measurement ... yet seemingly required to exist if the claims about electromagnetic and light waves were to be true. (1993, p. 149)

Admittedly, the notion of human nature is not the most central explanatory notion in the research programs alluded to earlier, but it is not explanatory idle either. As we will see later, the notion of human nature is an instance of a particular type of explanatory notion that has its place in science.

Second, as Paul Griffiths has noted,[3] laypeople often embrace some notion of human nature, and this notion is likely to be influenced by a flawed folk-biological conception of development (Griffiths, Machery, and Linquist 2009; Linquist et al. 2011). It would probably be difficult to eradicate laypeople's flawed notion of human nature. A simpler course of action consists of developing and promoting a successor notion of human nature that takes into consideration evolutionary and molecular biology.

Finally, and most important, the essentialist notion of human nature was supposed to fulfill several functions (see below). We need a notion to fulfill those functions that should and can be fulfilled. That notion is the successor to the essentialist notion of human nature.

The Nomological Notion of Human Nature

The "nomological notion of human nature" (Machery 2008) does not fall prey to the objections usually raised against the essentialist notion of human nature. According to the nomological notion, human nature is the set of properties that humans tend to possess as a result of the evolution of their species.[4] Being bipedal is part of human nature

---

[3] At the 2010 meeting of the Philosophy of Science Association in Montreal.
[4] One may object that evolutionary forces causally influence *all* traits. For discussion, see section 3.2 of Machery (2008).

because most humans are bipedal animals and because bipedalism is an outcome of the evolution of humans. The same is true of fear reactions to unexpected noise and the capacity to speak.

According to the nomological notion of human nature, the properties that constitute human nature are neither separately necessary nor jointly sufficient for belonging to the species *Homo sapiens*: Some human beings have no fear reaction to unexpected noise, but they are human all the same. In fact, so understood, human nature does not propose conditions for belonging to *Homo sapiens*; it merely describes what human beings tend to look like by virtue of evolution. Furthermore, the properties that constitute human nature need not be distinctive of human beings; rather, they can be shared by other species: For instance, fear reactions are found in many species. It is an empirical question – one that cannot be answered *a priori* – whether a candidate property such as the capacity to smile or a moral sense belongs to human nature. Science, not literature or philosophical thinking, holds the keys of human nature. As Bloom put it (2013, pp. 3–4), "We can explore our moral natures using the same methods that we use to study other aspects of our mental life, such as language or perception or memory." Finally, human nature changes with the evolution of our species. It is likely that the current human nature is to some extent distinct from the human nature of our ancestors 100,000 years ago. For instance, a dark skin color may have been part of human nature until the recent evolution of lighter skin pigmentation (e.g. Wilde et al. 2014).

The nomological notion of human nature is ecumenical about the nature of the evolutionary processes that create human nature. The traits that are part of human nature can be adaptations, by-products of adaptations, outcomes of developmental constraints, or neutral traits that have come to fixation by drift or that result from a founder effect.

According to the nomological notion of human nature, polymorphic traits are not part of human nature when they are not widely shared among humans in general. If males and females have different evolved mating psychologies, the properties of male and of female mating psychologies are not part of human nature. One could object to this exclusion on the grounds that having, say, a bimodal mating psychology is an important characteristic of our species and that the notion of human nature should reflect it, but it turns out to be impossible to exclude any trait from human nature once this type of bimodal trait is included in it (Machery 2012).

It is an empirical question whether human nature is "thick" – whether humans have many properties in common because of the evolution of *Homo sapiens* – and if human nature is thick, this is a *contingent* fact that calls for an explanation. Adaptations vary across subgroups of a given species when these live in different environments for a sufficient amount of time and when gene flow between them is limited. Breeds of dogs illustrate this phenomenon. Behavioral (e.g. reproductive strategies in pygmy swordtail) and physiological (e.g. body size polymorphism among elephant seals) polymorphisms (due to frequency-dependent selection or to sexual selection) can be common in a species. The norm of reaction of a given species also can be such that different subgroups living in different ecological niches have very different phenotypes, even if there is little genetic variation across groups (e.g. the morphology of the arrowleaf). For these three reasons, it is possible for conspecifics to have fewer properties in common than the members of other species, and the nature of a given species can be thinner than the nature of another species. The extent to which human nature is thick remains an unsettled question among evolutionary behavioral scientists, and some scholars highlight the plasticity of human phenotypes (e.g. Sterelny 2003, 2012). To illustrate, behavioral ecologist Eric Alden Smith (2011, p. 326) writes that "[i]t would be going too far to say that it is the nature of humans to have no nature. But the kernel of truth in that statement is that our species has extraordinary capabilities for generating behavioral diversity independently of underlying genetic variation."

The nomological notion of human nature stands in sharp contrast with the Aristotelian conception of human nature. For Aristotle, human nature stands in a *causal* relation to the properties shared by humans: Human nature causes and causally explains why human beings tend to have the properties they tend to possess (see, in particular, Lennox 2001). Walsh (2006, p. 430) summarizes recent scholarship on Aristotle's essentialism as follows:

Aristotle's essentialism, then, should be seen as an explanatory doctrine, rather than a taxonomic one. Organismal natures play a teleologically basic role in explaining why organisms have the traits they have and why they resemble one another in the ways they do. Natures do not play a role in demarcating natural kinds united by the common possession of structurally identical features.

In contrast, according to the nomological notion, human nature is not a cause; it consists of properties that are caused by various evolutionary processes.

The nomological notion of human nature has several virtues. First and foremost, it does not fall prey to the objections raised against the essentialist notion of human nature, and it is compatible with evolutionary, developmental, and molecular biology. (I won't repeat the arguments for this claim here: see Machery 2008.) Thus it meets the first condition of adequacy for a successor notion of human nature put forward in the introduction.

Second, it offers a better explication of what evolutionary behavioral scientists typically mean by "human nature" than the essentialist notion of human nature. Consider, for instance, Richerson and Boyd's quote earlier: the questions they are interested in "ultimately ... are questions about human nature," and "[t]he answers must be thought in the long-run processes that govern the interactions of cultural and genetic evolution in our species." It is clear from this quotation and from the rest of their work that "human nature" does not refer to a set of properties that defines membership in *Homo sapiens*; rather, when Richerson and Boyd use "human nature," they have in mind the properties that are common among humans, and they explicitly tie these properties to evolutionary processes. More generally, many evolutionary behavioral scientists are too sophisticated to conceptualize human nature along essentialist lines.

For all these virtues, one might worry that the nomological notion of human nature fails to meet the second condition of adequacy put forward in the introduction: it is not robust; it fails to fulfill the functions that the traditional notion of human nature was meant to fulfill. The remainder of this chapter addresses this concern.

Traditional Functions of the Notion of Human Nature

What is the point of having a notion of human nature? What does this notion do for us? This section reviews the functions the notion of human nature was traditionally supposed to fulfill.

The notion of human nature is meant to tell us what human beings are like; it tells us what properties human beings possess. Following Samuels (2012), I will call this function the "descriptive function." Bloom's use of the notion of human nature in the quotation provided at the beginning of this chapter illustrates this function. The essentialist notion of human nature fulfills the descriptive function because human nature is conceived as a set of properties that are separately necessary and jointly sufficient for being a human being: These properties are what human being are like. If it is part of human nature, conceived along essentialist lines, that human beings must have a capacity for speaking, as Descartes held, then what a human being is like involves being able to speak.

A second traditional function of human nature, which I will call the "taxonomic function," is to draw a line between human beings and other animals. Human beings, *and only them*, possess a human nature. Antony (1998, p. 75) puts the point as follows: "The first [idea], deriving from Aristotle, is that natures should be, in some sense, 'definitional.'"[5] The definition of a class determines what belongs to this class and what doesn't. If bachelors are defined as unmarried males, this definition distinguishes who belongs to the class of bachelors (e.g. Kant and Beethoven) and who doesn't (e.g. Barack Obama). The essentialist notion of human nature fulfills the taxonomic function because human nature is conceived as a definition.

A third traditional function of human nature, which, following Samuels (2012) again, I will call the "causal-explanatory function," is to explain causally why human beings are the way they are. Antony (1998, p. 75) puts the point as follows: "From Locke, we get the second idea of essence as an underlying explanatory structure."[6] People appeal to human nature when they want to account for a particular behavior. Because they are more salient, behaviors or characteristics people disapprove of, such as jealousy and greed, are more likely to be explained in terms of human nature, but behaviors people approve of can also be so explained. An Aristotelian approach to human nature would naturally fulfill this function because, as we saw earlier, it identifies human nature with the cause of the typical characteristics of human beings.

---

[5] I doubt Antony's historical ascription of the taxonomic function to Aristotle is accurate. For relevant discussion, see Lennox (2001) and Winsor (2003, 2006).

[6] I am skeptical of Antony's historical ascription of the causal-explanatory function to Locke.

Similarly, the essentialist notion of human nature often took the essence of human beings to explain human beings' typical characteristics.

When it fulfills the causal-explanatory function, human nature can be ideological. It can provide an erroneous and misleading explanation of some unequal and contingent states of affairs, presenting them as if they were not the product of human activities. As Foucault put it in his dialogue with Chomsky (Chomsky and Foucault 2006, p. 43): "Doesn't one risk defining this human nature ... in terms borrowed from our society, from our civilization, from our culture?" Unsurprisingly, human nature has long been put to use for oppressive ideological purposes, a point feminist philosophers have clearly established (e.g. Jaggar 1983; Antony 2000).

Traditionally, the notion of human nature was also meant to circumscribe human beings' behavioral flexibility, a function I will call the "limitation function." This idea takes at least three forms. First, human nature can determine what is possible and what is impossible. In contrast to most birds, human beings cannot fly without technological assistance: it is not part of human nature, traditionally conceived, to fly. The essentialist notion fulfills the limitation function so conceived because a human being must possess the properties that define human nature. A second take on the limitation function is less concerned with distinguishing what is possible and impossible for human beings than by distinguishing what human beings do well and what they do poorly: people can be successful at doing what is part of human nature but are bound to do what is beyond human nature poorly. In *Observations on the Feeling of the Beautiful and Sublime*, Kant (2011, p. 49) wrote that "whatever one does contrary to nature, one always does very poorly." The essentialist notion of human nature also fulfills the limitation function so understood because many of the typical properties of human beings are taken to be caused by the properties that humans must have to be human. Third, if one could somehow succeed at overcoming the limitations imposed by human nature, this would come with severe costs, an idea that is present in some contemporary appeals to human nature. Thus Wilson asserted in *On Human Nature* (1978, p. 47) that "human nature is stubborn and cannot be forced without a cost."

The limitation function of human nature can easily acquire political and social overtones. When it fulfills this function, human nature tells people that there are things that they should not attempt to do because doing them is simply impossible, because they can't be done well, or because doing them would be costly. Including something in human nature thus can prevent people in general or particular groups from attempting to modify or eliminate it. In particular, this inclusion can prevent members of oppressed groups (the working class, racial minorities, or women) to attempt to overcome the boundaries society imposes on them. It is no accident that human nature has often been appealed to in order to maintain the oppression of women. Thus, in *Observations on the Feeling of the Beautiful and Sublime*, Kant (2011, p. 36) wrote that "[l]aborious learning or painful grubbing, even if a woman could get very far with them, destroy the merits that are proper to her sex." A fifth traditional function of the notion of human nature is to provide norms bearing on the permissibility or worth of actions, character traits, behaviors, or ways of life. Some dispositions (e.g. homosexuality) have often been judged to be wrong because contrary to nature. Antony (1998, p. 65) refers to the "normative ... connotations carried by the notion of 'the natural.'" As an illustration of this normative function of the notion of human nature, Rousseau wrote in *Emile* (1979, p. 327): "Do you wish always to be well guided? Then always follow nature's indications. Everything that characterizes the fair sex ought to be respected as established by nature." The normative function of human nature can naturally acquire political and social overtones, as illustrated by Rousseau's quotation, and as a result, human nature can be used for oppressive purposes.

In contemporary philosophy, the notion of human nature has been used extensively for normative purposes, in particular, by Neo-Aristotelians like Foot (2001) and Thompson (2008).[7] Foot writes (2001, p. 24)

[A] moral evaluation does not stand over against a statement of a matter of fact, but rather has to do with facts about a particular subject matter, as do evaluations of such things as sight and hearing in animals, and other aspects of their behavior. Nobody would, I think, take it as other than a plain matter of fact that there is something wrong with the hearing of a gull that cannot distinguish the cry of its own chick, as with the sight of an owl that cannot see in the dark. Similarly, it is obvious that there are objective, factual evaluations of such things as human sight, hearing, memory, and concentration, based on

[7] See also Setiya (2012, chap. 4).

the life form of our own species. Why, then, does it seem so monstrous a suggestion that the evaluation of the human will should be determined by facts about the nature of human beings and the life of our own species?

The nature of a given species determines what a well-functioning member of this species looks like, which allows us to assess whether a particular conspecific is in some way deficient. With respect to human beings, human nature determines what a well-functioning human looks like in the theoretical and practical domains, and it is part of human well functioning to behave rationally.

Space prevents discussing the normative function of the notion of human nature and its relation with successor notions of human nature with the required care. Suffice it to say here that the most rudimentary and least sophisticated appeals to the notion of human nature (of the kind: "If doing *x* does not belong to human nature, it is wrong") are indefensible and that a candidate successor notion of human nature should not attempt to fulfill them but that there may be room to put the notion of human nature to normative uses.

In the remainder of this chapter I will examine whether, and in which sense, each of these functions should be fulfilled and whether the nomological notion of human nature can fulfill those that should. I begin with the descriptive function.

Does the Nomological Notion of Human Nature Fulfill the Descriptive Function?

The nomological notion of human nature fulfills the descriptive function; indeed, it was developed to fulfill it. The crucial insight this notion builds on is that there is something it is like to be a human being and that the notion of human nature is used to refer to what it is like to be a human being. Bipedalism is a characteristic of human beings, and it is part of human nature.

The nomological notion of human nature fulfills the descriptive function in a distinctive manner. First, it includes only part of what human beings are like because for a trait to belong to human nature according to the nomological notion of human nature, it is not sufficient that it be characteristic or typical of human beings; it must also have an evolutionary etiology. Even if everybody became a fan of K-pop, this musical taste would still not be part of human nature because its distribution would not be explained evolutionarily.[8] Second, in contrast to the essentialist notion of human nature, it does not focus on traits that are unique to human beings. Many traits that are typical of human beings because of evolutionary processes are shared by primates or by mammals. These two distinctive features are valuable. The first one endows the description of human nature with some stability: fast social and cultural processes do not change human nature. The second highlights the fact that in many respects human beings are like other primates and other mammals and that a description of what human beings are like that overlook this fact, as traditional notions used to, is deficient.

Because the nomological notion of human nature fulfills the descriptive function, human nature has predictive power: this allows scientists and laypeople to make probabilistic predictions about how people are going to behave in particular situations. This is line with the use of the notion of human nature in the sciences. For instance, Gintis (2008, p. 1346) makes the following prediction about a behavioral-economics game: "Because the four subjects are strangers, the standard view of human nature suggests that there will be zero contributions."

Other Candidate Successor Notions of Human Nature

Not all candidate successor notions of human nature currently on offer fulfill the descriptive function as well as the nomological notion. According to Ramsey's life-history trait cluster, human nature includes all the possible life-history trajectories of all the possible human beings (2013, p. 987): "If one were to take all of the possible life histories that form the basis for individual nature, and then combine them, one would possess the set of life histories that forms the basis for human nature." Importantly, in his view, human nature is not composed of traits such as altruism, moral sense, or color-vision relations but rather of relations between traits, for instance, the relations expressed by "If a child lives in a scarce environment, he or she will reach puberty earlier" and by "If a child grows up a in violent environment, he or she will not trust people" (Ramsey, personal correspondence). The relation

---

[8] To clarify, by "evolutionary explanation," I only have in mind explanations referring to organic evolution and not any explanation that appeals to Darwinian principles. In particular, a universal taste for K-pop could plausibly be explained in cultural-evolutionary terms.

between the set of all possible individual life histories and these conditional traits could bear some clarification. It is unclear whether conditional traits or the possible individual life histories are the parts of human nature in this view; if the latter is the right interpretation, then the conditional traits are causal generalizations grounded in human nature. Furthermore, it is not fully clear why human nature could not ground nonconditional traits such a color vision.

Be that as it may, Ramsey's (2013) account of human nature fails to fulfill the descriptive function in a satisfactory manner. Of course, in a sense, Ramsey's candidate successor notion of human nature does fulfill this function: it characterizes what human beings are like. Indeed, it characterizes all the ways all human beings could be! Yet, in another sense, the descriptive function is left unfulfilled: to describe all the ways the members of a kind could be is a poor answer to the question, "What are the members of this kind like?" A proper answer would provide information about the typical or diagnostic properties of the kind members.[9] In addition, Ramsey's notion of human nature seems to have little predictive power. Because every phenotype that a human being could have belongs to one of the life histories included within human nature, on this notion one cannot justifiably infer that a human being is likely to possess a trait from the fact that this trait belongs to a life history included within human nature. The kind of appeal to the notion of human nature illustrated by Gintis' quotation earlier in this chapter thus would have to be abandoned if the life-history trait cluster account of human nature were the best successor to the essentialist notion.

Ramsey would likely respond that the life-history trait cluster account has the resources to address these two concerns because it leaves room for distinguishing different kinds of associations between traits grounded in human nature. Some life-history patterns (i.e. associations between traits or conditional traits) will be much more common than others, and the notion of human nature will identify those as such. As Ramsey put it (2013, p. 989, emphasis added), "[B]ehaviors occupying a particular region [of a human nature space] are *core* features of human nature,

---

[9] This objection applies equally well if one conceives of human nature as composed of nonconditional traits or as grounding conditional traits. Even if the latter is the case, human nature poorly describes human beings if it grounds every conditional trait that could be true of every possible human being. The same is true of the other concerns expressed later.

while those in other parts of the space are less central." As a result, it will be possible to appeal to human nature to predict how a human being would behave, as Gintis does. For instance, the set of possible life histories could ground the conditional that people are unlikely to be altruistic when they interact with strangers, which would allow us to predict people's behavior in the experiment described by Gintis.

This response is not entirely successful because it concedes that one cannot justifiably infer that a human being possesses some trait by appealing merely to human nature, as Gintis does. The inference schema Gintis seems to be embracing would have to be replaced by the following one:

> $P(s[\text{conditional trait}])$ is a "core" property grounded in human nature.

> A human being is likely to possess property P.

While there is no inconsistency in this proposal, the life-history trait cluster account of human nature involves more reform than the nomological account in this respect: in some contexts, it must replace the notion of human nature with the notion of core human nature.

The Taxonomic Function

The nomological notion of human nature is not meant to fulfill the taxonomic function and indeed does not fulfill it. Properties that are part of human nature are not necessary for being a human being; indeed, many of them are common among primates and mammals (e.g. fear reaction) either because they are homologous (e.g. smile in chimpanzees and human beings) or analogous (e.g. color vision in human beings and some bird species) traits. Together they are not modally sufficient: an organism could possess all the properties in human nature without being a human being.

The fact that the taxonomic function is left unfulfilled should not be chalked up against the nomological notion of human nature because it should not be fulfilled. What makes a human being human consists merely of the fact that his or her parents were human (Hull 1986). The distinction between human beings and other animals is a genealogical matter, and it does not depend on properties possessed by human beings and other animals.

The nomological notion of human nature embraces this lesson from evolutionary biology and separates the effort to characterize human nature from the attempt to distinguish human beings from other animals. Aliens that would be indistinguishable from human beings would have a human nature without being human because they have no ancestor in common with any human.

Does the Nomological Notion of Human Nature Fulfill the Causal-Explanatory Function?

The causal-explanatory function seems to be a challenge for the nomological notion of human nature. In contrast to the Aristotelian approach to human nature, human nature is not viewed as a cause; rather, it is constituted by the outcomes of various evolutionary processes. Human nature is not the cause of, for example, bipedalism and color vision but is rather a set of outcomes or effects that includes bipedalism and color vision. Thus it is unclear how the notion of human nature could underwrite a causal explanation of human beings' characteristics (e.g. when one says, "It is in human nature to be jealous"). Indeed, Samuels (2012, p. 3) has argued that "[t]hough this conception [i.e. the nomological notion of human nature] fares quite well in capturing many of the traditional theoretical roles of human nature, there are some central roles that it will not readily play. Specifically, it will not play the traditional taxonomic and causal-explanatory roles of human nature."

It may be tempting to give the causal-explanatory function the same treatment as the taxonomic function, that is, to dismiss it. Perhaps human nature is not an explanatory notion, and perhaps, then, we should simply stop appealing to human nature for explanatory purposes. It is indeed not clear that the notion of human nature is used explanatorily in the evolutionary behavioral sciences. However, the more traditional functions we dismiss, the weaker the claim that the nomological notion of human nature is really a notion *of human nature*. Furthermore, if human nature plays no explanatory role, then it is dubious that it is an essential part of the evolutionary behavioral sciences; instead, it may be a notion that could be eliminated without loss. Thus, if at all possible, we should find an explanatory role for the notion of human nature.

Admittedly, we often explain *explananda* by identifying their causes, in large measure because explanation is often just a means for intervention. We want to change some phenomenon (e.g. to cure some disease), and understanding the causes of this phenomenon (e.g. identifying the virus causing this disease) allows us to fulfill this goal. Nonetheless, *explananda* need not be causes. Structural explanations of phenomena do not appeal to causes (e.g. Garfinkel 1981), nor do mathematical explanations. The proposal, then, is that human nature can be a causal-explanatory notion despite not being a cause.

To see what kind of explanatory notion human nature is, we need to consider what we may call "etiological kinds": a kind *K* is an etiological kind if and only to belong to *K*, it is necessary to have a given etiology. All the members of *K*, and sometimes only them (if the etiology is sufficient to belong to *K*), share this etiology. The class of adaptations in evolutionary biology is an etiological kind: traits are adaptations if and only if their distribution is explained by natural selection. The class of psychosomatic diseases is another etiological kind: a set of symptoms (e.g. abdominal pain) is a psychosomatic disease if and only if it is caused by the patient's mental states (anxiety, etc.).

Etiological kinds are at home in lay explanations: for instance, laypeople often classify syndromes as psychosomatic diseases, but what is important here is that they also have a good scientific standing, as illustrated by the class of adaptations. Science welcomes etiological kinds because of their connection to causal explanation, as will be argued in the remainder of this section.

Membership in an etiological kind is explanatory, and one can explain why the members of an etiological kind possess a particular property by classifying this member under its etiological kind. For instance, one provides some explanation of why John has painful abdominal cramps by classifying his pain as a psychosomatic disease; one also explains at least partly the distribution and functional organization of jealousy by classifying jealousy as an adaptation (e.g. Buss et al. 1992). How does classification in an etiological kind explain? Such classification is explanatory because it involves endorsing a particular explanatory sketch: by classifying a particular as an instance of an etiological kind, one asserts that this particular has a specific kind of etiology that accounts causally for its relevant features. Classifying John's abdominal cramps as psychosomatic is to assert that his cramps have a particular etiology: They causally result from his having some particular mental states. Typically, the etiology one endorses by classifying in an etiological kind is not fully specific, and this is why one typically endorses only an explanatory sketch, that is, an incomplete explanatory account that specifies, at least in part, the type of information needed to provide a more complete explanation (Hempel 1965). The explanatory sketches associated with etiological kinds can be fleshed out to provide more complete explanations of the *explananda*. For instance, the explanation of John's abdominal cramps is more complete when the nature of the mental states causing the cramps is specified: when John's anxiety is identified. Classifying in an explanatory kind also involves rejecting possible explanatory sketches. The psychosomatic classification of John's cramps involves rejecting the possibility of explaining it as a stomach flu; classifying jealousy as an adaptation involves rejecting any attempt to explain it as a mere by-product of another adaptation.

Etiological kinds are not only explanatory, but they are also causally explanatory, and classifying in an etiological kind is causally explanatory despite failing to involve explicit reference to any cause. Such classification is causally explanatory because one commits oneself to a particular form of causal explanation of the *explananda* and to the falsity of competing forms of causal explanation. Classification in etiological kinds is causal explanation without causes.

We can now see why human nature is causally explanatory. Human nature, as characterized by the nomological notion, is an etiological kind: all the properties of human beings that are included in human nature have the same etiology in that they are the outcomes of evolutionary processes. As is the case with other etiological kinds, classifying a trait as belonging to human nature is thus to endorse a particular explanatory sketch: it is to assert that this trait is a proper target of an ultimate explanation. Naturally, the details of this ultimate explanation are left unspecified, and the classification in human nature merely indicates the kind of explanation that is correct. The trait can be an adaptation, the product of drift, a by-product of an explanation, and so forth. Identifying the causal process of which the trait is an outcome turns the explanatory sketch into a more complete explanation. Furthermore, classifying a trait into human nature asserts that other explanatory sketches are erroneous: in particular, it is to assert that a trait is not to be simply explained as the product of cultural forces.

Mere sociological or sociocultural explanations are taken to be inadequate. For instance, to say that bipedalism is part of human nature is to assert that it is correct to search for an ultimate explanation of bipedalism, one that appeals to the historical changes of homologous traits as a result of various evolutionary forces. To say that jealousy is part of human nature is also to assert that it would be incorrect to explain this trait merely in terms of reinforcement or merely as a result of cultural schemas taught during childhood.

The nomological notion of human nature fulfills the explanatory function of human nature after all. While human nature is not a cause, classifying a trait as part of human nature is to endorse a particular causal-explanatory schema for this trait – one that asserts that this trait has an evolutionary etiology – and to insist on the inappropriateness of attempts to provide a causal explanation of this trait merely by nonevolutionary schemas. Admittedly, the nomological notion of human nature fulfills the explanatory function in a distinctive manner, which differs drastically from the manner Aristotelian notions of human nature satisfy it, but it does fulfill it.

Other Candidate Successor Notions of Human Nature

The causal-explanatory function is largely left unfulfilled by Ramsey's life-history trait cluster account of human nature. Every possible trait belongs to some life history included within human nature, so asserting that a given trait is due to human nature provides no information at all. In this respect at least, human nature is not explanatory.

It is also useful to compare how the nomological notion of human nature and Samuels' (2012) causal-essentialist conception fulfill the causal-explanatory function. Samuels (2012, pp. 2–3) proposes that "human nature is a suite of mechanisms that underlie the manifestation of species-typical cognitive and behavioral regularities." These mechanisms, which are constitutive of human nature, causally explain the properties that are typical of human beings. As a result, the causal-essentialist notion of human nature straightforwardly fulfills the causal-explanatory function of human nature; no need to make a detour through the explanatory significance of etiological kinds. Samuels views his account as superior for this very reason.

However, the causal-essentialist conception of human nature fulfills the causal-explanatory function at the cost of either failing to fulfill the descriptive function or redefining it. As I understood it, a successor notion of human nature fulfills the descriptive function of human nature if and only if identifying what constitutes human nature amounts to describing what human beings are like. So understood, fulfilling the descriptive function of human nature involves identifying traits such as color vision, jealousy, bipedalism, pedagogy, smile, and incest avoidance. But if human nature consists of mechanisms – as Samuels proposes in order, in part, to fulfill the causal-explanatory function so successfully – then identifying the constituents of human nature leaves the descriptive function of human nature unfulfilled. Color vision, jealousy, bipedalism, and pedagogy are not mechanisms; they are the phenomena the mechanisms Samuels has in mind explain. So either fulfilling the descriptive function does not consist of describing what human beings are like or the causal-essentialist conception of human nature fails to fulfill this function.

The Limitation Function

Strong readings of the limitation function are implausible, and no successor notion of the essentialist notion of human nature should attempt to fulfill them. According to the nomological notion, human nature does not delineate what is possible and what is impossible for human beings. The traits that are part of human nature are only typical of human beings, which means that some human beings do not possess them. In addition, modifying or even removing a trait that is part of human nature need not be difficult; it can be done successfully, and there need not be any cost. It may be part of human nature to like a salty diet, but the taste for salt is malleable (Henney, Taylor, and Boon 2010). So, if the limitation function is to be fulfilled, some weaker reading has to be the target.

In this final section I will argue for a probabilistic reading of the limitation function: the traits that constitute human nature are likely to be difficult to modify (contrast with Antony 1998, p. 80). This probabilistic version stands in contrast to the claims that it is impossible to modify human nature, that it is always costly to do so, and that one could never do it fully successfully. The existence of malleable traits is thus compatible with this probabilistic reading of the limitation function. This reading appeals to the notion of difficulty, which I propose to understand in three different ways. First, epistemologically: we are unlikely to know how to modify the traits that constitute human nature. Second and third, instrumentally: modifying these traits may require much social and educational engineering, and modifying them may have undesirable consequences.

Before explaining how the nomological notion of human nature fulfills the limitation function so understood, let's look at an example. In the early decades of Israel, left-leaning kibbutzim socialized child rearing (Golan 1958; Rapaport 1958; Beit-Hallahmi and Rabin 1977; Aviezer, Sagi, and Van Ijzendoorn 2002). Instead of being educated by their genetic parents, children were raised communally: they slept together and apart from the parents, the whole community took care of their needs (goods, clothing, medical care, etc.), and parents were responsible for all children of the kibbutz. While child rearing varies across cultures and times, the form of collective sleeping found in kibbutzim was unique: Caregivers and children slept separately, and the night watchperson rotated. This social organization prevented a plausible component of human nature, extensive bonding between caregivers and children, to occur. In line with the probabilistic reading of the limitation function, this social organization did not last: parents' involvement with their own children increased, and home sleeping progressively replaced communal sleeping.[10]

There are at least three reasons why the traits that constitute nature are likely to be difficult to modify. First, human nature, understood along the lines of the nomological notion, includes many traits that are not specific to human beings, such as inbreeding avoidance, color perception, and rough-and-tumble play. Some of these traits are homologous to traits shared by other primates, mammals, or vertebrates (e.g. smiling in human beings is homologous to smiling in chimpanzees), some of them just analogous (e.g. color vision in human beings is analogous to color vision in bird species). The development of the former traits is unlikely to be sensitive to educational practices or to culturally variables aspects of the environment. These traits have a long evolutionary history, and they first emerged in species whose ontogeny was very different from human ontogeny, which relies heavily on learning and pedagogy. It is thus plausible (although not necessary)

---

[10] The causal explanation of the decline of communal sleeping and, more generally, child rearing is not entirely clear. Broader social changes in Israel certainly played a role, too.

that their ontogeny is not influenced by the kind of social factors at play in education. As a result, it is unlikely to be sensitive to the type of factors that we think of manipulating and know how to intervene on when we are trying to influence people's characteristics. Thus we are unlikely to know how to modify them.

Emotional expressions provide a good example. Some emotions, known as "basic emotions" (Ekman and Friesen 1971; Ekman 1993), are plausibly universal and are expressed by distinct facial expressions found in all cultures (see, however, Nelson and Russell 2013; Hassin, Aviezer, and Bentin 2013; Gendron et al. 2014). For instance, across cultures, the disgust face includes a wrinkled nose, lowered eyebrows, slightly opened mouth, and raised corners of the mouth. These basic emotions and many of their expressions are not specific to humans but have rather a long phylogeny, as Darwin (2002) already remarked. We thus would expect the expression of emotions to be largely insensitive to educational and cultural variables. Research confirms this expectation (Ekman and Friesen 1969; see Safdar et al. 2009 for more recent discussion). In particular, Ekman and Friesen (1969) observed that while different cultures have different rules (known as "display rules") governing the proper expression of emotions, these rules do not prevent people from expressing their basic emotions by their typical facial expressions; rather, they lead people to override quickly these expressions in public settings. Japanese culture includes norms against the expression of emotions such as anger; Japanese express anger in the typical manner (lowered nostrils, lips, chin, brow and brow ridge, and raised cheekbones and mouth) when they are alone but quickly override this facial expression in a public setting.

The second reason why traits that constitute human nature are likely to be difficult to modify is that some of them are likely to be canalized against changes in the factors we know how to influence. A trait is canalized with respect to some aspects of the environment if and only its development is not influenced by variation in these aspects of the environment. Note that canalization is never an absolute property (i.e. traits aren't canalized *simpliciter*), but it always needs to be relativized to particular environmental aspects (Griffiths and Machery 2008). When the traits that are constitutive of human nature are adaptations, they are likely to be canalized against variation in cultural and social environments if their ontogeny could be easily disrupted by such influences. While the acquisition of syntactic competence does depend on hearing some language during childhood (e.g. Curtiss 1977 on Genie), it happens in a very wide variety of conditions, suggesting that syntax acquisition is largely canalized against variation in linguistic input. Canalized traits are likely to be difficult to change in the first and second senses distinguished earlier. We do not know how to intervene on their development, and in at least some cases such intervention probably would require extensive intervention. Consider, for instance, what it would take to teach a child to speak, as her native language, a version of English that would not have the hierarchical treelike structure of human languages.

The third reason why traits that constitute human nature are likely to be difficult to modify is that many are likely to be generatively entrenched. A trait is "generatively entrenched" if the development of other traits depends on its development (Wimsatt 1986). Because some of the traits that constitute human nature have a long phylogeny (e.g. some emotions such as anger), they may well be generatively entrenched. If they are, disrupting their development would also disrupt the development of other traits that depend on them, which is likely to have unfortunate consequences.

To wrap up, the limitation function has clear political and social overtones, which explains why the notion of human nature has often been misused in the past, and we should be particularly careful when we appeal to human nature to identify some limitations of human beings. That said, this section has defended the plausibility of Wilson's earlier quotation: because many traits that constitute human nature have a long phylogeny, they are likely to be insensitive to education and cultural factors, the kind of factors we know how to manipulate. Other traits are likely to be canalized against the variation in the kind of factors we can influence. And modifying traits that belong to human nature may have widespread unexpected consequences for other aspects of a human life. In all these cases, changing the traits that are part of human nature is likely to be difficult.

Conclusion

An intense theoretical effort is ongoing to reconstruct the notion of human nature in light of progress in evolutionary biology and genetics, and the nomological notion of human nature is one of the candidate successor notions to the discredited essentialist notion of human nature. Its credentials are excellent, it is immune to the objections raised against the essentialist notion, and it is robust. In its own way, it fulfills some of the traditional functions the notion of human nature was meant to fulfill.

11 | A Postgenomic Perspective on Sex and Gender

JOHN DUPRÉ

Introduction

Gender is the central concept of a thriving and diverse area of philosophy. The different roles, rights, responsibilities, and so on allocated to men and women by various societies raise basic questions for ethics and political philosophy. Assumptions about these differences have been argued to have important influences on metaphysics, epistemology, science, and more. Such topics are loosely encompassed within feminist philosophy, though, of course, this is not an isolated domain of inquiry but one that claims major relevance to all the fields just mentioned. Sex plays an obvious role in all this: differential roles in reproduction are typically central to the justification of the differences in male and female social roles that are called in question by feminists generally and feminist philosophers in particular. This, however, is a matter of what might be called "surface sexual difference." A central thesis of "second wave" feminists of the 1960s and beyond was that the differentiated social roles and statuses of men and women were contingent and mutable. Different societies, and societies at different times, articulated gender in quite different ways. Surface sex, a set of generally easily accessible biological differences, was the basis on which gender roles were assigned, but these roles were in no way determined by sex. This view specifically opposed a long-standing and continuing tradition that claims that much of gender difference is natural, growing inevitably out of underlying sexual difference. Gender, in this latter view, is seen as an expression of nonobvious ("deep") sexual difference.

This chapter is based on a public lecture given as the Diane Middlebrook and Carl Djerassi Visiting Professor of Gender Studies at the University of Cambridge. I am grateful to the Centre for Gender Studies for giving me this opportunity, to its director Jude Browne for making the stay so pleasant, and to Carl Djerassi for his very generous gift that made this visiting position possible. I have also benefited greatly from comments on an earlier draft by Juliet Mitchell and on several drafts by Regenia Gagnier. Finally, I am very grateful for support from the European Research Council, Grant SL-06034, which also contributed to this work.

Half a century ago, feminist scholars worked hard to establish this distinction between sex and gender, precisely to distinguish the biological differences between men and women (sex) from the cultural differences that many assumed were just as much a part of nature (e.g. Rubin 1975; Unger 1979; Fausto-Sterling 1985; the distinction was introduced by Stoller 1968). From the beginning, however, the distinction has been controversial, even among feminists; prominent recent critics have included, for example, Judith Butler (1990). The central concern has been that the distinction tends to reify gender as something as real as, but just different from, sex. Not only does the concept of gender threaten to occlude the importance of interactions between gender and race, ethnicity, class, and other important social divisions, but more generally, it obscures the uniqueness and diversity of individual women.[1] As will become clear by the end of this chapter, I am sympathetic to these concerns about gender. But the concept does remain valuable at least for analyzing biological accounts of the development of human differences, specifically, as is my aim in this chapter, for assessing the interaction between internal and external influences in the development of human differences. If in the end the intricacy of interaction between biology and culture is so great that the distinction between sex and gender can be shown to have no ultimate ontological import, it may be best seen as a ladder that we have climbed and are now in a position to kick away. If so, though, we should be very clear why we are doing so.

The sex/gender distinction is, at first pass anyhow, intuitively clear. Sex is a biological distinction grounded in reproductive physiology. Most people – though the word "most" is very important – have a reproductive physiology and, later on, secondary sexual characteristics such as breasts or facial hair that clearly distinguish them as male or female. Gender, on the other hand, refers to behavior that is characteristic of members of one sex or the other in a particular society. In most societies, men and women wear different clothes, but not in all societies and not the same clothes wherever they are differentiated. This is an aspect of gender. The most important aspect of gender is where it interacts with the division of labor. All societies allocate different tasks to different people, and thinkers such

---

[1] A classic example of such criticism was the response to Betty Friedan's classic, *The Feminist Mystique* (1963), often credited with launching the second wave of feminism. Critics objected that Friedan, in objecting to the confinement of women to the domestic sphere, was essentializing the experience of middle-class white women while failing to notice that poor women (disproportionately minorities) were already in work, often in the homes of middle-class white women and enabling them to pursue nondomestic careers (Hooks 1984).

as Adam Smith have made a compelling case that this division of labor is the foundation of the economic success of human societies. In all or almost all societies, this division is more or less gendered. Some work is considered appropriate for men; some for women. An array of differences in status, pay, responsibility, and much else follows from these differences in allocation of work.

My aim in this chapter is not, however, to contribute to the extensive and important body of work that has explored the social and political ramifications of gender difference. Rather, it is to look at the biological and ultimately ontological foundations of the sex/gender distinction. This is where the biophilosophy, which is the focus of this volume, has an important part to play, for while public debate continues as to whether gender is ultimately an expression of deep sex or rather, as feminists have assumed, a politically mutable feature of the organization of society, this should no longer be an issue in light of advances in biology over the last half century. The biological determinism of the former perspective is no longer scientifically defensible. The reasons for this, and the exploration of its implications for philosophical understanding of the human and the social, offers a paradigm for what I take to be the role of biophilosophy. And nowhere are these implications more telling than for the understanding of sex and gender.

Essentialism

Sex and gender have often been understood in terms of essences. "Essentialism" is a doctrine both about language and about the world. We cannot speak without dividing the world into kinds. When I tell you that the cat is on the mat, I convey information by distinguishing the thing on the mat from all the countless kinds of animals and for that matter nonanimals that might have been there, and I distinguish the thing it is on from all the rugs, carpets, rocks, logs, and so on that the cat might have been on but isn't. But what, philosophers have asked since antiquity makes something a member of the cat kind rather than the dog kind, the badger kind, and so on? Does the world in some way divide things up for us, and do our words register these naturally given divisions? Essentialism answers both these questions in the affirmative.[2] And if this is right, the question for the philosopher, or

---

[2] An influential contemporary version of essentialism holds that real kinds are demarcated by essences but that only physics and chemistry are likely to contain

the scientist, is to discover what are the real divisions determined by nature.[3]

Classic versions of essentialism deriving from the ancients were famously criticized by John Locke. Like other contributors to the scientific revolution of his time, Locke thought of the natural world as ultimately composed of nothing but atoms moving in the void. If things had essences, they must be determined by the structure and relations of these atomic parts. But since, as he famously remarked, we lack microscopical eyes, such essences were inescapably beyond our reach, and there was no reason to believe that the ways we divide up the world at our own gross macroscopic level correspond to any reality at the microscopic level. However, many have concluded that Locke's pessimism was premature. We may still lack microscopical eyes, but we do have electron microscopes, high-throughput gene sequencers, and even atomic tweezers. So our ability to correlate the observable world with an underlying reality is very considerable and growing. Essences, it appears, are back within our grasp. Every day the complete genome sequences of more organisms are announced. Are these perhaps contributions to a growing library of essences?

Biology at a less rarified level has been a fertile breeding ground for essentialism. Anyone who enjoys the outdoors is likely to be struck by the distinct kinds of organisms that are encountered in the wild. There are foxes and rabbits, dandelions and oak trees; one rabbit is pretty much like another, very different from a fox, and there are no intermediate hard cases. Yet a wider spatial view and, especially, reflection on evolutionary history tell us that if we look a little further in time or space, there are always intermediates and always hard cases. Not so many million years ago there was a common ancestor of the fox and the rabbit. If we could go back in time observing all the ancestors of the rabbit until we reached that common ancestor and then forward again through the ancestors leading up to the fox, we would have a more or less smooth series of intermediates leading between these two so very different animals. With a lot more time, we could do the same thing for ourselves and a mushroom.

---

real kinds (Ellis 2001). In some respects, this view is congenial to the antiessentialism that I shall defend, but only because of the explicit denial that essentialism has immediate relevance to the kinds of kinds with which I am concerned.

[3] For more detail on these issues, see Bird and Tobin (2012).

A similar point can be made with respect to space. A striking example is provided by ring species. Where I write, in the United Kingdom, the herring gull (*Larus argentatus argenteus*) and the lesser black-backed gull (*L. fuscus*) are two very familiar and quite distinct species, not known to interbreed. Yet it appears that if we track round the globe at roughly the same latitude, there exists a series of gradually diverging species each member of which is capable of interbreeding with the next but at the end of which are the herring gull and the lesser black-backed gulls. Here we find exactly the phenomenon just described in a temporal context recapitulated in space.[4] In biology, it appears, distinct kinds are not given to us by nature but rather by our local and limited perspective on nature. So, to return to the main topic, our natural intuition that men and women are essentially different kinds distinguished by distinct inner natures should be treated with caution. We should at least look very carefully at what exactly the differences – and similarities – really are.

From Essence to Process

Let me turn, then, to the broadest of the categories relevant to the present topic, male and female. It is easy to imagine that these are biological universals, fundamental to the reproduction of living beings. Nothing could be further from the truth. The vast majority of organisms do not have sexes at all. This includes the single-celled organisms that constituted the living world for 80 percent of its history and remain by far the most common organisms but also many so-called higher animals and plants. Many plants, though they engage occasionally in sexual reproduction, generally reproduce asexually. And many organisms have more than two sexes or breeding types.[5]

Even among organisms that reproduce only sexually and that have only two sexes, sex can be fluid. Many reptiles become male or female in response to environmental conditions, such as the temperature at

---

[4] This classic example has recently been called into question, and it has been suggested that the relations in question are considerably more complex (Liebers et al. 2004). This does not affect the general point that populations of organisms exhibit gradual change over space and that terminal members of such graded populations, or clines, may be very different from one another.

[5] Nanney (1980) describes a protozoon in which seven distinct mating types can be distinguished.

which their eggs incubate. Some fish change their sex in midlife: as the position of dominant male becomes vacant in a group of bluehead wrasse (*Thalassoma bifasciatum*), the largest female turns into a male. These examples bring me to the main biological idea underlying this discussion. Organisms in general and sex in particular must be understood developmentally.[6] And development, at least for complex multicellular organisms, is not something predetermined "in the genes" but a process of interaction between the developing organism and its environment.[7]

A philosophical corollary of this thought is the following: organisms are not things but processes. This is an ancient distinction, often associated with the famous remark attributed to the Greek philosopher Heraclitus, "No man ever steps in the same river twice, for it's not the same river and he's not the same man."[8] The only constant, for Heraclitus, was change. Modern opinion has tended to embrace the alternative opinion of Democritus, that ultimately there was nothing but atoms – unchanging things – in the void. Indeed, a version of atomism was a central plank of the scientific revolution of the sixteenth and seventeenth centuries in the West and has tended to remain a default assumption of most thinking about science (except, perhaps, for the last hundred years, by physicists). But this view, understanding the world as ultimately composed of unchanging things, has not served biology well. A process, unlike a thing, is maintained by change. A chair can sit in an attic for decades doing nothing but still remain the very same chair. Organisms, by contrast, maintain themselves by doing all kinds of things – metabolism, cell division, and so on. An animal that does nothing is a dead animal. The integrity of a process is maintained not by the constancy of its temporary parts but by their causal connections. Our paradigm of a human tends to be of an average-age adult, but that is no better or worse than thinking of a child or, for that

---

[6] Rather wonderfully, the bluehead wrasse just mentioned also illustrates this idea in a quite different way: young blueheads, but not older fish, will often serve (work?) as cleaner fish.

[7] Detailed defense of this assertion is well beyond the scope of this chapter. For extensive biological details on the developmental interaction between organism and environment, see Gilbert and Epel (2009). For a philosophical discussion of modern understandings of genes and genomes, see Griffiths and Stotz (2013).

[8] I must admit that this famous quote is in fact a bit problematic, for, of course, if a man and a river are processes, then the man and the river may well be the same processes at different times.

matter, a fetus or an old lady. Biologically, what is fundamental is a life cycle: what makes parts of a life cycle stages of the same life cycle is not having the same properties at different times but relations of continuity and causality between stages. The whole need not be held together by constant, still less essential, properties.

Sexual Differentiation

So let us now look at the processes through which differentiated sexes in humans develop.[9] Whereas we tend to analyze a thing into its parts, a process is naturally analyzed into stages. Needless to say, perhaps, in neither case can we assume that the divisions are clear or unambiguous. However, the following provides a sufficiently clear series of stages for present purposes.

1. **Chromosomal sex.** Most women have two X chromosomes, and most men have an X and a Y chromosome, and they originated from a fertilized egg with those chromosomes. The word "most" is very important, however. First, not all humans have either an XX or an XY genotype. There are people with XYY, XXY, and XO chromosomes (or karyotypes), of which the first are generally assigned a male gender, and the last two are generally treated as female. Second, for various reasons, including now elective reassignment, later stages in gender development do not always coincide with chromosomal sex.
2. **Fetal gonadal sex.** By 12 weeks, most fetuses have embryonic gonads, irreversibly committed to becoming either testes or ovaries. The development of testes appears to be triggered by a gene on the Y chromosome, the product of which binds to a gene on chromosome 17 and triggers a cascade of events involved in the production of testes. A different sequence of genetic events pushes the as yet undifferentiated gonad in the direction of becoming an ovary. The Y chromosome gene just mentioned is known as the *Sry* gene, which stands for "Sex Reversal on the Y chromosome," echoing the

[9]   This chapter, and especially the present section, is deeply indebted to the work of biologist and gender theorist Anne Fausto-Sterling. Her *Myths of Gender* (1985) pioneered biologically informed criticism of purportedly scientific accounts of gender difference, a project developed in new directions in *Sexing the Body* (2000). The outline of the stages of sexual differentiation here closely follows her *Sex/Gender* (2012).

curious idea, dating from Aristotle, that being female is a default. The persistence and untenability of this idea are noted by two experts on the relevant genetics: "The discovery that gonads develop as ovaries in the absence of the Y-chromosome (or, more specifically, the *Sry* gene) supported the prevailing view that the testis pathway is the active pathway in gonad development. However, as Eicher and others have emphasized, the ovarian pathway must also be an active genetic pathway" (Brennan and Capel 2004, citing Eicher and Washburn 1986). Of course, if the *Sry* gene is indeed the relevant "switch," it might equally well be described as preventing ovary development. In neither case is the ensuing genetic cascade fully understood.

3. **Fetal hormonal sex.** As the gonads develop, they begin to produce their characteristic mix of hormones. The reproductive system, under the influence of these hormones, begins to differentiate toward characteristically male or female physiologies. Again, this depends not only on the production of hormones but also on the proper functioning of receptors that recognize these hormones. So, for example, occasionally XY fetuses carry a mutation that hinders androgen recognition and produces children born with highly feminized external genitalia. If everything follows the standard path, however, this leads us, finally, to genital sex.

4. **Genital sex.** These are the standard criteria used to distinguish the sex of babies at birth.

The process of fetal differentiation, then, is complex and multifactorial. While most babies will be born either with an XY genotype and typical male physiology or an XX genotype and female physiology, there are many ways in which these typical outcomes can be derailed. It is no surprise that there are a significant number of atypical outcomes, sometimes described as "intersexed," now more often said to exhibit "disorders of sex development," though one may wonder whether describing atypical development as "disordered" constitutes progress.

The next crucial point in human development is, of course, birth. This is the point at which the wider community decides whether a baby is a boy or a girl. In the cases where this decision is difficult, standard medical practice has been to attempt to adjust the baby to one or other of the standard kinds. This often involves surgical reshaping of the external genitalia and treatment with hormones. The *exhaustive* division of people into two sexes is not a reflection of how things are in the world but of a social policy that everyone must be assigned to one or other of these categories. Very recently, some countries, including Germany, Australia, and New Zealand, have allowed babies to be registered at birth as of indeterminate sex, though this move is highly controversial and has been criticized by some advocates for intersex people as maintaining a fixed and determinate set of categories.

Gender Differentiation

Though techniques of fetal surveillance such as ultrasound may rapidly be changing this, to a rough approximation, gender begins at birth.[10] And the countless institutions that enforce gender require that it be decided on which side of this fundamental dichotomy every individual falls. On endless forms we must say whether we are male or female – a question generally framed as a request for our sex, though, more accurately, it should ask for our gender. As noted earlier, however, in some places this dichotomy is being challenged, and the effects of this on the gendered organization of social life are as yet impossible to predict.

At any rate, development moves on. For most of us this continues to follow physiologically one of two fairly well-distinguished paths of sexual differentiation, though with wide variations in detail and with a few more along the way joining the ranks of those whose sexual development differs substantially from either norm. While the typical differences radiate out into many other parts of physiology, the further these are from the core reproductive systems, the less this difference will be sharply dichotomous and the more it will become statistical and overlapping. The average upper body strength of men, for instance, is greater than that of women, but there are many women with greater strength than many men.

[10] From the point of view of development, we should not, with due consideration to its significance and sometimes traumatic nature for the mother, see birth as a cataclysmic turning point. The baby is little more independent from the mother, for instance, than it was before birth, though it may derive its nutrition from a different part of her anatomy (though certainly being born is traumatic and a serious struggle from the baby's point of view, and the world is a very different place from the uterus. Thanks to Juliet Mitchell for reminding me of this!).

Of central importance in our species, social and psychological development also takes off, with an enormous range of external factors
impinging on the developmental process, many of which are relevant
to the continuing bifurcation of the population into the socially con-
doned male and female kinds. Boys and girls are differentially hugged,
given dolls or guns, pink toys or blue toys, and taught the intricacies of
the gendered division of people. By three, children more-or-less well
know that they are boys or girls and know many of the behaviors, likes,
and dislikes that are expected of them as such. These systematic differences in behavior are elaborated in distinctive ways through the life
cycle. Most men and women continue to dress differently, to choose
different leisure activities, and most important, to do different kinds of
work, both in the labor market and in the home. The nature of these
differentiated pathways certainly has changed over time, though not
always in the ways feminist activists have hoped. As is often
observed, increasing participation by women in the labor market has
tended to be concentrated in less-well-paid employment, and when the
same employment, pay for women is still always lower; increased male
participation in domestic work has not been commensurate.

Explaining Gender Difference

There are certain purported explanations of gender difference that have
particularly attracted scientific attention. One of these has been the
exploration of differences in male and female brains, a tradition that
goes back at least to the nineteenth century (Cahill 2006; for penetrat-
ing criticism, see Fine 2000). Because, it is often said, brains cause
behavior, such research is often seen as a search for a fundamental
cause of behavioral difference. An even more fundamental cause may
then be sought in the genes if, as many also suppose, genes explain the
properties of brains.

In parallel with the investigation of genetic and neurologic differences between men and women has been the search for evolutionary
explanations of gender difference. Here attention has focused on
realms of behavior that are seen to be especially significant for
evolutionary success, most notably mate choice and parental investment (Buss 1999). The familiar central argument is that because
women invest far more in a pregnancy than men – eggs are bigger
than sperm, and gestation takes a lot longer than copulation – they

will be more concerned to optimize the chances of success for any reproductive endeavor. This is taken to imply that women will have evolved to be very careful about whom they mate with, looking at least for the best genes on offer and, if possible, for a little help in rearing the offspring. Men, however, need only make a minimal investment. The evolutionarily rational strategy is to fertilize as many women as possible and trust that some offspring will make it to maturity. As sociobiologists like to remind us, the potential reproductive success of a male is almost limitless. It is said that 10 percent of the male inhabitants of what was once the Mongol Empire are descended directly from Genghis Khan, approximately 16 million individuals, or one man in 200 in the entire human population (Zerjal et al. 2003).

These differences in reproductive strategy are the starting point for evolutionary speculation, but their implications are seen to ramify far more widely. It is natural for women to monopolize childcare and domestic work, given their evolved concern to invest in their children; inevitably, they have less time for the outside world of work. Perhaps the need to compete with other men in the labor market – and ultimately thereby for access to women – will require cognitive capacities unnecessary in the differently demanding home environment. At least, evolved cognitive capacities are likely to be different.

These stories fit together into a broader picture that understands gender difference – or here we might as well just say sex difference – in an impressively integrated way. Natural selection placed different pressures on our male and female ancestors; these resulted in the selection of different genes, which are expressed in different brain structures; different brains cause different behavior. Let us call this the "biological big picture."

I think almost everything is wrong with the biological big picture (for more details, see Dupré 2001, 2012, especially chapter 14). Here, however, I will concentrate on one set of pivotal players in the story, genes. Genes, in the big picture, cause organisms to have particular properties, in this case properties of their brains that make them, for instance, keen on spreading their seed as widely as possible. Such properties make the individuals that exhibited them evolutionarily successful, and the genes that cause them are selected. But can genes really do this job?

Genes and Genomes

The science of genetics took off in the early twentieth century with the work of Thomas Hunt Morgan and collaborators on the fruit fly, *Drosophila* (Kohler 1994). This work was the study of the inheritance of difference. Some flies have red eyes, some white. When a red-eyed fly mated with a white-eyed fly or another red-eyed fly, what proportion of the offspring had red or white eyes? Morgan and colleagues bred and counted many thousands of flies and their differentiating traits, and the results of this work were interpreted in terms of the seminal insight that an individual had two sets of genes, one from each parent, who, in turn, contributed half their genes to each offspring. Entities such as genes for red or white eyes were thus inherited from parents, and these interacted in specific ways. For example, the red-eye gene is said to be "dominant" because a fly with a red-eye gene from one parent and a white-eye gene from the other will have red eyes. This kind of work, describing the transmission of genes bearing specific traits, is often referred to as "Mendelian genetics," honoring Gregor Mendel's pioneering work on peas fifty years earlier.

Morgan's work made fundamental contributions to the advance of genetics, and Mendelian genetics still plays an important role in areas of medicine and agriculture. But Mendelian genetics is now a very small part of genetics or, as some now prefer to say, genomics. This is so because Mendelian genes turn out to be a very minor part of genomes (Barnes and Dupré 2008). Most genes[11] cannot be correlated with a particular feature of an organism. Those that can are generally defects that make a gene nonfunctional. Consider the familiar example from human genetics, blue eyes. Blue eyes reflect the failure to make melanin in the iris. One functioning gene will suffice to produce melanin, so the brown gene is dominant. The blue-eye gene is not really a gene to make blue eyes but a defect in the gene that makes eyes brown.[12] And, of course, single-gene diseases such as cystic fibrosis or Huntington's disease, to which Mendelian models still apply, are unsurprisingly caused by dysfunctional genes.

---

[11] I will assume, for the sake of argument, that it's even useful to think of genomes as divided into genes. This assumption, however, is increasingly debatable (see Barnes and Dupré 2008; Griffiths and Stotz 2013).

[12] Eye color, like most relations between genotype and phenotype, is really much more complicated, but the simple story will serve for present purposes.

What Mendelian genetics most crucially leaves out is *process*. While no one doubts that there is a process that leads from the zygote or embryo to the adult, talk of genes for this or that trait allows us to ignore it and thereby allows us to ignore all the further factors that are necessary for this process to occur and all the different outcomes that interactions with these factors may make possible. This omission meshes with a related perspective on evolution.[13] Natural selection, it is sometimes said, cares only about the outcome, and if a gene for outcome X is selected, then somehow or other outcome X will appear at the proper time. Development – the process – can be black boxed. We know what goes in and we know what comes out. We don't need to worry about what happens inside the box.

One might have supposed that this lacuna would have been filled with the development of molecular genetics that followed the iconic discovery by Crick and Watson[14] of the structure of DNA, by then recognized as the genetic material. But, in fact, though this did lead to the discovery of some fundamental processes, notably the way in which sequences of nucleotides, constituents of DNA molecules, could determine the production of particular proteins, the main functional molecules in living systems, processes of development were still not closely integrated into genetics.

One reason for this was that many geneticists continued to think (or anyhow speak) in terms of genes for this or that feature of the phenotype. Of course, they were well aware that when one spoke of a gene for high intelligence or a gene for homosexuality, this did not provide the whole causal story. Many other genes – and much else besides – would be involved in the pathway from the gene to the trait it helps to cause. However, the genome as a whole was still seen as providing the complete code, recipe, or blueprint for the organism. The recipe was susceptible to minor changes, no doubt, as witnessed by the variability observable in actual individuals. The variations could be understood in terms of Mendelian genes that caused molecular differences, which, in turn, changed the probabilities of particular outcomes. Both the standard pattern and the variations from the pattern could be seen as determined by the genes, and there was no pressing need to take the developmental processes out of their black boxes.

---

[13] A perspective best known in the work of Richard Dawkins (1976).
[14] And Maurice Wilkins and Rosalind Franklin.

Within this framework, sex determination was a paradigmatic Mendelian system in which, perhaps unsurprisingly, the Y chromosome was dominant. Being female resulted from having two copies of the recessive X gene.[15] As with other Mendelian systems, the differences between individuals, the XX and XY "phenotypes," were taken to be both explained and caused by the genetic differences.

Counterposing this model with the complexity of the process of sexual determination sketched earlier begins to reveal the problems with the black-boxing strategy. Though there are typical developmental trajectories for embryos with XX and XY chromosomes, there are many ways in which individual developmental histories can diverge from this. Other genes, such as the binding site for the transcript of the *Sry* gene, determine whether the Y chromosome has its typical effect. And, as will be explained later, the activity of genes is frequently influenced by environmental factors. A strict and exhaustive dichotomy of outcomes is enforced at birth rather than supplied by nature.

The development of gender differences after birth may seem closely parallel to the development of sex differences: there are two standard, typical developmental trajectories. While there are anomalies – tomboys, transvestites, homosexuals, and so on – there is a typical path of development toward, let us say, heterosexual, promiscuously inclined men competing with one another in various workplaces and marketplaces and heterosexual, preferentially monogamous women gossiping pleasantly with one another while taking care of the children and the home. These are the stereotypes implied by popular models of the evolutionary elaboration of sex roles in reproduction. It is admitted that many contemporary societies have moved some distance from these stereotypes, opening the workplace to women and domestic labor to men, and are increasingly tolerant of those outside the main pathways of gender normality. But this, it is often added, is always with some difficulty, requiring a battle against the tendencies laid down by nature. We can try to get more women to be physicists or philosophers or men to do the housework, but we are fighting against their intrinsic

---

[15] An important anomaly in the system is that only XX and XY pairs are capable of mating. This curious feature underlies Fisher's (1930) famous argument for why, under most circumstances, XX and XY genotypes will be equally common.

nature. Nature, here, is the innate tendencies of the genes, as selected by millions of years of evolution.[16]

But nature, or genes, do not work like this. There are no genes dedicated to heterosexuality, the love of big machines, or good housekeeping that need to be diverted from their natural trajectories. There is a genome that, given a specific sequence of surrounding circumstances and subject to a certain amount of unpredictable noise, produces an adult individual with certain characteristics and dispositions. Change the environment and you may very well change the outcome.

So what is a genome? We often think of genomes as sequences of the letters C, G, A, and T that form a code; and sequence can be a very useful thing to know about a genome. Technologies from molecular phylogeny, the genetic exploration of evolutionary relations, to forensic genomics, the identification of criminals by the material they leave at crime sites, depend on the comparison of genome sequences. But there is a great deal more to a genome than its sequence. Considering that the chromosomes in a human cell measure about 2 meters and the diameter of a cell is of the order of 100 micrometers, there is an obvious question of how the genome can be made to fit. In fact, it is not just stuffed in any old how but exquisitely coiled and folded. Moreover, the details of this folding, or condensation, are crucial to what the genome does. To put it simply, to be expressed, a gene or a section of the genome must be accessible to the transcription machinery, and condensation implies that most of it is not accessible. The shape of the genome changes constantly, and so does, partly in consequence, its activity. And these changes are brought about by other molecules in the cell responding to many features of the wider system and even environmental influences far beyond. The study of these changes is part of the science of "epigenetics," the exploration of chemical and physical changes to the genes or the genome, how they occur in response to a wide range of external causes, and what are their effects. Paradigmatic and detailed work here is on the development of behavioral dispositions in rodents

---

[16] In their more general theoretical statements, evolutionary psychologists are usually careful to distance themselves from genetic determinism and note that actual outcomes depend on a range of environmental inputs. This then raises a problem in how to understand their more empirical work aimed at demonstrating that the phenotypes predicted by evolutionary speculation are indeed found in human populations. These phenotypes must at least be understood as typical or default developmental outcomes, even if environmental accidents sometimes derail them from this default tendency.

(Champagne and Meaney 2006; Champagne et al. 2006), but there is also a growing body of research on the way human physiology or psychology responds to developmental influences in ways that are mediated by changes to the genome.[17]

The crucial point is this: we have been encouraged to think of the genome as something static and fixed, a program or recipe that guides or directs the development of the organism. This is quite wrong. It is important that gene sequence is very stable, because the genome is indeed a repository of information about possible protein structures, but the genome does not itself say what is to be done with that information. The application of genomic information occurs as part of a process in which the genome is a dynamic participant and that is highly sensitive to a range of external influences.

Back to Gender

So what does all this tell us about gender? Gender is a bifurcated developmental process that tends to lead to two distinct suites of characteristics that are mapped onto the typical physiological states male and female. These processes are not inscribed in the genes: nothing is; they result from an array of molecular, physiological, and environmental factors coordinated reliably to produce certain typical outcomes. The fact that they are not written in the DNA does not mean that we can change them at will. Developmental processes tend to be very stable for good and obvious reasons. Indeed, life would be impossible if there were not developmental processes that fairly reliably reproduced in offspring the characteristics of parents. Parents not only provide genomes, but they also provide for their offspring with the sequence of environments that channel development in the typical direction. This may be no more than providing exactly the right place to deposit an egg, or it may involve creating a complex built environment such as a bird's nest, a beaver dam, or a termite mound.[18] It will often also involve imparting behavior through imitation or other kinds of

---

[17] For an overview of the significance of recent advances in epigenetics, see Meloni and Testa (2014).

[18] For the importance to evolution of so-called niche construction, of which such environmental modifications are examples, see Odling-Smee, Laland, and Feldman (2003). The central role of this process in human evolution and development should be self-evident.

training, and the training imparted will typically be that to which the parent, in its development, was exposed.

Humans have taken the complexity of these developmental processes far beyond anything else in the natural world. The environments in which we place our children have reached a bewildering complexity, parenting is an often frighteningly difficult skill, and socially provided institutions from maternity wards to universities are designed to contribute to the development of our offspring. Because so much of the developmental matrix in which humans grow is constructed by us, it follows that we have unparalleled abilities to change the developmental trajectories of our children. I do not say that it is simple to change these institutions, still less that it is easy to tell what will be the consequences of changes that we make, but I do say that it is possible. Feminist scholars have for decades been pointing to the variety of gender systems found in different places and at different times and inferred that the presence of a particular system is always contingent. Their critics, committed to a biologically grounded view of gender development, have claimed that this diversity is largely illusory. But given the view of development I have just presented, there is no reason to suppose that things are not as they so clearly seem. The institutions and norms surrounding gender development have diverged in different places, and over time, and the gender system has changed, too.

Let me finally take up the idea just mentioned of norms. Gender is, of course, thoroughly norm ridden. We teach our children how boys and girls, men and women, ought to behave and often that they ought to behave differently from each other. The importance of norms, and many central points of the foregoing discussion, can be nicely illustrated with a brief consideration of the issue of homosexuality. Homosexuality is, of course, a huge problem for the very prominent kind of biological determinism, or at any rate biological causality, inferred from reflections on evolution. Prima facie, at least, homosexuality seems a poor strategy for maximizing one's reproductive success. Sociobiologists and evolutionary psychologists have battled manfully (I use the word advisedly) with the problem. Perhaps ur-homosexuals worked very hard at raising their nephews and nieces, and the genes for homosexuality that they had some chance of sharing with these young relatives were thereby favored. This is, of course, nonsense, not least because there are no genes for homosexuality or, perhaps better, there are so many genes for homosexuality – genes that in more or less subtle

ways affect the probability of becoming homosexual in specific environments – that it would be better to say there were none. It is also the
worst kind of "just so" story: beyond the fact that it might possibly
explain an anomaly in a dominant system of ideas, it has no evidence
going for it at all.

Being gay, lesbian, or straight is a developmental outcome.[19] Like all
human developmental outcomes, it results from a complex interaction
between internal, including genetic, and external causes. Crucially, the
latter are partly normative. Liberal societies do not, I think, now
mandate heterosexuality, though no doubt they favor it, but they do
mandate a dichotomy. One is one thing or the other. When men or
women after decades of heterosexual marriage take up homosexual
relations, it is generally said that they have discovered that they were
gay or lesbian. Their marriages are discovered to have embodied a gross
failure of self-knowledge. Teenagers who feel attracted to members of
their own sex agonize over whether they are gay or whether this is some
passing anomaly of desire. As with sex, this dichotomy is not an
immediate problem for the many people who have no doubt on
which side of the line they fall. And the suggestion that the division is
a normative one is often unwelcome to the unambiguously homosexual, who understandably feel that a quasi-biological dichotomy is
a solider ground for defending their lifestyle than a normative dichotomy. However, since the pioneering studies of Alfred Kinsey over sixty
years ago (Kinsey et al. 1948, 1953), it has been quite clear that in terms
of the behavior generally supposed to define these categories, people lie
on a spectrum, with many engaging in sexual activities with members
of their own and the opposite sex at various stages of their lives.
Nowadays it is common to distinguish not only straight, gay, and
lesbian people but also bisexual, transgendered, and queer –
a category that is best defined by its refusal to accept a category.
No doubt there are many strata of society in which heterosexuality
remains normative, but it is increasingly clear that maintaining this
norm, and the normativity of the dichotomy between straight and gay,
will be difficult as a growing number of people refuse to accept it.
Actual developmental histories produce mixed and diverse objects of

---

[19] What follows here has an obvious debt to Michel Foucault (1979 [1976]). I also
continue to follow Anne Fausto-Sterling (2012).

sexual desire. Sexuality, very possibly, is leading the way where even sex may eventually follow.

A final striking perspective on the ontogeny of desire, the developmental process that leads to the preference of one object of sexual desire over another, is provided by the much-debated issue of pornography. Prominent feminists have suggested that pornography, or certain forms of pornography, may promote violence against women or normalize various demeaning treatments of women. This may well be so. Psychiatrist Norman Doidge (2007) provides a compelling and disturbing argument that pornography can, at any rate, radically reshape sexual desire. He describes patients becoming increasingly addicted to pornography and simultaneously increasingly unable to become sexually excited by their live partners. He also describes the evolution of pornography from the relatively uncomplicated depiction of sexual intercourse to the growing menu of violent, abusive, or just plain bizarre genres currently available on the Internet. He even reports that consumers of Internet pornography may reach a state where they are sexually aroused not just by thinking about the activities performed in pornography but by thinking of the computer itself. Even if the simplistic evolutionary psychological stories about universal preferences for ideally curvy female figures (Singh 1993) proved true as statistical averages, they would be irrelevant for understanding the diversity and plasticity of desire. Desire, it appears, is almost indefinitely malleable and can be shaped in the most unexpected ways.

Conclusion

Let me conclude. The picture I have sketched is one in which both male and female sex and male and female gender point to the most typical outcomes of developmental processes, but outcomes from which many individual trajectories diverge. At birth, or perhaps sooner, as prenatal surveillance becomes more and more routine, the male/female dichotomy of sex is normatively enforced, with medical intervention common in response to atypical individuals. This dichotomy is then the basis for a more systematically normative enforcement of dichotomous gender development. While it is still commonly supposed that both stages of this process are largely determined by genes, the growing understanding of the complexity of human development, and the deep entanglement of internal and external influences that development involves, make this kind of genetic determinism wholly implausible. An essentialist perspective on sex or gender is disastrously misguided.

So what should we make of the sex/gender distinction with which I began this chapter? Sex is an important biological concept, and it is, of course, central to human reproduction; gender is a diverse and malleable superstructure erected socially on this biological base. Nevertheless, there are reasons, in the end, not to make too sharp a distinction between the two. The distinction between male and female sexes is important but not wholly sharp. There are many individuals who fall in the gap between these two kinds, and there is much to be said for relaxing the normative requirement of sexual dichotomy. Moreover, sexual differentiation is no more immune to external, especially epigenetic, influences than are other aspects of physiological development. These influences may well include aspects of gender so that the system of gender differentiation may act causally on the physiological articulation of sex. Though I think that the distinction between sex and gender will continue to be pragmatically useful, most fundamentally, it may be better to think of sex/gender as one seamless axis of differentiated development. But, of course, this is not the pair of predetermined developmental tramlines imagined by genetic determinists; rather, we should see broad and well-trodden pathways within a much wider range of more esoteric possibilities, perhaps ever widening as we increase our tolerance of difference. Those whose sex/gender development lies some way from these pathways should be welcomed, not least as reminders of the flexibility and open texture of the human developmental process. If there is a boundary between sex and gender, it is a moving and slippery one. But no problem with that. That's what biological – and social – boundaries are like.

12 | *Biophilosophy of Race*

LUC FAUCHER

This is a chapter on Biophilosophy. The term "Biophilosophy" echoes its better-known cousin, "Neurophilosophy" (with a capital *N*). The latter discipline emerged in the 1980s as philosophers immersed themselves in neurosciences. Neurophilosophy originated from the desire of philosophers to establish a richer and denser connection between philosophy and neuroscience and since has given rise to two different subdisciplines: philosophy of neuroscience and neurophilosophy (with a lowercase *n*). "Philosophy of neuroscience" is a branch of the philosophy of science that is devoted to problems raised by the concepts, methods, and theoretical claims at the heart of the neuroscientific enterprise. Examples of the philosophy of neurosciences include questioning the capacity of neural imagery (FMRI) to provide the proverbial window into the brain (Klein 2010) and assessing the validity of the inferences from disorders to specific modules in neuropsychology (Machery 2014). "Neurophilosophy" (with a lowercase *n*) is a branch of philosophy that uses neuroscientific resources to shed light on traditional philosophical concepts or problems or to consider new philosophical problems that arise from recent development in neurosciences, and for this reason, it comprises many different projects. One of the aforementioned projects is the refining of folk concepts so as to align them with current scientific knowledge, a process that can lead to the elimination or radical redefinition of the concepts (Churchland 1986). Another such project is the use of neuroscientific knowledge to provide solutions to, or reframings of, age-old philosophical problems, such as the problem of determining responsibility for our actions or the roots of moral judgment (Roskies 2010; Greene 2014).

Like its cousin Neurophilosophy, Biophilosophy (with a capital *B*) comprises two distinct projects: philosophy of biology and biophilosophy (with a lowercase *b*). "Philosophy of biology" has been traditionally interested in understanding and discussing concepts used in biology, such as the concepts of selection, fitness, adaptation, function, or species (e.g. Neander 1991; Rosenberg and Bouchard 2003). By contrast, "biophilosophy" (with a lowercase *b*) has sought to use the resources of biology to shed light on traditional philosophical concepts or problems, such as the concept of human nature (Machery 2008).

In this chapter, I seek to illustrate two kinds of projects (similar to projects that are pursued in neurophilosophy) that can be undertaken by biophilosophers (lowercase *b*), namely, the elimination of a folk concept (and perhaps its replacement by a new concept) and the description of the evolutionary origin of a capacity or a disposition. Accordingly, this chapter will be divided in two sections. In the first section, I illustrate how contact between philosophy and biology could result in the refinement or elimination of some folk or proto-scientific concepts. More specifically, I will present arguments cited in debates concerning the existence of "race." Contrary to what some might have expected (given what might be referred to as the "consensus of the nonexistence of biological races" that has flourished in biology and social sciences from the 1960s onward[1]), "race" as a concept is currently making a comeback in biology, mainly due to the impetus of genomics. This is what some have labeled the "genomic challenge to the social construction of race" (Shiao et al. 2012). I will argue that the notion of "race" used in this context – and that some think receives validation from this body of work – differs in many (and important) respects from what is taken to be the "folk" notion of race. Biophilosophy then faces a challenge: should the term "race" be retained even though the entities to which it refers do not have the properties attributed to them by folk notions, or should we simply drop the term? As I will show, part of the answer involves normative stakes – considerations of a sort that have not often been taken into account in neurophilosophy when proposing the elimination or pruning of a concept. Reflections in biophilosophy thus shed light on factors other than empirical adequacy, which should be taken into account when deciding the fate of a concept.

In the second section of this chapter I will discuss the question of whether evolution could leave, or in fact has already left us, with a set of

---

[1] This consensus is expressed, for instance, by Omi and Winant (1994), who write that "we have now reached the point of fairly general agreement that race is not a biological given but rather a socially constructed way of differentiating human beings" (p. 65).

domain-specific mechanisms devoted to thinking about people as belonging to different racial groups. In that section I will demonstrate that despite the fact that it is unlikely that we have inherited mechanisms specialized in racial cognition, other mechanisms such as mechanisms specialized in ethnic cognition might have been coopted to produce the distinct way we think about races. I will argue that knowledge of these mechanisms is crucial to understanding and eradicating racism. This section will illustrate how contact with biology can inform us about the way we should think about phenomena like racialism and racism.

The Genomic Challenge and Race

The answer to the question of whether race exists is unfortunately not as straightforward as it seems, mostly because the meaning of "race" has varied and still varies greatly (for a historical perspective, see Hudson 1996). According to many, a substantial part of the debate on race relies on our folk concept of race (hereafter race$_f$). In that context, one question concerns the existence of races$_f$ possessing the features that the folk attribute to them. In order to answer this question, two things need to be done: first, we must specify the features that the folk attribute to races$_f$; second, we must see if there is anything in the world that corresponds to races thus described.

Another question at the border between biophilosophy and philosophy of biology concerns the use of the concept "race" in science (hereafter race$_s$). For some people (e.g. Andreasen 1998, 2004; Kitcher 2003), there is a perfectly legitimate use of the concept of race in biology (even if that concept does not capture some or most of the features that the folk generally attribute to race). Thus, in recent years, we have experienced heated discussion concerning the existence of races in biology itself. Does biology need a concept of race? Should biology get rid of the concept of race and replace it with another one? Some have argued that we should conserve the folk concept of race$_f$ with minimal modification (Sesardic 2010); others favor its elimination or its ontological deflation (Zack 2002; Spencer 2014; Hardimon 2012).

The Folk Concept of Race (Race$_f$)

There are debates concerning exactly how the folk think of races, and it has become obvious recently that more work should be done to reveal the content of that concept because it is possible that different people attach different beliefs to the same concept (Condit et al. 2004). Until recently, researchers mostly attributed the following conception of race to the folk, and it is on the basis of this conception that the debate has been conducted.[2] Human races are social groups in which

1. Individuals share a number of physical and psychological features that are specific to their group and that they do not share with any other group;
2. The fact that they exhibit these features is explained by the presence of an underlying and unobservable cause, an "essence";
3. The possession of this essence is necessary and sufficient for membership in the group; and
4. They share these features in virtue of a biological mechanism that ensures the transmission of the racial essence from generation to generation.

On this view, the folk think about races somewhat like certain philosophers have said we think about natural kinds in chemistry. According to these philosophers, something is a piece of gold if and only if it has a certain microstructure. If something does not have this microstructure, it is not gold. This microstructure explains the observable features of gold. Similarly, someone is of Race$_f$ X if and only if he or she has a certain unobservable essence. If someone does not have this essence, then he or she is not a member of Race$_f$ X (even though he or she might try to pass for an X and might even succeed at doing so). This is in a nutshell what Mallon (2013) calls "racial essentialism," which he claims is the result of a default disposition of the human mind (see also Machery and Faucher 2005a).

It is against this kind of essentialism that some of the first criticisms of the race concept were addressed. The idea that there is more genetic variability inside racial groups than between them, which makes it possible that an individual inside a racial group might be more similar, on the genetic level, to another in another group than with one of his or her own group, was a serious blow to racial essentialism (Lewontin 1972; Brown and Amelagos 2001). Another blow came from the observation that classifications based on different phenotypic traits

---

[2] See, for instance, Zack (1998), Feldman and Lewontin (2008), and Appiah (2006).

do not overlap perfectly (Brown and Amelago 2001; Diamond 1994). For instance, skin color and lactose tolerance are not correlated in the way that folk racial essentialism requires them to be. If the race concept requires racial essentialism, these observations force one to conclude that race$_f$ should be done away with. Nothing in the world has all the attributes that folk races are supposed to have; races are thus fictions – and damaging ones at that.

Recently, some (Glasgow et al. 2009; Spencer 2013) have challenged the claim that the folk concept of race is essentialist, in the sense described earlier, and that our "ordinary" concept of race is not essentialist but instead makes reference to geographic ancestry and visible physical features (Hardimon 2003, 2012). Some have considered visible features to be the only "essential" feature of race (Glasgow 2009, 2011). The dispute concerning the meaning of "race," as held by the folk, has forced philosophers to be clearer about what they are talking about when they talk about "race" and has pushed them to inquire about the right methods for investigating the question of what, exactly, "race$_f$" is supposed to be.

The meaning of "race" that has been central to the philosophical debate about the existence of race has mainly been a historically important one, a conception of race that philosophers take to underlie racist projects of the past and the present (e.g. Appiah 1996). For instance, the link between this conception of race and racist projects is clear in Zack's (2003) project of getting rid of the concept of race in order to get rid of racism (see also Kelly et al. 2010). But even if this conception has been historically important, it is not necessarily the one that people generally have now.

This leads us to the question of how best to capture what people think that race is. One possibility is to make use of conceptual analysis, on the assumption that philosophers somewhat share a concept of race with other folk and that a way for philosophers to study folk concepts is to examine their own intuitions. This approach has increasingly been challenged in philosophy in general (Knobe 2007) as well as in the particular case of race. A number of philosophers have turned away from the method of conceptual analysis and have rather endorsed the use of empirical methods to establish their claims. In order to access and evaluate folk concepts, these philosophers use literature in psychology (Condit et al. 2004; Gelman 2010) and anthropology (Astuti et al. 2003; Hale 2015) or perform their own experiments (Glasgow et al. 2009; Shulman and Glasgow 2010; Machery and Faucher, unpublished).

The least that one can say on the basis of a consideration of this literature is that the picture of folk concepts that emerges from it is far from clear: some (Glasgow et al. 2009; Hale 2015) argue that not all the folk have an essentialist conception of race, while others (Gelman 2010) think that they have one (but not necessarily all the time). Others, like Condit et al. (2004), argue that the folk have an inconsistent theory of race and that their essentialism applies mostly to physical traits but not always to nonphysical ones.

The fact that the picture of what the folk think about race is not clear should make one pause when discussing the question of the validation or nonvalidation of the concept of race$_f$ by biology. While some folk conceptions are obviously not candidates for scientific validation, others (a purely geographic and ancestral conception) might be. It is with this cautionary statement in mind, that we can now turn to the new forms of biological racial realism.

One Form of Biological Racial Realism

"Biological racial realism" is the idea that human races$_f$ exist in nature, objectively – that is, independently of human interest – and that they are genuine biological kinds.[3] On this view, there are patterns of biological variation among human populations that correspond more or less to human races$_f$ (see, for instance, Risch et al. 2002). These patterns are therefore not arbitrary inventions and are authoritative – that is, they not only exist but also are a scientifically fruitful way to categorize human populations. As Risch and his colleagues put it, "from both an objective and scientific (genetic and epidemiological) perspective there is a great validity in racial/ethnic self-categorizations, both from the research and public policy points of view" (2002, p. 1).

The debate about biological racial realism comprises two questions that should be kept distinct. One concerns the existence of races$_f$ (under

---

[3] Biological racial realism is committed to more than the *usefulness* of biological categories. Indeed, some have defended the idea that racial categorizations have biological effects on some people thus categorized and, for this reason, that races have an objective biological reality and can be useful for prediction or explanation (e.g. Kaplan 2010; Fausto-Sterling 2008). See Spencer (2012) for an illuminating discussion of this issue.

one interpretation of what races$_f$ are), and the other concerns the existence of races$_s$ in biology, independently of the question of whether races$_s$ validate races$_f$. As I stated earlier, answering the first question is a perilous enterprise given that we are not certain of the content of the folk concept (or concepts) of race$_f$. The second question is debated mainly inside philosophy of biology. In recent years, a plethora of theories of race have been offered, most of which do not claim to capture much of our folk conceptions: examples of such theories include races as lineages (Templeton 1998), races as clades (Andreasen 1998, 2004, 2005), races as ecotypes (Pigliucci and Kaplan 2003), and races as structured populations (Spencer 2014). At least some of these authors (Andreasen 1998, 2005; Piglucci 2013) have defended the view that their concepts of race are different in important ways from the folk concept. Piglucci goes as far as to say that "races (in the folk sense) do not really exist" (2013, p. 4). Others, like Spencer, completely drop racial folk concepts (which he finds multifarious and logically inconsistent) while holding that folk race names (or at least the ones that are used in the US Census forms) refer to biologically existing entities.[4]

In the following I will present the concept of race that has probably been the most discussed in the philosophical literature recently: the concept of races as "genetic clusters."

Races as Genetic Clusters

One source of biological racial realism is the studies focused on a population's genetics, measuring genetic variation patterns using genetic markers (Wilson et al. 2001; Risch et al. 2002; Rosenberg et al. 2002, 2005; Rosenberg 2011). To understand the results of these studies, first consider two undisputed facts about the genetic structure of human populations. There is a widespread agreement in biology that human genetic variations are geographically structured in two ways. First, the greatest genetic variation occurs within Africans,

---

[4] In recent philosophical literature on the topic, only Sesardic (2010, p. 344) comes close to proposing a concept of race that is not radically different from the "historically important" folk concept (only dropping the essentialist claim). I will not be discussing Sesardic's proposal in the following, partly owing to limitations of space and partly because it is considered almost unanimously to be a dead horse (see, for instance, Hochman 2013; Puglucci 2013; Taylor 2011).

the variation in non-African populations being a subset of African diversity or new variants of it. As Bolnick puts it, "From a genetic perspective, non-Africans are essentially a subset of Africans" (2008, p. 73). As Long and Kittle (2009) imaginatively observed, if a malevolent being were to wipe out a local group, the loss in terms of genetic variation would be different if that group were from some part of Africa than if it were from Papua New Guinea. Second, human populations roughly follow a clinal pattern of genetic variation, which means that "[p]opulations are most genetically similar to others that are found nearby, and genetic similarity is inversely correlated with geographic distance" (Bolinick 2008, p. 72). It "roughly" follows a clinal pattern because geographic distance does not smoothly predict genetic variation "because of chance aspects of reproduction and population growth, plus factors like mountains, deserts, bodies of water, long-range migration, and religion and other cultural proscriptions produce deviation from simple gradual variation over space" (Weiss and Fullerton 2005, pp. 166–7). So both physical and social-cultural obstacles to the gene flow can create sharper genetic variations between local groups.

Almost everyone in the debate on race accepts – and should accept – that "human populations differ in the frequencies of particular alleles, and that these differences are not uniform or random but follow patterns associated with the ease and historical frequency of the gene flow (as well as with local selection pressures)" (Kaplan 2011, p. 1). It is precisely these structures that genetic studies try to uncover by examining frequencies of polymorphic sequences of randomly chosen DNA (single-nucleotide polymorphisms [SNPs], or haplotypes, microsatellite loci, copy-number variants [CNVs], or *Alu* sequences), sometimes with as few as twenty randomly chosen genetic markers, but typically with more.[5] These studies have revealed the existence of clusters, that is, groups of individuals that share some variations in frequencies of alleles at particular loci.[6] For instance, studying about 4682 alleles

---

[5] The optimal number of markers is around 200 unselected polymorphic sequences, with fewer being required when using microsatellites (Risch et al. 2002).

[6] Concerning the nature of these differences, it is important to note that "[b]ecause most alleles are widespread, genetic differences among human populations derive mainly from gradations in allele frequencies rather than from distinctive 'diagnostic' genotypes. Indeed, it was only in the accumulation of small allele-frequency differences across many loci that population structure was identified"

from 377 autosomal microsatellite loci[7] in 1056 individuals from fifty-two worldwide populations with a model-based clustering algorithm (a program called *STRUCTURE*), Rosenberg et al. (2002, 2005) were able to group these individuals into sets of clusters (called "*K* clusters," the number of which is chosen in advance). When the number *K* of clusters was set at two, the partition seemed to show the migration event from Africa into the rest of the world. At *K* = 5, the split was among Africans, Eurasians, East Asians, Oceanic populations, and Native Americans.[8] That is, at *K* = 5, *STRUCTURE* partitioned the sample into groups roughly corresponding to vernacular racial divisions.

Is It Appropriate to Talk About Race in Biology?

Though researchers like Rosenberg describe the groups discovered via this method simply as "clusters," some have identified them with races. For instance, Risch and his colleagues write that "[e]ffectively, these population genetic studies have recapitulated the classical definition of races based on continental ancestry – namely African, Caucasian (Europe and Middle East), Asian, Pacific Islander ... and Native American" (2002, p. 3). Others, like Spencer (2014), hold a somewhat similar position, saying that these population clusters are the referents of the racial terms that are used in the US Census. However, one thing is sure; talking about race in this context is not talking about race$_f$. As Feldman et al. (2003) observe, it is one thing to use polymorphic sequences of DNA to determine the geographic origin

(Rosenberg et al. 2002, p. 2384). To put it differently, the method does not use "racial genes" but a combination of small statistical differences of allele frequencies at many loci to create a genetic profile of particular human groups and uses this profile to infer the geographic origin of a subject.

[7] Microsatellites are highly variable; for instance, Rosenberg and colleagues (2005) report a mean number of distinct alleles per locus of 11.94 (Rosenberg 2011, p. 663). Some of these alleles have the same frequencies across regions, while others have substantial differences in frequencies. According to Rosenberg and colleagues, 46.6 percent of all alleles are present in all geographic regions, while 7.53 percent are present in only one region; more than half of them (56.89 percent) are found in Africa (but their frequency is typically low). By collecting small amounts of allele-frequency variation across many loci, it is possible to make inferences about individual genetic ancestry from these markers.

[8] Some, like Bolnick (2008) and Serre and Pääbo (2004), think that these results are due to sampling biases. See also Pääbo (2003, p. 410).

of someone; it is quite another to answer the question of the fraction of variability that we can find inside of geographically isolated populations. As we saw earlier,[9] the data are clear concerning the latter question: there is more variability within "racial" groups than between them. Knowing the geographic origin of an individual's ancestors does not allow the precise prediction of that person's genotype[10] or the set of phenotypic properties that person will exhibit. Races are not names for genetically homogeneous groups of individuals. Nor are the borders of race discrete: Ethiopians or Latinos cannot be ascribed to one and only one of these "races." Finally, this conception of race doesn't say much about the physical appearance and nothing about the psychological capabilities of members of these putative races (Coop et al. 2014). So, on all counts except for the fact that members of a particular race share a geographic origin, races$_s$ are different from races$_f$.

Should we call these clusters "races"? Many reasons have been invoked to say that we shouldn't. Some of these are semantic: for instance, Glasgow (2003), commenting on Andreasen's proposal, cites the fact that the concept of "genetic clusters" differs both in intension and in extension from our ordinary concept of race, so to use "race" to designate the genetic clusters is to talk about something else.

Recently, Spencer granted the fact that the intension of the term "race" as a genetic cluster is different from that of the folk concept, but he proposed that race terms should be understood as "proper names" and invoked Kripke's (1980) theory of proper names to support the view that all we should care about is the fact that our terms refer or not, and we should not be concerned with their intentions. But Spencer uses the names for races mentioned in the US Census, and one might argue that it is not at all clear that these are the names for races that people use, even though they know how to use them. For instance, Condit and her colleagues (2004, p. 258) demonstrated that people in the United States use inconsistent principles to identify races: they sometimes use continental origin (such as when one is talking about Caucasian, African American, etc.), sometimes language (Latino), sometimes national origin (Cuban, Japanese), and sometimes regional grouping (South Asian). Folk conceptions of race seem to be much more inclusive than the official US classification. Saying that races exist

---

[9] Rosenberg et al. (2002, p. 2381) recognize this variability.
[10] See Rosenberg (2011, p. 673).

because the terms used in the US Census might be used to refer to genetic clusters has an unacceptably stipulative aspect. Why should one give precedence to the US Census categories and not to folk race categories? The fact that one can find clusters that correspond to US categories does not show that these clusters have a special biological significance, in contrast to other clusters (Maglo 2010).

Some other reasons not to equate genetic clusters with races that are invoked by philosophers are methodological. Kaplan (2011) observes that if it is possible to regroup individuals according to their continental origins, it is also possible to regroup them at a lower scale that we do not identify as continental races. For instance, Novembre and colleagues (2008) showed that it is possible to assign an individual to particular geographic regions *inside* Europe by using the same techniques as Rosenberg used. It is thereby possible to distinguish a Portuguese from a Swiss German and even, inside Switzerland, a Swiss German from a Swiss Italian. Genetic clustering can produce many different clusters at many different levels and assign individuals to ethnic groups that do not correspond to races. Why, then, should we privilege continental races more than other clusters at lower or higher levels?[11] As Kaplan observes:

So we cannot in good faith say that, for example, people with recent ancestors from Africa form a race (Blacks) because they are more likely to share alleles with other people so identified, without recognizing that the same fact holds for populations we do not normally identify as races (e.g. people with recent ancestors from Spain and Portugal). Nor does it make sense to think of human "races" as biological entities when it is clear that knowing the location of the ancestors of someone provides at best very weak predictive power with respect to a particular assortment of alleles they will have, and hence that members of the same "race" are not particularly likely to share any particular features (genetic or otherwise). (Kaplan 2011, p. 3)

There is so much structure in the population's genetic variation that the decision to consider race only the continental clusters is arbitrary, and continental clusters do not provide much inferential power, which is something that the folk (as well as many biomedical researchers) have expected from racial membership. Being told that someone is African or Latino does not tell you as much about what kind of important

---

[11] This is what Hochman (2011) has called the "grain of resolution problem" (for a similar conclusion, see Gannett 2005).

genetic variation (e.g. genetic variation underlying certain diseases) you
might expect to find as being told that the person is Bantu or Cuban.
As Feldman and Lewontin put the point, "[l]ines of ancestry, rather
than genetically arbitrary racial categories, can provide much accurate,
biologically interesting, and potentially medically useful information"
(2008, p. 99; see also Tishkoff and Kidd 2004; Bamshad et al. 2004).[12]

The fact that races-as-genetic-clusters is not inferentially powerful
poses a problem for people like Risch, who claim that the major reason
to isolate such clusters is that they would provide precious tools for
healthcare research or public policy (for more reason to doubt the
interest of races as genetic clusters for epidemiology, see Larusso and
Bacchini 2015). Others, like Spencer (2012, 2014), argue that real (or,
as he prefers to say, "genuine") categories do not need to be inferentially useful as long as they are epistemologically useful. According to
Spencer, Rosenberg, and others, it is sufficient that we have detected
"real (enough) patterns" in populations, patterns that are not
accounted by geographic distance alone. These patterns are thus epistemologically useful in a respectable research program in biology,
even if they do not offer much ground for inferences in other domains.
Still, one might wonder why this deflated version of race should be of
any *special* interest for biologists in contrast to finer-grained clusters.
As I mentioned earlier, the fact that at $K = 5$, the clusters correspond
more or less to groups in the US Census is not a proof of the special
interest of that level for biologists.[13] Neither semantic objections nor
methodological ones have been decisive in the debate about the existence of races$_s$, and I will leave these arguments aside and consider
arguments of a different nature, what Mallon (2006) called "normative
arguments."[14]

---

[12] Risch et al. (2002, p. 6) seem to recognize that continental clustering does not
      produce very inferentially powerful categories when talking about the hemo-
      chromatosis gene mutation *C282Y* that has a frequency of less than 1 percent in
      Armenians and Ashkenazi Jews but 8 percent in Norwegians, all groups being
      considered to be Caucasians. In other of Risch's papers, the story is less clear.
      Peralta et al. (2009) link African ancestry with higher coronary artery risk.
      Though, when read carefully, it is not the degree of African ancestry per se that is
      measured, but the degree of Yoruba ancestry. The results hold in the United
      States because a large number of African Americans have ancestors that came
      from West Africa.
[13] For a similar argument, see Hochman (2014).
[14] See also Maglo (2010), who talks about "axiological empiricism."

Ethical or pragmatic reasons might lead one to argue for abandonment of the term "race" to refer to the groups isolated by genomics. This sort of consideration is invoked by Montagu, who perspicuously observed that "[i]t is simply not possible to redefine words with so longstanding a history of misuse as 'race', and for this, among cogent reasons, it is ill-advised." As Simpson has said, "There … is a sort of Gresham's Law for words; redefine them as we will, their worst or most extreme meaning is almost certain to remain current and tend to drive out the meaning we prefer" (1962, p. 923; see also Kaplan 2014). As we will see in the next section, in the case of races as genetic clusters, there is some good empirical evidence that gives us reason to think that Simpson's remark about Gresham's law might prove to be validated once again.

Looking into the Heads of Racialists Through an Evolutionary Psychological Lens

In this section I will turn to a second project encompassed by the "biophilosophy" (lowercase *b*) of race. First, I want to illustrate how an evolutionary approach can contribute to our understanding of racialist cognition.[15] Second, I want to indicate how the psychological research that I use to explain racialism can point to some potential problems with the "genetic clusters" conception of race.

According to a social constructivist view, the ideas that one has about the taxonomy of the human species or the stereotypes that one entertains about a racial group are induced entirely by instruction or through imitation. This is what Mallon and Kelly (2012) call "social constructionism about the representations." So, not only is racial thinking social-historically local (because the racial identity of an individual can vary through time, place, and culture), but the *content* of racial representations is determined solely by culture and might change drastically over time or between cultures.

In a series of papers co-authored with Edouard Machery (2005a, 2005b), I rejected this picture. Machery and I argued that there is good

---

[15] According to a distinction often drawn in the literature on races, "racialism" refers to the belief in the existence of races, while "racism" refers to the negative evaluation of individuals based on the fact that they belong to some races. "Racialist cognition" thus refers to the psychological mechanisms that underlie the way people think about racial groups.

reason to think that folk representations of racial groups are constrained by an innate, evolved, domain-specific cognitive mechanism. By the same token, we also argued against an idea that underlies much of traditional social psychology: the idea that all social groups are cognitively equal, that is, that the same cognitive processes that we apply when thinking about firefighters or fans of Kurosawa's movies are also applied when we are thinking about ethnic or racial groups. But, as Kurzban and Neuberg (2005) point out, this "conceptualization of 'group' – most simply, two or more individuals who influence one another – is likely inadequate."

For example, the intergroup relations literature within social psychology has focused on groups in a nonspecific way, as implied by general terms such as in-group favouritism and out-group homogeneity. This literature implies that relations between members of different genders groups, families, ethnic groups, work teams and college majors operate similarly: a group is a group is a group. In contrast, we believe that it is important to recognize that there exist qualitatively different types of groups that the mind treats differently from one another. (p. 654)

According to Kurzban and Neuberg (2005), from an evolutionary point of view, we should rather expect humans to exhibit "discriminate sociability" (p. 653). This view also has been gaining currency recently in psychology. For instance, Prentice and Miller (2007) remarked that all human categories (by which they mean all social categories) "are not created equal in the mind's eye"; indeed, as the continued, "[S]ome are essentialized: They are represented as having deep hidden, and unchanging properties that make their members what they are" (p. 202). This way of representing a class of individuals is called "psychological essentialism." Using previous work by Haslam's team (Haslam, Rothschild, and Ernst 2000), Prentice and Miller (2007) argued that social categories are not all treated the same way: indeed, some categories, such as gender, ethnicity, race, and physical disability, are strongly essentialized, while categories such as interests, politics, appearance, and social class are only weakly essentialized. Though the tendency to essentialize might vary among individuals inside a culture, they concluded that "people show a robust tendency to essentialize gender, racial and some ethnic categories across cultures and subcultures" (p. 202).

In the following I want to defend the view that a biologically informed psychology can shed light on the psychological mechanisms that underlie racialist cognition. Accordingly, I will describe my proposal that racial cognition is the product of an "evolved ethnic cognition mechanism." This evolved mechanism[16], I argue, biases us toward a particular sort of psychological essentialism when thinking about some social groups. So not only do we not think the same way about different kinds of social groups in essentializing some of these groups while not others, but it is also the case that the manner in which we essentialize them varies depending on the type of group. Here I agree with Barrett, according to whom, "there are *a priori* [evolutionary] reasons to expect the existence of multiple kinds or modes of essentialism, because a single set of essentialist assumptions is unlikely to produce valid inferences for all kinds, from non-biological substances such as water and gold to biological kinds such as predators" (2001, p. 10). Since we have reason to expect different forms of essentialism to apply to different parts of our physical and social world (e.g. in the case of the social world, essentialism about race is different from essentialism about sex or age; the first is a form of "lineage essentialism," while the others do not need the idea of inheritability or vertical transmission), it is important to be specific about the types of inferences that one supposedly can draw about a particular species, if only because these may allow different testable predictions.

Machery and Faucher's Previous Proposal

The proposal that Machery and Faucher defended in previous papers (2005a, 2005b) was that there exists an innate, evolved, domain-specific psychological mechanism that explains that despite local variations in conceptions of races or of certain social groups (such as

---

[16] Because most of the research I will mention in this section involves the concept of "essentialization", I will use it as well to describe the position that Machery and myself defend (2005a, b), but only for the sake of uniformity, as we have qualms about the way this concept is described in the literature. I will use the term "essentialization" to refer to a set of *reasoning patterns*. People who essentialize think that for some categories the properties of category members are inherited and stable. They are prone to generalize these properties to category members and they refer to category membership to explain the possession of these properties. However, contrary to the view of those psychologists who attribute essentialism to subjects, I do not think that people have to believe in inner, maybe unknown, properties that define the identity of categories and explain the possession of observables properties in order to display these reasoning patterns.

castes, for instance), a common conceptual core is found in many unrelated cultures in the way people think about these groups. That mechanism would have evolved to allow life in the large cooperative groups that have characterized our evolutionary adaptive environment.[17] These groups would have posed a certain number of *sui generis* problems (among them, coordination problems) that could only be solved by a new kind of psychological mechanism: a mechanism for ethnic cognition. Racialist concepts would be the result of the functioning of that mechanism in certain types of social contexts.

Gil-White (1999, 2001a, 2001b, 2005) proposed that our ethnic cognition is an exaptation of a module governing folk-biological generalizations (Atran 1998; Sousa et al. 2002). In other words, our ancestors came to represent ethnic groups as if they were biological species. According to Gil-White, our folk biology module would have been applied to ethnic groups because ethnic groups and biological species share many important properties. Ethnic groups are characterized by a set of stable norms that are transmitted culturally to biological descendants (Boyd and Richerson 1985; Henrich and Boyd 1998), and different ethnic groups have different norms (Richerson and Boyd 2001). A consequence of this is that individuals who belong to the same ethnic group tend to behave in similar ways, while individuals belonging to different ethnic groups tend to behave differently. Among other things, interactions between individuals belonging to different ethnic groups might not have been as profitable as interactions between individuals belonging to the same group because different cooperation norms in each ethnic group generate interaction costs (Gil-White 2005). For this reason, ethnic borders have often corresponded to the limits of social interaction, in particular, for marriages. Thus marriages and reproduction have probably more often than not been endogamous. Finally, ethnic groups distinguished themselves from each other by using ethnic markers so as to avoid a cost that would result from interacting with people who do not share the same norms (McElreath et al. 2003). Our ancestors displayed signs marking their ethnic membership (clothing, scarification, dialects, accents, etc.) and paid attention to them (see, for instance, Kinzler and Spelke 2011). Ethnic groups

---

[17] The formation of such large groups (that are called "ethnies" or "tribes" in ethnology) is exclusive to humans and requires certain psychological mechanisms to ensure their internal cohesion (see Dubreuil 2010).

thus have four properties in common with biological species: ethnic group members have a distinctive appearance because of ethnic markers, individuals who belong to the same ethnic group behave in similar ways because of the norms they share, ethnic membership is transmitted by descent, and reproduction is typically endogamous. Gil-White concluded that the use of our folk biology mechanism to think about ethnic groups was an exaptation rather than merely a mistaken activation of a module because to consider ethnic groups as biological species was adaptive. To put the point differently, representing ethnic groups as if they were biological species might have been a good rule (from the point of view of natural selection) – even if it is bad science – because it allowed our ancestors to make inductive generalizations about other ethnic groups on the basis of limited contact and thus allowed possessors of such mechanisms to avoid interactional costs. Most important, conceptualizing ethnic groups like biological species would presumably have the number of interactions between individuals of different ethnic groups, in particular, exogamous marriage (for an example, see Regnier 2015, who also offers a very interesting story about how a social group came to be essentialized).

Building on Gil-White's work, Machery and Faucher favored the hypothesis that racialized groups and certain other groups (e.g. castes) act as triggers for an ethnic cognition module, which is based on the folk biology module. Indeed, the physical features that define racial membership, even if they are constructed in some ways, are similar to ethnic markers. However, the triggering of the ethnic cognition module by race is a *misfiring* because races are not ethnic groups. People labeled as being members of the same race can be members of different ethnic groups, even though race can be ethicized. Racial policies may cause those who are racialized to conform to, or to be seen as conforming to, norms that are passed from one generation to another. Racial policies and the resulting norm sharing might make it more likely that marriages will be endogamous, producing even more norm homogeneity in the group and creating even more distance between the racialized groups, making it less likely that racialized groups will interact with each other, even if they inhabit the same physical locale. Thus it is in general not adaptive to conceptualize races as ethnic groups. As a consequence, our naive biology has not been exapted for racial cognition.

But given this view, how are we to explain the fact that ethnic groups and racialized groups are not always essentialized?[18] I proposed that the acquisition of ethnic and racial concepts is under the control of two kinds of factors: (1) an innate mechanism that I label the "ethnic acquisition device" (EAD) that provides a default core concept of ethnies (which includes the particular form of essentialism that I described earlier) and (2) contingent social and cultural factors, which give particular and local content to ethnic concepts and can neutralize or counter the essentialist bias so that ethnies or races are not essentialized.[19] Explaining how these factors can interact would exceed the limits of this chapter, but it is important to bear in mind that according to the view presented here, positing that essentialism is part of EAD does not necessarily lead to the formation of explicitly essentialistic ethnic and racial concepts. So, for instance, Astuti (1995) has shown that members of the Vezo of Madagascar have a folk theory of race that is not essentialist, although Vezo children seem to have one that is (Kanovsky 2007). This precision is important because recent anthropological research by Moya and Boyd (2015) indicates that ethnic thinking is not always accompanied by essentialism, and those authors conclude that "different cognitive mechanisms underlie several functionally distinct ethnic phenomena" (2015, p. 2) – for example, stereotyping, in-group loyalty, intergroup hostility, and essentialism. I agree with this view insofar as it claims that some unrelated (to ethnic thinking) cognitive mechanisms are involved in supporting essentialist thinking about ethnic and racial groups and that some groups do not think about other ethnic groups in terms of their having a distinctive essence. But I explain this departure from essentialist thinking by proposing that the EAD comprises an essentialist component or bias that can be, in some cases, countered or overridden by other, nonessentialist biases.

The Role of Language in Essentialization

As we saw earlier, not all types of social groups are essentialized, and among those that are essentialized, not all instances of them are

[18] See, for instance, Hale (2015), who studies villagers of Yapatera in Peru, who apparently do not essentialize race.
[19] As they can also work in the other direction to strengthen the bias; more on this later.

essentialized in every culture or at every point in history. So there must be a means to transmit information concerning which kinds of groups, in any given sociopolitical context, *should* be essentialized.[20] At the time of my initial proposal, I suspected that one way through which information about which group to essentialize was conveyed by means of language. In this connection, I was impressed by a paper by Gelman and Heyman (1999), in which the authors demonstrated how the use of a common name ("carrot eater") compared to a verb ("is eating carrots") led children to view a property as more stable and more likely to persist without parental encouragement than otherwise would be the case. More recent work suggests that another lexical structure plays an important role in the transmission of social essentialism: "generics" (Rhodes et al. 2012; Leslie 2014; Leslie, forthcoming).

Generics are statements of the following form: "Tigers are striped," "A cheetah runs fast," "Pitt bulls have an aggressive nature." As Rhodes et al. (2012) stated that generic statements are generally understood as "communicating non-accidental generalizations" (p. 13527); that is, they are understood as communicating something that is likely to be true of the members of a category in general, so it accommodates the fact that there are counterexamples or that most members of the category do not possess the property. To illustrate this using Leslie's (forthcoming) example: the statement "Mosquitoes carry the West Nile virus" is regarded as true even if very few mosquitoes actually carry the virus.

In a series of cleverly designed experiments, Rhodes and colleagues have shown that generics play a role in the transmission of essentialist beliefs about a social category. In a nutshell, here is what they did: they showed pictures of a new category of people, Zarpies, to four-year-old children. That category was diverse relative to known essentialized groups (e.g. Zarpies are of different races, genders, ages). Pictures of characters were accompanied by a text that was either formulated using generics ("Zarpies eat flowers") or nongenerics ("This Zarpie eats flowers" or "This one eats flowers"). They then asked questions to the children to measure to what extent participants "(1) expect properties associated with the new category to be innate

---

[20] There might also be cues that are used by children and adults that lead them to essentialize a group. Among these cues are those that increase the saliency of a group through competition or threat or segregation (see Bigler and Liben 2007; Martinovic and Verkuyten Ercomer 2012; Plante et al. 2015).

and inevitable (*inheritance* items), (2) expect the properties attributed to a single category member to extend to other category members (*induction* items), and (3) view category membership as causing/explaining the development of typical properties (*explanation* items)" (p. 13527). Their results indicated that generics strongly increased the odds of essentialist responses from participants (in some cases, doubling the essentialist responses relative to comparison conditions).[21]

In another related experiment, parents were introduced to Zarpies in terms that either led them to hold essentialist or nonessentialist beliefs. They were then given a book with pictures of Zarpies without text and asked to talk through the book with their child. Here Rhodes and colleagues measured the use of generics by parents via interaction with their children. The results indicated that the use of generics more than doubled when parents hold essentialist beliefs about Zarpies. They authors also discovered that parents were producing more negative evaluations in the essentialist conditions than in nonessentialist ones. Taken together, the results of these experiments indicate that generic language "can facilitate the transmission of social essentialism from one generation to another" (p. 13529). However, hearing a generic statement about a category does not necessarily induce someone to hold essentialist beliefs. Sometimes information about a category blocks essentialist inferences – for instance, if someone says that people sharing an area phone number are "snobs," one might not infer that this is an essential property of that group of people because one knows that area codes do not demarcate a natural kind. For this reason, Leslie (2014, p. 217) holds that essentialism is a "default assumption" when hearing generics about members of a novel kind. Leslie makes another interesting point concerning what she calls "striking property generalizations" (generalizations like "Sharks eat bathers" and "Muslims are terrorists"). She points out that contrary to positive properties, it takes very few instances of a negative action performed by very few individuals of an essentialized category to cause others to conclude that members of that category either typically have the property to perform the action or are typically

---

[21] Applying the same procedures to adults, they found an even stronger effect. According to these researchers, it makes sense that adults respond thusly to generics given "that learning about the social world continues over a lifetime – and new social categories may be encountered for the first time in adulthood" (p. 13527).

disposed to it. Also, in cases involving social categories, the asymmetry is apparently much stronger in the case of out-group members than in-group members. Leslie states, "[I]t is not hard to see the evolutionary benefits of such a disposition, since the costs of under-generalizing such information are potentially huge" (forthcoming, 4–5).[22] But if there are evolutionary advantages to such a tendency, they are also detrimental effects for the life in society (e.g. in promoting stereotypes of certain groups). Solutions to the problems posed by generics are not obvious because parents are not always conscious of using linguistic forms that induce generalizations. To make matters worse, Leslie (2014) notes that category-wide statements such as "all" or "most" are often interpreted as generics by adults, while children interpret statements quantified as "some" as generics.

Let me conclude this section by making two points. First, in the case of thinking about animal categories, generic language has been thought to be only one factor (and a weak one) leading to the development of essentialism (Rhodes et al. 2012). In contrast, Rhodes and her colleagues think that "social essentialism develops more slowly, more selectively, and with cultural variation, suggesting that cultural input plays a much more important role" than in the case of animal categories (Rhodes et al. 2012, p. 13527). This might be seen as inconsistent with the modular view of racialist cognition that I proposed earlier. However, I do not think that this is the case. Think of sex preferences. According to Lieberman's evolutionary theory of incest (Lieberman et al. 2007), one type of information (but not the only one) concerning who is a potential sexual mate and who is not is coresidence during the early years of development. Being raised with someone is a reliable-enough cue that this person is related to you and is therefore not a potential mate. According to Lieberman, males take more time to reach this conclusion than females. One hypothesized reason for this is that according to parental investment theory, the cost of concluding erroneously that a potential mate is not related to one is less for males than it is for females. The decision to consider a person as a potential mate thus demands the consideration of information from the environment. This is not in contradiction with the thesis that the mechanism in

---

[22] Schaller and Neuberg make a similar claim: "[L]ike other evolved biases in person perception, the psychology of threat detection is characterized by a tendency toward overgeneralization: many people who pose no threat whatsoever are implicitly assumed ... to pose some potential threat" (2012, p. 14).

charge of determining potential mates is modular. We do not need to *learn* which information to attend to or how long to attend to this information in order to decide who is a potential mate. A similar story might be told for animal and social essentialism. While animal essentialism is adaptive in every context and might be used quite early in development, social essentialism is not adaptive in every context and does not come into play as early, one might think, as animal essentialism. It is likely that in the environment of evolutionary adaptation, social interactions with other ethnic groups were (1) infrequent prior to later stages in life and (2) monitored by parents until later in life. So it makes sense that the mechanism in charge of determining which social categories are essentialized kicks in later in life and is more open to modification in light of social information.

Second, as I stated earlier, I agreed (with Gelman and later Rhodes et al.) with the proposal that some aspects of language might act as cues to indicate which categories are to be essentialized and therefore that language might provide a social vehicle for the transmission of essentialist beliefs. But my theory commits me to something more than this. In my view, not all forms of essentialism are alike. Essentialism about ethnies or races is not is importantly different from essentialism about gender. The former involves a form of what might be called "lineage essentialism," while the other does not. That is, in the case of ethnies or races, one acquires essential properties from one's parents, while this is not the case for gender. As far as I know, neither Gelman nor Rhodes and her colleagues explain this feature of essentialist cognition, while I explain it by saying that my thinking about ethnies or races is the product of an exaptation of the folk-biology module (i.e. my thinking about a human's group follows more or less the lines of my thinking about biological species[23]).

Essentialism, Stereotypes, and Prejudices

The adoption of essentialist beliefs concerning some social groups is not without effects on the adoption of stereotypes and prejudices. Indeed, in a series of papers, Haslam and his colleagues (among others,

---

[23] Though not exactly, because culture can introduce some variation in ways of thinking, such as hypodescent, for instance – that is, the fact that in a mixed couple, one would inherit the identity of the parent of the socially inferior group.

Haslam et al. 2000, 2006; Bastian and Haslam 2006, 2007; Haslam and Whelan 2008) have showed that individual differences in the adoption of essentialism predicted stereotype endorsement in general. This effect was not restricted to negative stereotypes (Bastian and Haslam 2006, p. 234). Endorsement of negative stereotypes depends, among other things, on the fact that the essentialized group is seen as highly entitative (uniform). For instance, the category "male" is believed to refer to an essentialized group (once you are a man, you are a man for the rest of your life), but males comprise a group that has a low degree of entitativity (there is a significant degree of perceived variability among men). However, being Jewish or being black is often seen as being part of an essentialized group that has high entitativity (i.e. members of the group are thought to be similar to each other). According to Prentice and Miller (2007), it is the latter groups that tend to be negatively evaluated.

These results linking stereotypes with essentialism are confirmed by work with children by Pauker and her colleagues (2010) showing that the ability to perceptually discriminate between racial groups is not sufficient to explain the fact that the older children get, the more likely they are to acquire stereotypes about (other) racial groups but that both racial saliency (the fact that racial differences or racial labels are used for certain tasks) and racial essentialism (about other groups) contribute to children's acquisition of such stereotypes.

In a widely cited paper, Williams and Eberhardt (2008) also have shown that the impact of an essentialist conception of race is not only cognitive (in that it influences the adoption of stereotypes) but also motivational. In a series of studies, they found that individuals who adopt an essentialistic racial conception (because they were primed by exposure to information concerning the possibility of the genetic validity of racial categorization) tend to "understand racial inequalities as natural, unproblematic, and unlikely to change" (p. 1034). They also found that adopting such a conception makes them "less motivated to change racial inequities, but also less concerned with and moved by such disparities. At the interpersonal level, [they showed] that those with a biological conception of race maintain friendship networks that are less racially diverse, have less desire to develop friendship across race lines and are less interested in simply sustaining contact with a person of another race than those with a social conception of race" (Williams and Eberhardt 2008, p. 1034).

An important issue related to the problem of determining the benefits and costs of retaining discourse about race in science is highlighted by studies related to those about essentialism. These studies investigate genetic essentialism, the "tendency to infer a person's [or a member of a group] characteristics and behaviors as based on their perceived genetic make-up [or the perceived shared genetic makeup of his or her groups]" (Dar-Nimrod and Heine 2011, p. 802). Genetic essentialism is a variety of psychological essentialism where the essence is postulated to be genetic. As in previous studies on essentialism that I have discussed, when such genetic essentialism is primed (Jarayatne et al. 2006; Keller 2005), it tends to make people more likely to adopt stereotypes about other groups and to increase the perceived similarity of group members and to accentuate group differences. Moreover, priming genetic essentialism in people who chronically hold a belief in genetic determinism also has the effect of increasing prejudice and group biases (Keller 2005, p. 697).

Phelan and her colleagues have uncovered an effect that should be taken into consideration when talking about race and genes. In a series of experiments, they asked subjects to read three vignettes that they called the "Social Construction Vignette," the "Race-as-Genetic-Reality Vignette," and the "Backdoor Vignette." The "Social Construction Vignette" presents race as socially constructed without biological reality. The "Race-as-Genetic-Reality Vignette" presents race as a concept validated by genetic research. Finally, the "Backdoor Vignette" presents information concerning the link between a feature of a racial group (a disease that disproportionally affects this racial group) and a genetic variant that is present in that group. According to the authors, the difference between the "Race-as-Genetic-Reality Vignette" and the "Backdoor Vignette" is that the second "clearly endorses the idea of genetically based racial difference in a serious health outcome but makes no statements about more general genetic differences between racial groups" (Phelan et al. 2013, p. 174). Next, the investigators measured the level of belief in essential racial differences. For instance, they asked people to rate statements such as the following: "Although Black and White people may be alike in many ways, there is something about Black people that is essentially different from White people" (p. 188). They discovered that the "Backdoor Vignette" elicit about the same level of belief in essential racial differences as the "Race-as-Genetic-Reality Vignette," which is much higher than the "Social Construction Vignette." They also found similar results with regard to measures of implicit racism and social distance from blacks. All in all, this suggests that establishing a link between a feature of a race and a gene is enough to prompt an essentialist attitude toward race with all the cognitive and motivational features that accompany racial essentialism usually. This should be weighed in the balance when deciding to state, as Risch does, that a conception of race is validated by genetics or (like Spencer) that racial terms have a reference (see earlier). Discussing race in this manner might lead, as Montagu predicted, to essentialism about race.

So far I have argued that there is a mental mechanism for producing ethnic concepts that may use language as a cue and that once a group is essentialized, stereotypes are more likely to be acquired about it, and in certain conditions, prejudices are likely to increase as well. I will now focus on the part that emotions might play in the acquisition and triggering of particular types of prejudices and behaviors.

Emotions

In this last subsection I will explore another body of work that sheds light on the psychological mechanisms underlying racialist cognition. This is work related to the specificity of emotions and behavior toward prejudiced groups, and it rests on an evolutionary "sociofunctional" theory of emotions. I will first explain the theory and then present some of the empirical results explained by it.

In psychology, prejudice has been traditionally thought as a "general undifferentiated phenomenon" (Schaller and Neuberg 2012, p. 10) – for instance, as an "unreasonable negative attitude" (Fishbein 1996, p. 6) or an "antipathy based on a faulty and inflexible generalization" (Allport 1954, p. 10). Thus an essential part of the definition of prejudice is the reference to the valence of attitudes of an attitude directed toward members of a group. Theories that explain why we are prejudiced against this or that group are, for the most part, undiscriminating in that their predictions ascribe different valence-based attitudes (negative or positive) to different groups but without precision concerning the *content* of those attitudes. But "why," Neuberg and Cottrell ask, "do gay men often evoke disgust and desires to distance one's school-aged children from them, whereas Native Americans often evoke pity and desires to establish community outreach programs for them, and African Americans often evoke fear and desires to learn new self-protection techniques?" (2006, p. 163). By contrast, some authors, inspired by the evolutionary theory, proposed that prejudice is more textured than what is suggested by traditional psychological theories.[24] This view has been put forward by Cottrell, Neuberg, and Schaller in a series of papers that adopted a "threat-based" framework for understanding prejudice (Cottrell and Neuberg 2005; Neuberg and Cottrell 2002, 2008; Schaller and Neuberg 2012).[25] This model is based on a "sociofunctional" theory of emotions; according to the functional approach to emotions, each emotion (fear, disgust, anger, embarrassment, etc.) is a specific response, or a set of coordinated responses, to a specific problem. For example, psychophysiologist Robert Levenson (1994, p. 123) characterized emotions as follows[26]:

Emotions are short-lived psychological-physiological phenomena that represent efficient modes of adaptation to changing environmental demands. Psychologically, emotions alter attention, shift certain behaviors upward in response hierarchies, and activate relevant associative networks in memory. Physiologically, emotions rapidly organize the responses of disparate biological systems including facial expression, somatic muscular tonus, voice tone, autonomic nervous system activity, and endocrine activity to produce a bodily milieu that is optimal for effective response. Emotions serve to establish our position vis-à-vis our environment, pulling us toward certain people, objects, actions and ideas, and pushing us away from others.

The sociofunctional approach to emotions is a version of the functional approach. In contrast to other functional approaches, which focus more on the problems posed by the physical environment (e.g. avoiding toxic food or avoiding predators), it highlights the problems that we encounter in our social lives. The problems posed by members of other groups to members of our own group are among these problems. According to the sociofunctional approach, some emotions are adaptive responses to those problems. So emotions evoked by a particular ethnic or racial group should correspond to the problems – more specifically, threats – that this group is seen as posing. Because perceived threats posed by different groups are themselves different (threat of interpersonal hostility, threat of contamination, threat of being

---

[24] For a different approach taking a similar stance, see also Fiske, Cuddy, and Glick (2006).
[25] See also Tapias et al. (2007).   [26] See also Keltner and Haidt (1999, 2001).

cheated out of valuable resources, threat to group shared values, etc.), they elicit different evolved psychological processes that produce qualitatively different responses, and because different threats demand different responses (e.g. contamination demands avoidance of contact with someone, while hostility demands either fear or preparation to fight), it is expected that different groups seen as posing different threats will activate different specialized psychological processes. In a nutshell, different groups are expected to have different "prejudice profiles" (Schaller and Neuberg 2012, p. 10). Because a given ethnic or racial group might be seen as posing several problems, it might evoke a combination of responsive emotions. This account also allows for differential responses to subgroups within these groups, which may be seen as posing different threats and thus may themselves evoke different emotions. Finally, because vulnerability to threat is likely to change over time, the mechanisms that are responsible for processing information about the out-group and producing adaptive responses to the threat they are seen as posing should exhibit "functional flexibility" (Schaller and Neuberg 2012, p. 15).

These predictions are supported by the results of experiments run by Cottrell and Neuberg with white subjects in Vancouver. When white subjects were asked what types of emotions African Americans, Asians, and members of First Nations evoke, it was found that African Americans evoked mostly fear (but also disgust and anger), members of the First Nations pity (and some anger), and Asians envy. Notably, each group evoked an equivalent measure of prejudice (i.e. whites were as negatively prejudiced against each group); thus levels of prejudice that are superficially similar may in fact conceal a striking diversity of emotions. Cottrell and Neuberg also examined how subjects perceived the problems posed by these racial groups. They found that African Americans were seen as posing problems for property, health, reciprocity, social coordination, and security. Other groups (such as Asians) were seen as posing problems for the economy and the dominant values, but not for social coordination or reciprocity.[27] They also showed that there is a link between perceived threat, emotions, and

[27] Other studies (which I will not discuss) also claim that we are disposed differently toward men than women of an outgroup, at least in regard to threats to our physical safety. This difference may be grounded in the difference of the degree of danger that men have represented in our evolutionary past compared to women (see McDonald et al. 2011).

behavioral dispositions (see Cottrell et al. 2010) so that, for instance, "People who are viewed as a threat to physical safety (e.g. African Americans) elicit not only fear but also inclinations to learn new self-defense strategies and to increase police patrols" (Schaller and Neuberg 2012, p. 11).

Finally, Shaller and Neuberg (2012) report the effects of manipulating the feeling of vulnerability in subjects (e.g. by placing them in a darkened room or making them watch a horror movie).[28] They found that this manipulation increases the activation of specific stereotypes related to danger about a group (like "murderer" or "rapist") but not unrelated negative stereotypes (like "lazy" or "ignorant").[29] They also found that it has an effect of increasing particular prejudices against specific groups: for instance, manipulating the degree of perceived vulnerability to illness has an impact on xenophobic attitudes toward immigrant groups that the subjects were not familiar with but not to other groups of immigrants that were familiar to the subjects (Faulkner et al. 2004). Similarly, asking people who are chronically germ adverse to wash their hands with antibacterial wipe before performing a task resulted in lower levels of prejudice against immigrants, overweight people, and people with physical disabilities (Huang et al. 2011).

These studies indicate that it is a mistake to think of racial prejudice as a single, unitary phenomenon motivated by only one kind of emotion. An evolutionary theory of prejudice helps us to reject what we called elsewhere a form of "psychological monism" about racial prejudice, whereby "hate" or some other indistinct, negatively valenced emotion is the only affective phenomenon mentioned in the explanation of prejudice (Faucher and Machery 2009). It also promotes a situated psychology of racial prejudice by showing how social and cultural factors, such as the perception of a particular threat posed by

---

[28] There are also studies showing a link between disposition toward a particular emotion and prejudice toward outgroups whose stereotypic traits are likely to elicit this emotion (Tapias et al. 2007).

[29] Darkness amplified prejudicial beliefs about danger-relevant traits (trustworthiness and hostility) but did not much affect beliefs about equally derogatory traits less relevant to danger (Schaller and Conway 2004, p. 155). See also Dasgupta et al. (2009), who demonstrate that specific emotions increase implicit bias against any group deemed to be relevant to those emotions (inducing disgust increases implicit bias against any group by whom we are disgusted), but see Correll et al. (2010) for a different view.

an out-group or feelings of vulnerability, influence both the content of prejudices and their activation.

If one wants to understand and eradicate particular forms of prejudice, it is crucial to study the workings of particular emotions (such as fear, disgust, anger, envy, pity) and the context of their activation because it is likely that a single strategy will not inhibit all forms of prejudice at all times. As Schaller and Neuberg (2012) put it: "[A]n intervention that mitigates fearful reactions to someone who looks like one of 'them' (rather than one of 'us') may be entirely ineffective in ... reducing resentment toward someone else's status as a welfare recipient ... and when a target group is characterized by features that connote multiple kinds of threat, no single strategy – no matter how thoughtfully designed – is likely to be completely effective" (p. 44).

Conclusion

In this chapter I discussed two projects in the biophilosophy of race. The first concerns the existence of race. I demonstrated that when examining the question of the existence of human races, one must distinguish between the question of the existence of $race_f$ and the question of the existence of $race_s$. I have shown that even though the question of the existence of $race_f$ once seemed to have been settled, the question of the exact content of $race_f$ brought forth in recent discussions has blurred the contours of the debate and reopened it. The question of whether $race_f$ exists cannot be settled until the content of $race_f$ is settled empirically. The debate about $race_s$ is still raging. Philosophers of biology invoke semantic and methodological considerations in an attempt to settle the debate, and I have argued that normative considerations should be invoked as well.

The second biophilosophical project that I considered concerns the mind of the racialist. I have argued that the adoption of an evolutionary perspective on racialist psychology allows researchers to give a more nuanced and fine-grained view of phenomena such as essentialism and prejudice. Though this is a project that is not as well known to philosophers as the first, it is nonetheless an important one. If one wants to understand and eradicate racism and other racial harms, one will have to take into account (among other things) the specificities of the racialist mind. I argue that an evolutionary psychological approach may provide the best way of doing this.

13 How Philosophers "Learn" from Biology – Reductionist and Antireductionist "Lessons"

RICHARD N. BOYD

Philosophers have learned lots of lessons from biological theories and concepts, some of them good and some bad. Here are some important examples.

Antireductionist Lessons

Nonreductionist Physicalism and Multiple Realizability. Logical empiricists "rationally reconstructed" materialism as "physicalism" – all phenomena can be deductively subsumed under "fundamental laws of physics." By Craig's lemma, this implies that if some complex item or state of affairs is physicalistically acceptable, it must be describable in the vocabulary of "fundamental physics."

That's wildly implausible about pains, so materialist philosophers adopted an "identity theory" formulation of materialism: it would be enough for "pain" to be definable in neurophysiological terms; perhaps pain = C-fiber firing. (Why C-firings counted as physical was left unexplored; that's a topic for later.) This was still a reductionist-sounding conception, but then real biology intruded. Sometimes after brain injuries, a different part of the brain becomes able to take over the psychological function of the damaged part. Moreover, psychological states are not realized in exactly the same structures in all species. This taught philosophers that some physical states are "multiply realized" and that realization (pun) helped to launch "functionalist" and other nonreductionist treatments of materialism.

Coming Full Circle. Roughly continuous with nonreductionism about mental phenomena are related nonreductionist treatments of, for example, causal properties, dispositions, semantic relations, moral categories, social and economic categories, genders, and so forth. Basic idea: complex phenomena need not be grounded in some very simple way in smaller phenomena but may be realized in somewhat hetero-geneous aggregations of smaller things. Once you get the idea, more biological examples come to mind: species, taxa, organisms, popula-tions, gene complexes, and so forth. Philosophers of biology (and biologists) have explored these and related phenomena using antireduc-tionist resources initially developed in the philosophy of mind. What began as philosophers learning from biologists has come full circle to inform approaches to theoretical biology and the philosophy of biology.

Reductionist Lessons

Natural Selection Accounts of Biological Function. I have in mind accounts according to which the biological function of an organ, behavior, signaling systems, and so on must be an effect such that the structure in question was established or sustained by natural selection because it produced that effect.

   "Evolutionary" Approaches to Moral Psychology and Issues About "Human Nature." I have in mind here efforts by philosophers to address philosophically relevant psychological questions by relying on contemporary "evolutionary psychology."

Strategy

I will defend *and* extend the antireductionist lessons and critique the reductionist ones. Here is how the arguments will go:

The Metaphysics of Buffered Aggregations. The antireductionist lesson from biology is that almost all nonmicroscopic phenomena are (aspects of) "buffered aggregates," where the "buffering" stabilizes their explanatorily important causal profiles. This conception underwrites an "accommodationist" conception of kinds and things (etc.) and of reference.

   Rethinking Materialism. *Why are C-fibers physical?* The accommodationist conception permits us to formulate materialism in a thoroughgoingly nonreductionist "compositionalist" way, thereby underwriting nonreductionist approaches in biology, psychology, and metaphysics generally.

Resisting the Reductionist Impulses. Materialism is, *in some sense*, a "reductionist" doctrine. The accommodationist and compositionalist conceptions show why defending materialism does not require defending anything like syntactic or conceptual reductionism or any distinctly reductive approach to human psychology.

Let's begin with some metaphysics.

A Process Theory of (Almost) Everything

Return with us now to those thrilling days of yester year when "analytic functionalism," "psychofunctionalism," and "central nervous system state identity theory" competed for the loyalty of physicalists; when "contingent identity" was being challenged by "metaphysical necessity." The "identity theory" was challenged along two dimensions. One involved multiple realizability. The other (due to Kripke) focused on "metaphysical necessity." If M names a mental state and P is a physical description of a physical state, and if both are "rigid designators," then $M = P$ will be, if true, true in all possible worlds. On the plausible view that P refers to something physical in every possible world in which it refers, it follows that if the identity theory is true, then it is true in all possible worlds. But "philosophical intuitions" tell us that this is not so. So, from our armchairs, we can settle the scientific question of materialism about the mental!

I have a preferred response (Boyd 1980): (1) Materialism is not best understood as an "identity theory." Materialism about some class of phenomena is the claim that they and their causal powers are *composed of* unproblematically physical phenomena and their unproblematically physical causal powers. "Multiple realizability" shows that this does not imply "identity" in the reductionist sense. (2) Still, "M is physical" *does entail* "M is identical to some physical phenomenon" (namely itself). (3) But "M is physical" does not entail "M is physical in all possible worlds." Much of this chapter is about the metaphysical and semantic underpinnings of this approach and their relation to how philosophers learn from biologists (and vice versa).

The Case of C-Fibers. Okay, but suppose that (in humans) the only realizations of pain are C-fiber firings, or whatever. Would pain then be associated with a "physical state" in the way anticipated by reductionists? When reductionist philosophers wrote about "physical states" they had in mind something like the "exact total physical states" (for the Newtonian case, a specification of the exact mass, position, and velocity of every particle and likewise for later physical theories). The multiple realizability (of C-fibers themselves!) shows that there is no "reduction" in that sense. How about a token C-fiber firing, call it "CFF"? Suppose that there is an exact function from times to exact physical states that characterizes the token C-fiber firing process during its duration. Does good methodology require that we posit the "token identity" of CFF with the process corresponding to *exactly* that exact function? Do the causal powers of CFF on which we rely in psychological or neurophysiological explanation depend on its *exact* physical state? No, the relevant causal properties of CFF would be the same even if the ratio of ^{12}C to ^{13}C (isotopes of carbon) in the subject's most recent meal had been slightly different and thus ever so slightly altering the physical states manifested in CFF. Similarly, the C-fiber itself would have been the same C-fiber under those conditions.

This is unsurprising from a biological point of view. Biological systems persist and reproduce only because their important structures and causal functions are to some extent buffered against even more serious changes in organisms and their environments. C-fibers are dynamic stabilities of temporally extended molecular processes. C-fiber firings are dynamically buffered structured processes.

Of course, this is true of other sorts of biological entities and properties. One interesting fact about the "species question" is that many of the answers involve buffering: they treat species as defined by the buffered evolutionary persistence of some sort(s) of structure. They differ about what sorts of structure and what sources of persistence are involved, but there is agreement that species are (manifested in) some sort(s) of evolutionarily buffered processes (see, for example, Mayr 1969, 1970; Hull 1978; Boyd 1999; Magnus 2011). A similar view seems to be emerging about homologies (Wagner 2001; Rieppel 2005a, 2005b) and, perhaps, about higher taxa (Rieppel 2005b; Boyd 2010b).

Other Stuff

Biological individuals, properties, relations, and so on are something like processes buffered in ways that preserve biologically relevant causal profiles. Is that peculiar to biological phenomena? No: (almost?)

all causally efficacious entities are buffered composites of smaller entities. Rocks, rivers, tables, and whatever are not sets or mereological
sums of particular atoms or molecules. Their molecular constituents
vary over time, sometimes very little, sometimes a lot, but they are held
together – buffered – by cohesive forces. Their relevant causal profiles
are similarly buffered against some changes in their internal structure
or environment.

The causal efficacy in particular instances of (almost all?) entities and
their causal powers depends on ongoing buffering processes at the time
of their causal interactions with other entities and their causal powers.
The potential window-breaking capacities of a baseball are
sustained – buffered – under ordinary circumstances by ongoing stabilizing bonding processes between its constituent molecules that sustain
its shape, hardness, resiliency, and so forth. But in any particular case in
which a baseball breaks a window, the molecular constituents of the
baseball break the window by *continuing* (somewhat differently) to
be bonded to each other. The token buffering processes that constitute
the baseball are part of the token process of window braking (for the
best exposition of this idea, see Earley 2008; for biological taxa,
see Boyd 2010b and Rieppel 2005b).

To a very good first approximation,

1. (Almost?) All causation has dynamic aspects.
2. (Almost?) All correct causal explanations involving macroscopic
   phenomena *reflect* facts about dynamic-less macroscopic causal
   processes and the persisting causal profiles that they sometimes
   sustain and stabilize. That is true because
3. "Static" macroscopic states, persisting macroscopic entities, their
   persisting macroscopic properties … *just are* causally sustained
   structural stabilities in dynamic interactions of (more nearly) microscopic causal processes. The causal effects of a composite X are
   brought about by its components *interacting together* to sustain the
   existence of X and of its macroscopic causal profile.

General Lesson from Biology. (Almost) Everything is process-like.

Accommodationism

Causally efficacious things are buffered aggregates, and their causal
powers are those that are underwritten by the constitutive buffering

processes. Our successes in scientific explanations (and predictions and guesses [Boyd 2010a]) depend on our being able to deploy concepts, instruments, and terminology that somehow "latch onto" those aggregates and their causal profiles. Of course, individual entities have lots of different buffered causal profiles: some particular wolf is an instance of *Canis lupus*, of Mammalia and, in some places, of "top predator," and so on. We individuate kinds of entities in ways that reflect buffering processes that sustain their *discipline-and-interest-specific* causal profiles. *Canis lupus* is individuated in ways that reflect not just the buffering processes that stabilize the biologically relevant features of individual wolves but also the evolutionary buffering processes (such as reproduction within lineages of wolf populations and stabilizing selection) that underwrite relevant stabilities in the causal profiles of wolves over (some range of) evolutionary time; similarly, for Mammalia. In the case of "top predator," their distinctive causal profile is sustained (buffered) across taxa not only by the buffering mechanisms that stabilize individual top predator species but also by the sorts of stabilizing selective processes that operate on top predator populations even when they are in very different taxa.

The appropriate metaphysics and semantics to go with this conception can be provided by the "accommodationist" conception of natural kinds and natural kind terms (Boyd 2010a). Here is a simplified version: let M be a disciplinary matrix, and let $t_1, \ldots, t_n$ be the natural kind terms deployed within the discourse central to the inductive/explanatory successes of M. Then the families $F_1, \ldots, F_n$ of properties provide definitions of the kinds referred to by $t_1, \ldots, t_n$ and determine their extensions, just in case

1. *Epistemic access condition.* There is a systematic, causally sustained tendency – established by the causal relations between practices in M and causal structures in the world – for what is predicated of t_i within the practice of M to be approximately true of things that satisfy F_i, $I = 1, \ldots, n$. In particular, there is a systematic tendency for things on which t_i is predicated to have (some or most of) the properties in F_i.[1]
2. *Accommodation condition.* This fact, together with the causal powers of things satisfying these explanatory definitions, causally

---

[1] Think of predicating t_i of some expression a as predicating "has a as a member of t_i."

explains how the use of $t_1, \ldots, t_n$ in M contributes to accommodation of the inferential practices of M to relevant causal structures. It explains whatever tendency there is for participants in M to identify causally sustained generalizations, to obtain correct causal explanations, or to obtain successful solutions to practical problems. (For more, see Boyd 2010a, 2010b.)

Homeostatic Property Clustering

Sometimes the accommodation condition requires that the a natural kind be defined by a naturally occurring "clustering" of properties with the consequence that (1) the kind lacks precisely defined membership conditions and, sometimes, (2) the properties in the defining cluster vary over time and/or space. Biological species are paradigmatic homeostatic property clustering (HPC) natural kinds. It is an intended consequence of the HPC account that participation in the relevant clustering mechanisms is, often or always, part of the definition of an HPC kind (for further discussions, see Boyd 1999).

"Mind Dependence," "Relativity," and "Reality" of Natural Kinds

In one sense, accommodationism makes natural kinds discipline-relative social constructions. Does that make them "unreal" or ontologically suspect? No. Objects typically have lots of different causal properties, some relevant and others irrelevant to the prediction or explanation or practical use of any given class of phenomena. So we classify them into kinds in ways that (if we get things right) reflect the sets or clusters of causal powers by which they contribute to produce the sorts of effects we are interested in. The causal powers in question, their clustering (if they constitute a HPC), their causal efficacy, and the effects they produce are all perfectly real. The fact that some effects interest us diminishes the ontological standing neither of those effects nor of the kinds that contribute to causing them. This remains true even in those cases where the effects of interest are partly caused by our classificatory practices. Only if it were claimed that human classificatory practices determined the effects of interest in some noncausal, spookily metaphysical way would issues of "reality" arise, but the accommodationist conceptions makes no such claim (for details, see Boyd 2012).

Extending Accommodationism

The similarities between the HPC conception and the buffering-process conception is obvious: for HPC natural kinds, the processes underwriting homeostasis are just the sort of buffering processes that the buffering conception posits. But not all natural kinds are HPC kinds: for some, there is no indeterminacy of boundaries, nor are their definitive causal powers merely homeostatically shared. There is no indeterminacy in the extension of " ... is a $^{12}C^{16}O_2$ molecule," nor do $^{12}C^{16}O_2$ molecules differ in the causal powers they can contribute to producing chemical effects. Nevertheless, $^{12}C^{16}O_2$ molecules exist and have the causal powers they do because of buffering processes that tend to preserve their chemically relevant structures.

So the proposal that all or almost all things and properties are buffered aggregations is substantial extension of the HPC component of accommodationism. The fact that species, higher taxa, kinds of economic organization, and so on are stabilized by HPC processes provides no reason to believe (or to doubt) that molecules are buffered processes whose causal powers are underwritten by buffering. *But they are, as are (almost?) all natural phenomena.*

Biology's Metaphysical (and Semantic) Lesson for Philosophy

Recognizing multiple realizability of mental states helped philosophers of mind to articulate nonreductionist materialism. The credibility of nonreductionist materialism and an appreciation of the role of stabilizing processes have fed back into less reductionist approaches to biology. I urge here that the lessons from this feedback have broader scope: (almost) all natural phenomena exhibit the process-mediated buffering so familiar from biological systems!

Rethinking Materialism and Its Evidential Basis

Back to C-fiber firings. Suppose that we did know that all (human) pains were C-fiber firings. Why would that show that pains are physical processes?

Philosophers have been drawn to two different approaches to understanding materialism:

1. *Reductive approaches.* Materialism says that all concepts and laws are conceptually/syntactically reducible to laws and concepts of "fundamental physics." This is the *antimetaphysical* "rational reconstruction" offered by logical positivists. It gave rise to mind-central nervous system "identity theory," even though central nervous system states such as C-fiber firings are not identical to fundamental physical states as logical empiricists understood them.
2. *Aggregative approaches.* Materialism is the doctrine that (*a*) + (*b*): (*a*) all things/properties/capacities/forces are *aggregates* of very small *physical* things and the *forces, fields, physical* properties associated with them, and (*b*) the *italicized* stuff above is okay: there is nothing mental, teleological, purposive, representational, or theological about the very small things in question, their causal properties, associated forces and fields, or about how they aggregate (Wilson 2006).

Which to Choose?

Insofar as scientists have confirmed materialist conceptions in biology and elsewhere, the aggregative approach captures their concepts and practices. No one thinks that the emerging biochemical understanding of genetics entails the possibility of the conceptual reduction of genetic and developmental laws and generalizations to the "laws of fundamental physics." The reductive approach cannot make any sense of even of "reductive" findings in science. The choice is obvious, *but* . . .

Evidence?

Okay, so *insofar as scientists have confirmed materialist conceptions*, what they have confirmed are doctrines about aggregation. But why should we think that they have confirmed *materialist* conceptions? Why not instead say that they have shown that lots of important effects are caused by the sorts of phenomena we ordinarily say are physical? One initial motive for identity theories was the conception that materialism would be confirmed by scientists who would determine exactly which physical phenomena (described in the "vocabulary of fundamental physics") were identical to, say, pains. If they have not done that, what is the evidence for materialism? Why believe that (*a*) and (*b*) are true of neurophysiological phenomena like C-fibers and their firings? There are two issues here.

First, it is possible to confirm that some phenomenon is aggregated from small things – atoms and molecules, let us say – and that the aggregation of their causal powers is *sufficient to produce* that phenomenon's characteristic causal effects without being able to identify the exact atomic constituents in question. Indeed, this is our situation with respect to almost all phenomena scientists study. If life forms are discovered on other planets, we will have (without further investigation) good reason to suppose that they too are composed of atoms and that their causal powers derive from the aggregation of the causal powers of their atomic constituents. Of course, our knowledge in such cases is highly theory mediated and involves considerations of overall theoretical unifications, but this is true of all scientific knowledge (Boyd 2010a). We are equally in a good position to be pretty sure (actually, as sure as science gets) that we do not *need* to invoke mental properties of atoms and their constituents.

Here is the second issue: the causal sufficiency of items satisfying (*b*) to produce, when aggregated, all the effects we know about does not, by itself, entail that there are not other, nonphysical factors at work. Perhaps aggregated physical phenomena are sufficient to cause all human behaviors, but still, perhaps some nonphysical mental factors also operate to help produce behaviors. Perhaps it is not necessary to posit vital forces to explain biological phenomena, but still, perhaps, there are some whose operation helps to cause biological effects. What is the evidence that there are no such nonphysical helpers?

The work of Jaegwon Kim (1993) has made this an important issue for philosophy of mind and metaphysics. According to Kim, what rules out unnecessary causes is a "causal exclusion principle" that rules out "causal overdetermination": positing two different causes for an effect when each of them would be causally sufficient to produce the effect. This would rule out nonphysical mental "helpers" all right, but (as Kim insists) it would rule out nonreductionist materialism as well because the causal sufficiency of microscopic physical causes to produce all natural effects (which nonreductionist materialists accept) would rule out the causal efficacy of macroscopic physical causes.

Here is a solution to this problem: nonreductionist materialism says that macroscopic things are real and that big composite things can cause stuff *by being* composite. Still nonreductionist materialists *do* need a nonredundancy principle: composite physical things and their macroscopic powers are okay; explanatorily unnecessary dualistic posits are not okay. Of course, there are methodological principles governing the acceptance of explanations at different "levels" – more or less microscopic ones, for example. As Sturgeon (1992) emphasizes, these principles do not rule out all cases of multiple or overlapping causal explanations. There are cases where such non-competing explanations are mutually corroborative: sometimes the credibility of a macroscopic causal explanation is enhanced by an explanation of how the (more nearly) microconstituents of a macroscopic cause could, in aggregated concert, (help to) cause some macroscopic effect.

Still, sound methodological practices do *sometimes* treat a microscopic explanation and a macroscopic explanation as mutually exclusive. If you posit (just) collision with a moving car as the cause of damage to a tree, and I, instead, posit (just) interactions with the microconstituents of a tornado, then we have offered incompatible explanations. Accepting yours excludes accepting mine *but not because of the size of the posited causal factors*. So the appropriate nonredundancy principle will rule out some but not all macroscopic causes. But then what is the relevant "causal exclusion principle"?

It is misleading to frame the issue as one of "causal" exclusion because that terminology encourages (even though it does not strictly entail) the idea that the principle in question can be elucidated by analyzing the concept of causation (think firing squads, windows broken either by baseballs or by atomic baseball constituents).

Thinking this way misleads because even in the most straightforward cases, where all the causes are physical, the principle in question is an *a posteriori* physical principle not accessible by conceptual analysis of ordinary causal notions. Instead, the relevant principle in the physical case is underwritten by the fact that adding a new nonzero force to a physical system produces a nonzero change in the acceleration of some object(s), as does adding new mass. In cases where the relevant particles and masses are held fixed, the right nonredundancy principle rules out saying that forces {Fi} produce effect E and so also do forces {Gj} *if and only* if the forces {Fi} are not composites of components of the forces {Gj}. If, for example, the forces {Fi} compose the forces {Gj}, then exclusion is not justified. Now, this is quite plausible, but it is not *a priori*. In classical systems with conservative forces, the nonredundancy principle is equivalent to conservation of total energy and mass.

So, when properly understood, the principle does not rule out positing causal efficacy *both of* a macroscopic object and its macroscopic properties *and of* its macroscopic components and their microscopic properties. Roughly, composite things and their composite causal powers do not compete with their components for causal efficacy *precisely because those components exercise the relevant causal powers by composing the composite object.*

What about applying the antiredundancy principle in cases where it is not taken as certain whether some or all of the systems under examination are physical – for example, when we are discussing the question of whether or not there are nonphysical mental causes? This is applying a methodological principle whose credibility rests on its connection to the *a posteriori* justified physical principle. There, too, talking about "causal exclusion" and focusing on conceptual analyses of "cause" and related concepts are prima facie unjustified. In such cases, too, the principle requires that posited additional causes have additional effects where, as in the classical case, a composite of already posited causes does not count as "additional."

Learning from Biology, Learning from Science

So far we have seen a reciprocal learning process involving philosophers and biologists and other scientists. There is an underlying message here: at least with respect to natural phenomena; broadly construed, metaphysical and epistemological questions are to be addressed by *a posteriori* methods continuous with those of the sciences rather than by *a priori* considerations. This is true with respect to understanding materialism, and it is true with respect to methodological principles about causal aggregation. Like all important methodological principles in science, they are justifiable *a posteriori* if at all (Boyd 2010a, 2012).

Resisting the Reductionist Impulse

Although materialism is *in some sense of "reductionist,"* a reductionist doctrine, what it says about some class of phenomena [i.e. (*a*) and (*b*) earlier] is not reductionistic in any of the usual senses. Nor does its confirmation regarding some phenomena or other ordinarily require their reduction, in any usual sense, to some other materialistically acceptable phenomena. It suffices that there be good scientific reasons to believe that (*a*) and (*b*) hold for those phenomena, and ordinarily, this does not require anything like a reduction.

But there are cases where, in order to justify a materialist conception of some class of phenomena, *something like* a reductive explanation does seem necessary. Functional adaptations to environments in biological species are the key example. Before Darwin and Wallace, it was unclear that any sort of purely materialist account of such adaptations was possible. Neither Darwin nor Wallace nor contemporary biologists have reduced the phenomenon of adaptation in any of the senses envisioned by logical positivists, but they did offer what we might call "reduction sketches" to show how adaptations could have purely physical causes.

Plainly sometimes such reduction sketches are necessary. Absent such a sketch for some phenomena, one could either tentatively reject materialism or decline to believe in the phenomena in question (or withhold judgment, of course). Perhaps the phenomena of consciousness currently require a reduction sketch (I do not think so, but that is a topic for another chapter). Here I will examine two recent cases where some philosophers have turned to evolutionary biology as a source of something like reduction sketches: the concept of biological function and the question of human social psychology, especially aspects of psychology relevant to issues in moral and social philosophy. In the first case, I will argue that insofar as a reduction sketch is needed, it is not provided by selected-effects conceptions of biological function (because they get the concept wrong scientifically) but is instead provided by largely nonbiological considerations about natural kinds. In the second case, many philosophers have sought to clarify questions about the moral and social motivations by relying on the deliverances of "evolutionary psychology." In such cases, I will argue that philosophers are misled by reductionist conceptions of biologists and psychologists who misunderstand the implications of evolutionary theory. Absent such misunderstanding, there remain philosophical conceptions about moral and social psychology that can be addressed by broadly scientific research. Indeed, moral and political theory might – like the rest of philosophy – be continuous with the empirical sciences (Boyd 2010a), but the relevant resources are not mainly evolutionary.

Selected-Effects Conceptions of Biological Function

According to a line of argument initially developed by Millikan (1984) and Neander (1991), an adequate concept of biological function must come with an associated concept of "malfunction." The function of vertebrate hearts is to circulate blood, but a nonfunctional heart – a *malfunctioning* heart – is still a heart because it is *supposed to* pump blood. All this normative-sounding language seems to many thinkers to require a reduction sketch in order to be naturalistically acceptable.

Selected-effects conceptions of biological function propose, roughly, that the function of some aspect of a biological system is a matter of its history. The function of A is F just in case ancestral A's (sometimes) produced F, and the emergence or persistence of A is explained by some A's being favored by natural selection because they produced F. Biological functions are evolved functions in this sense.

It is clear that this account does not fully capture the notion of biological function even in evolutionary biology. In cases of "exaptation" (Gould and Vrba 1982), some aspect A of organisms in a lineage once had function F, perhaps satisfying the proposed account, but then in some subsequent sublineage, survival and reproduction depended on A's underwriting some different function F'. One way that this can happen fits the selected-effects conception. In populations under some new conditions, selection might have favored genetic changes that led to A's underwriting F' so that there would be a natural-selection explanation for A's underwriting F' in subsequent lineages.

But in cases where developmental plasticity leads selection (see West-Eberhard 2003), individual organisms may respond adaptively to new conditions by changing their behaviors and physiology so that (prior to any action of natural selection) they operate so that their A's underwrite the function F' and thus ensure the survival of the lineage. Think, for example, about a situation where a specialized mouthpart A was established by selection in a lineage of organisms that initially were obligate feeders on some particular plant species P precisely because it facilitated feeding on that particular species. On any plausible account, its function was to facilitate feeding on P. If individuals in some isolated population in that lineage were faced with the local extinction of P's, it might happen that (without any changes in gene frequency) they were able to survive and reproduce because all or most of them responded by using their A's to feed on some quite different plant P', even though their A's were not especially effective for that purpose. Ordinarily, if this were to happen, we would expect that *subsequent* selection would result in modifications to A's in their descendants to better underwrite the function of feeding on P'. But even when that happens, the evolutionary scenario requires that selection favored (modifications of) A because A *already had* the *new* function of underwriting feeding on P'. Biologists' working notion of function is as much forward looking as it is historical.

So we have reasons to reject the selected-effects conceptions, but what about the normative implications of the concept of malfunctioning? There *is* something normative here, but the normativity is ordinary epistemic normativity, not anything peculiar to biology. We are interested in how biological systems survive and reproduce. As accommodationism predicts (and epistemic normativity requires), we need concepts of kinds and categories and relations that are accommodated to the causal structures and relations that underwrite survival and reproduction. In biological systems, the following is utterly commonplace (perhaps universal). In organisms in some lineage, there is a structure or behavior or some such phenotypic feature, A, such that (a) in many organisms in the lineage A has some effect F, (b) in some or all of these cases A's having F contribute to the survival and reproduction of the organism in question or its descendants, (c) A's are associated with distinctive developmental pathways, so A's can be individuated in ways independent of the production of F, (d) some A's do not produce F, (e) it may, but need not, happen that there is natural selection favoring some organisms whose *A*'s do produce *F* or produce *F* especially effectively, and (f) understanding (a) to (e) is important to understanding how some of the organisms in that lineage live and reproduce and how (or whether) the lineage continues. It's epistemically normative, if we're to understand the lineage's biology or evolutionary, history that we have ways to express the ways in which (a)–(e) are or have been or can be manifested. The natural kind/relation/etc. expressions "*A* has proper biological function *F*" and "*A* malfunctions or malfunctioned in organism (or population or whatever) y" are the linguistic expressions used to describe phenomena such as (a) to (f). They reflect the ways in which our concepts and inferences are accommodated to the relevant causal structures. The normativity here is entirely epistemic: sorting out the biological facts. We need not be "endorsing" *F* or its proper functioning. Analogously, a pacifist can describe an ICBM as malfunctioning without endorsing its function.

"Evolutionary" Psychology

Lots of philosophers have engaged with (human) evolutionary psychology. Some, such as Joyce (2005) and Street (2006), have a sophisticated understanding of the conclusions of the evolutionary psychology literature and have used its findings to defend their conceptions of human moral psychology and of what they each take to be deep evolutionary challenges to moral realism. Other philosophers (e.g. Buller 2005; Fedyk 2012; Richardson 2007) have offered important philosophical criticisms of contemporary evolutionary psychology. I will summarize in a moment the most powerful criticisms. What I will then do is to address an often-heard response to those criticisms: that the approach of evolutionary psychologists has led to the formulation of important fruitful hypotheses: hypotheses that have subsequently been supported by experimental evidence: evolutionary psychology has proven to be a valuable strategy in the "context of invention" whose proposals have fared well in the "context of confirmation."

I will offer and defend two replies. First, the "context of invention" versus the "context of confirmation" distinction is bogus. It ignores the crucial role of projectibility judgments in science. Thinking of evolutionary psychology as a harmless theory-invention strategy renders all but invisible the scientific flaws in current evolutionary psychology and the unwarranted bias toward reductionist and nativist conceptions that they underwrite. The second point concerns the idea that many of the theories commended to us by evolutionary psychology have been met with significant empirical confirmation. In addition to the unjustified nativist and reductionist methodological bias associated with evolutionary psychology there is an additional problem here. Many of the experiments designed to test evolutionary psychological theories have, I shall argue, the property that the nativist conclusions they are supposed to test imply that the experiments themselves are ill designed.

Let us begin with a summary of the inferential practices characteristic of mainstream evolutionary psychology[2] (for an overview, see Cosmides and Tooby 1997; not every mainstream evolutionary psychologist would agree with the details, but their ideas are very influential).

Contemporary evolutionary psychology is a research strategy that is grounded in the idea that findings from evolutionary theory provide independent constraints on theories of human developmental psychology so that some issues can, at least prima facie, be resolved by appeals to "predictions" from evolutionary psychology. To a very good first approximation, the central inferential patterns in evolutionary psychology involve (1) advocating an evolutionary scenario S regarding selection for a behavioral profile B in the environment of evolutionary adaptation according to which B was favored by selection because it served evolutionary function F and then (2) taking that scenario to "predict" that humans have (something very much like) an innate (or almost always learned) and relatively nonmalleable unconscious motive to achieve F (so that the propositional content of the motive approximates the posited evolutionary function). For an

---

[2] I summarize here what I take to be the most plausible currently popular approach. Importantly, it was the dominant approach among scholars such as Wilson and Barash, who practiced what they called "sociobiology" before the term "evolutionary psychology" was introduced. Other researchers, especially some human behavioral ecologists, understand evolutionary theory to predict that humans will exhibit (approximately) reproductively optimal behaviors even under conditions very much unlike the EEA (see Cashdan 2013). This cannot be a general prediction from evolutionary theory! If organisms tended to behave optimally outside the (supposed) EEA, this would be inexplicable via natural selection, and evidence would be provided for intelligent design. So I ignore such "optimality everywhere" approaches as not worth serious consideration.

even better approximation, add some inference patterns identifying motivationally altruistic social behaviors with those that are "altruistic" in the evolutionary sense of reducing individual fitness while contributing to the fitness of conspecifics. For an almost perfect approximation, add (often tacit) inferences from premises of the form "*B* has a biological/genetic basis" to "*B* is (something like) innate and relatively nonmalleable."

Here are two classic examples. The first is from early work of Wilson (1975) (note that he changed his mind later [Wilson 1978]) and Barash (1979).

1. *Premise:* In the EEA, altruism reduced individual fitness. (Note that this is not obviously right unless altruism in the motivational sense is identified with "altruism" in the special evolutionary sense.) It was established by kin selection because displays of altruism increased fitness of altruists' kin in ways that more than compensated for the reduction in individual fitness.
2. *Conclusion:* The (unconscious) motive of most altruism is a concern for one's kin or in-group, and altruism is thus associated with innate xenophobia.

The second comes from Daly and Wilson (1985). Here is the key argument: "Child rearing is a costly, prolonged undertaking. A parental psychology shaped by natural selection is therefore unlikely to be indiscriminate. Rather, we should expect parental feelings to vary as a function of the prospective fitness value of the child in question to the parent" (p. 253). Note that this conclusion is supposed to apply generally, not just in the specific environment of evolutionary adaptation. Recasting this argument, we get

1. *Premise:* Natural selection established in the EEA patterns of child rearing where people directed care toward children as an increasing function of the fitness value of the child in question to the caregiver.
2. *Conclusion:* Humans have an innate tendency (or at any rate a tendency that persists in a very wide range of environments) to have their (child care motivating) feelings vary as a function of the prospective fitness value of the child in question to the caregiver.

What is wrong with such inferences? There is almost a consensus among critics that evolutionary scenarios question are not subject to anything like the evidential standards prevailing elsewhere in evolutionary biology. But suppose that we accept scenario S. What about the inferences from S to nativist conclusions about human psychology? A crucial question concerns the range of psychological hypotheses compatible with S.

Behaviorally Based Scenarios. A scenario regarding selection for behavioral patterns in the EEA is *behaviorally based* if the selection narrative itself – including calculation of the effects on fitness of possible behavioral patterns – does not depend on any particular hypotheses about the proximate causes, psychological or neurophysiological, for the behaviors but only on the patterns having once been heritable traits under selection. To an extremely good first approximation, all the scenarios that figure in contemporary evolutionary psychology are behaviorally based.

Ultimate-Proximate Plurality Thesis. Almost always for a scientifically plausible behaviorally based ultimate hypothesis regarding a behavioral profile in the EEA there will be a great many different scientifically plausible proximate hypotheses each of which would explain the behaviors in question. Moreover, almost always many of these hypotheses will treat the behaviors in question as learned, others will treat them as arising from something like innate dispositions, and others will resist easy classification.

Behavioral Equivalence. Two psychological theories are "behaviorally equivalent" in an environment E just in case they predict exactly the same behaviors in E. *Key theorem:* for any behaviorally based scenario S, if two scientifically plausible psychological theories are behaviorally equivalent in the EEA, then they are equally compatible (or incompatible) with the selection narrative in S.

Conclusion. The characteristic inference from S to the conclusion that B is underwritten by something like an innate but perhaps unconscious motive to do F is unjustified. Lots of other scientifically plausible hypotheses are equally compatible with S.

Consider the scenarios described by Wilson and Barash, where kin selection favored a behavioral profile in the EEA where altruistic behaviors were disproportionately directed toward kin. What other scientifically plausible psychological hypotheses other than innate xenophobia might explain the posited behavioral kin bias in the EEA? There are many scientifically plausible ways in which the required differential responses might have been underwritten. Perhaps people responded differently to familiar people (who in hunter-gatherer societies would be more likely to be kin) than to others. Perhaps they responded to respects of similarity and difference in appearance, language, dress, smell, or to some combination of these factors. Perhaps the response to some of these factors was innate; perhaps for some it was a matter of social learning; perhaps for some a combination of both.

Rebuttal. The inference pattern is justified by the empirically confirmed finding that evolved behaviors are underwritten by something like instincts (the "massive modularity thesis"; see, for example, Caruthers 2006; Cosmides and Tooby 1997).

Response (1). In the 1940s and early 1950s, this conception of (nonhuman) animal behavior was widely accepted (see, for example, Tinbergen 1951). Beginning in the early 1950s (Lehrman 1953), this conception had been profoundly criticized by evolutionary biologists; it is now widely recognized that developmental and behavioral plasticity and learning play very important roles in evolution (West-Eberhard 2003). Lots of evolved adaptive behaviors, even in invertebrates, are learned.[3] So there is currently no justification for thinking that evolved behaviors must be underwritten by something like instincts (Fedyk 2014).

Response (2). Suppose that *B was* underwritten by something like an instinctual motive *M*. All the scenario requires is that *M* produce, in the EEA, the behavioral profile *B*. Lots of scientifically plausible "instinctual" motives could have produced *B* in the EEA. So no inference to a motive to accomplish *F* is justified.

These inference patterns reflect deep confusions about the evolution of behavioral repertoires and about the relationship between evolved behaviors and learning. Perhaps no evolutionary biologist would admit to accepting them if they were made explicit. Nevertheless, one cannot understand the literature unless

---

[3] The classic early paper here is Lehrman (1953). The majestic account of the role of developmental plasticity (including behavioral plasticity) in evolution is West-Eberhard (2003). See Fedyk (2014) for an overview. An interesting general evolutionary framework for integrating these behaviors with still other observations from human anthropology is proposed by Jablonka and Lamb (2006).

one engages with these pathologically defective inference patterns (Boyd 2001).

Context of Invention?

Might these inferences still be defensible as a fruitful theory-invention strategy? No. Here's why.

Projectibility, Confirmation, and Radical Contingency. Scientists choose, at any given time, between the answers to the questions they ask that are projectible at that time, where projectibility amounts to theoretical plausibility given (what they take to be) the best available science (Boyd 2010a). The contribution of projectibility judgments to the epistemic reliability of scientific methods is radically contingent: those methods are reliable exactly (and only) to the extent that the relevant background scientific conceptions are accurate enough that, often enough, when scientists investigate a question, an answer pretty close to the truth is among the hypotheses deemed projectible by current standards. So the inferences we're discussing contribute to the reliability of methods in psychology only if the background assumptions about the evolution of behaviors that underwrite them are pretty good. We have every reason to think that they are not.

But do not evolutionary psychologists contribute methodologically by proposing hypotheses that compete with other less nativist projectible hypotheses? They might do so if people thought of evolutionary psychology as "out-of-date, uninformed evolutionary speculation about psychology." Then there might be a contribution to "brainstorming" about human psychology. The actual practice is the very different. Among those influenced by evolutionary psychology, the hypotheses licensed as projectible by the inferences in question are understood to be all but predicted by evolutionary theory and thus to have a methodological priority that renders nonreductionist social learning hypotheses scientifically dubious. If nativist hypotheses *were* all but dictated by the best-confirmed theory in biology, *then* they should be granted this sort of methodological priority. But they are not, so they should not. The methodological effect of evolutionary psychology is to direct researchers' attention away from credible alternatives to nativist and reductionist hypotheses.

"Massive Modularity" and Experimental Design

We have already seen the following:

1. From a scenario *S* positing selection for behaviors *B* in the EEA with evolutionary function *F*, evolutionary psychologists characteristically infer that humans have something like an innate and pretty nonmalleable (but unconscious) inclination to achieve *F*.
2. This is unjustified. Any plausible psychological theory behaviorally equivalent in the EEA to positing that innate inclination is equally compatible with *S*.
3. Evolutionary psychologists who adopt the massive modularity hypothesis can, on this dubious assumption, conclude that evolved adaptive behaviors are usually underwritten by innate instinct-like modules, but *even then*, any module that would produce the same behaviors in the EEA is compatible with *S*; it would not have to be something like an innate inclination to achieve *F*.

Adopting the massive modularity hypothesis (MMH) fails to justify characteristic evolutionary psychology inferences; it also raises problems of experimental design. Many experiments presented by evolutionary psychologists are ill designed to detect the posited sorts of insulated modules. To see why, consider what is distinctive about such modules.

Consider Daly and Wilson's (1985) inference from a scenario about selection for child care behaviors to a nativist conclusion about preference for caring for closely related children. The only psychological conclusion actually dictated by the scenario is

1. Humans in the EEA had a developmental tendency which, implemented in the EEA, tended to lead them, as adults, to exhibit caring behavior of a sort that, in the EEA, was directed toward children who had some features or other correlated, in the EEA, with offspring-hood or with being the offspring of close relatives.
   How does this minimal hypothesis differ from their actual conclusion?:
2. Humans have, *in general* (not just in the EEA), a psychological tendency to *differentially care about their own or their relatives' children*.
   Or from positions (3) and (4), which might be behaviorally equivalent in the EEA?

3. Humans have, *in general*, a psychological tendency to *differentially care about familiar dependent children.*
4. Humans have, *in general*, a psychological tendency to *differentially care about children commended to their care by learned social norms and proximally available for emotional bonding.*

Answer: They posit different respects of computational integration of the psychological states that were responsible for child care in the EEA with other features of human psychology. The minimal position (1) is silent about integration. The other three posit different respects of integration. Position (2) predicts that biological parents of adopted children will tend to try to care for those children when they can identify them and that parents of adopted children will tend to have weaker child-care motives toward them than toward their biological children. According to position (2), child-care motives will tend to vary in response to information about biological relatedness *whether or not that sort of information was available in the EEA or was correlated in the EEA with biological relatedness.* Position (3) predicts that under a wide variety of conditions, the fact that a child became dependent and familiar to a potential caregiver will tend to produce child-care-related motives *even if the circumstances occasioning their dependence and/or familiarity are different from any present in the EEA.* Position (4) predicts that patterns of child care can vary substantially between different cultures as a result of differences in cultural learning *even when the learning takes place via processes not extant in the EEA (like books or television or newspapers).*

Call such respects of computational integration "rational integration." Of course someone who advocated position (2) or position (3) or position (4) would not be committed to the posited tendency being underwritten by fully rational processes, but she would tentatively expect potential caregivers to respond rationally to new information about biological relatedness or dependence or cultural norms.

It is exactly this sort of rational integration that the MMH denies about evolved modules. A defender of the MMH will not deny that the neurologic machinery that underwrites some insulted evolved module will be integrated *somehow* with the rest of the nervous system, but the whole point of modularity is that it cannot be expected to exhibit, *in response to stimuli different from those that triggered it in the EEA,* the sorts of behavioral responses that a more fully rationally integrated structure might be expected to exhibit. A modularized realization of the developmental tendency (1) *would not* be expected to respond, in a contemporary society, in such a way as to make child-care behaviors responsive to new information about genetic relatedness of potential caregivers to available children. In so far as we might have evidence favoring position (2) – *even if that evidence indicated that the tendency in question was innate* – it would be evidence *against, rather for,* an evolved care-for-related-children module.

How is this relevant to experimental design? Daly and Wilson (1985) do not offer experimental evidence; they offer recent data about actual instances of child abuse or child neglect. But many evolutionary psychologists do offer experimental evidence: they present human subjects with stimuli and examine their responses to see whether or not they fit the responses predicted by the theories they propose. For theories positing evolved modules, these experiments are often fatally flawed: the modularized hypotheses often predict the opposite of the responses that authors count as confirmatory. Here is why.

Suppose that theory *T* posits an evolved modularized psychological state, say, a desire for *C*. Suppose that a wide variety of experimental subjects are presented with stimuli *unlike those that operated in the EEA* but that would ordinarily be expected to activate a *nonmodularized* desire for *C* and that they respond as though such a desire has been activated. Suppose further that the experiments are so well designed that they provide evidence that a desire for *C* is something like a human universal. Would they provide evidence that humans have a *modularized* desire for *C*? No. *The opposite*. If behaviors appropriate to a *non*modularized desire for *C* were reliably elicited by such stimuli, that would be evidence *against* the hypothesis of an *insulated* module *even if* those behaviors provided evidence that a desire for *C* was an *innate* human universal. If you use the MMH to justify the claim that evolutionary theory predicts that adaptive human behaviors are underwritten by innate structures, then you need to be very careful about experimental design.

Consider the well-known studies of human mate choice initiated by Buss (1989). The Buss studies and many others involve asking subjects to respond to written descriptions of possible mates or romantic partners or to such descriptions associated with photographs (see, for example, Buss 1989; Brown and Lewis 2004). There are lots of methodological problems with such studies (see, for example, Hazan and Diamond 2000; Pedersen et al. 2002), but there are special problems of experimental design if the posited mate choice strategies are taken to be modular. No one thinks that mate choice in the EEA was mainly mediated by verbal descriptions or by pictures. It is plausible that the neurologic structures involved in mate choice would have evolved in our lineage long before language emerged and surely very, very long before pictorial representations of possible mates were available. It is entirely plausible that an evolved nonmodular (rationally integrated) innate mate preference (whenever it evolved) would be triggered by verbal descriptions of possible mates or by photographic presentations. It is (somewhat remotely) plausible that such preferences would be accurately reflected in responses to such highly culturally specific items as questions in psychological surveys.

What is not even remotely plausible is that *modularized* mate choice preferences would be triggered by verbal and pictorial representations of a sort that played little or no role in mate choices in the EEA or that answers to contemporary questionnaires would reflect them. Indeed, a modular conception of those preferences predicts the opposite.

The same problem infects the famous hip-to-waist-ratio studies (e.g. Singh 1993) that rely on outline drawings of women in bathing suits, a representational form familiar to comic book readers but probably not to early humans. Lesson: the results of many standard evolutionary psychology studies of evolved mate choice cannot provide evidence for *modularized* mate preferences.

How about the evolved psychology of morals? Haidt's famous experimental paradigm (see Haidt and Bjorklund 2008) examines subjects' verbal responses to verbally presented moral dilemmas. This, even though Haidt actually claims that his postulated modules evolved so as to responded, in fitness-enhancing ways, to *observed episodes of behaviors* (not descriptions) *and* even though his case for the modules he proposes rests in part on studies of (non-language-using!) great apes.

Finally consider the proposal of Cimino and Delton (2010) that humans have an evolved modular newcomer concept whose evolutionary function was to make people less vulnerable to free riders by making then more wary of newcomers to coalitions than of coalition members of longer tenure. Their paper reports several different and ingenious studies. Many involve presenting subjects with pictures of different members of an imaginary group, the "Ice Walkers." Each picture was accompanied with information about the tenure of the person as a member of the Ice Walkers. Each Ice Walker tenure length pair was associated with three sentences attributed to the imaginary Ice Walker in question. Subjects were asked, among other things, to remember which Ice Walkers said which sentences and to say toward which Ice Walkers they had various more or less favorable impressions.

Among the finding said to support the hypothesis of a modularized newcomer concept are these: the errors subjects made in matching Ice Walkers with sentences indicated that they to some extent classified Ice Walkers by length of tenure in the imaginary group, and they were more favorably disposed toward long-tenure Ice Walkers than toward those with shorter tenure.

It is (remotely) plausible that these experimental results show that an easily activated concept of tenure in social groups and a more favorable attitude toward those with longer tenure are human universals. It is (even more remotely) plausible that they provide evidence that such psychological features are innate. But insofar as they do provide such evidence, the concepts and attitudes in question would have to be integrated with other features of psychology such as the ability to recognize and understand pictures, the ability to understand stories about imaginary people, the psychological capacities, whatever they are, that underwrite memory for utterances by characters in stories, the psychological capacities, and social knowledge necessary to understand how to participate as a subject in a psychology experiment. So, insofar as the experimental results do provide evidence about the concepts and attitudes in question, they provide evidence *against*, rather than for, their modularity.

In all cases of experiments like these, there is a general worry about the extent to which subjects' responses reflect features of their psychology that would be exhibited in real life. This is worrisome enough, but suppose that somehow the worries can be overcome and the experimental results do (sometimes) provide evidence for real (perhaps unconscious and innate) desires, preferences, beliefs, and so on. Still there is no reason to believe that they provide evidence for the sorts of evolved modules posited by evolutionary psychologists who adopt the MMH. Indeed, the opposite. It is a consequence of how they are designed that, insofar as such experiments provide evidence for real psychological phenomena, those phenomena must be pretty well rationally integrated rather than modularized.

In cases involving the MMH, the emperor often has no data.

References

Introduction

Allen, C., and Bekoff, M. 1995. "Function, Natural Design, and Animal Behavior: Philosophical and Ethological Considerations," in Thompson (ed.), *The Oxford Handbook of Religion and Science*. Oxford: Oxford University Press, pp. 1–46.

Almeder, R. 1998. *Harmless Naturalism: The Limits of Science and the Nature of Philosophy*. New York: Open Court.

Bunge, M. 1979. "Some Typical Problems in Biophilosophy," *Journal of Social and Biological Structures* 2:155–72.

Clayton, P., and Simpson, Z. 2006. *The Oxford Handbook of Religion and Science*. Oxford: Oxford University Press.

De Carlo, M., and Macarthur, D. 2004. *Naturalism in Question*. Cambridge, MA: Harvard University Press.

Dennett, D. C. 2006. "Higher-Order Truths About Chmess," *Topoi* 25:39–41.

Flanagan, O. 2006. "Varieties of Naturalism," in Clayton and Simpson (eds.), *The Oxford Handbook of Religion and Science*. Oxford: Oxford University Press, pp. 430–52.

Gilson, E. 2009. *From Aristotle to Darwin and Back Again: A Journey in Final Causality, Species, and Evolution*. San Francisco: Ignatius Press.

Godfrey-Smith, P. 2014. *Philosophy of Biology*. Princeton, NJ: Princeton University Press.

Griffiths, P. 2014. "Philosophy of Biology," in E. N. Zalta (ed.), *The Stanford Encyclopedia of Philosophy* (Winter 2014 Edition). Available at: http://plato.stanford.edu/archives/win2014/entries/biology-philosophy/.

Kitcher, P. 1992a. *Freud's Dream: A Complete Interdisciplinary Theory of Mind*. Cambridge, MA: MIT Press.

1992b. "The Naturalists Return," *Philosophical Review* 101:53–114.

Koutrofinis, S. A. (ed.). 2014. *Life and Process: Toward a New Biophilosophy*. Berlin: de Gruyter.

Mahner, M., and Bunge, M. 1979. *Foundations of Biophilosophy*. Berlin: Springer.

Millikan, R. G. 1984. *Language, Thought, and Other Biological Categories: New Foundations for Realism*. Cambridge, MA: MIT Press.

Papineau, D. 1993. *Philosophical Naturalism*. London: Blackwell.

Rosen, M. 2012. *Dignity*. Cambridge, MA: Harvard University Press.

Rosenberg, A. 1996. "A Field Guide to Recent Species of Naturalism," *British Journal for the Philosophy of Science* 47:1–29.

Sulloway, F. 1992. *Freud: Biologist of the Mind*. Cambridge, MA: Harvard University Press.

Thompson, N. S. 1995. *Perspectives in Ethology*, Vol. XI: *Behavioral Design*. New York: Plenum Press.

Chapter 1

Dawkins, R., 2014. "Essences," Edge.org (Answer to Edge.org annual question: What scientific idea is ready for retirement?).

Dennett, D. C. 1971. "Intentional Systems," *Journal of Philosophy* 68:87–106.

1987. *The Intentional Stance*. Cambridge, MA: MIT Press.

1991. *Consciousness Explained*. Boston: Little, Brown.

2009. "Darwin's 'Strange Inversion of Reasoning,'" *Proceedings of the National Academy of Science USA* 106(Suppl. 1):10061–5.

2013. *Intuition Pumps and Other Tools for Thinking*. New York: W.W. Norton.

and Plantinga, A. 2011. *Science and Religion: Are They Compatible?* Oxford: Oxford University Press.

Fodor, J. 2008. "Against Darwinism," *Mind and Language* 23:1–24.

Hodge, J., and Radick, G. (eds.). 2009. *The Cambridge Companion to Darwin*. Cambridge: Cambridge University Press.

Hofstadter, D., and Sander, E. 2013. *Surfaces and Essences: Analogy as the Fuel and Fire of Thinking*. New York: Basic Books.

Kitcher, P. 2009. "Giving Darwin His Due," in Hodge and Radick (eds.), *The Cambridge Companion to Darwin*. Cambridge: Cambridge University Press, pp. 99–420.

MacKenzie, R. B. 1868. *The Darwinian Theory of the Transmutation of Species Examined*. London: Nisbet.

Magnan, A. 1934. *Les Vols des Insects*. Paris: Hermann.

McMasters, J. 1989. "The Flight of the Bumblebee and Related Myths of Entomological Engineering," *American Scientist* 77:164–8.

Pinker, S. 1997. *How the Mind Works*. New York: W.W. Norton.

Quine, W. V. O. 1969. *Ontological Relativity and Other Essays*. New York: Columbia University Press.

Raffman, D. 2005. "Borderline Cases and Bivalence," *Philosophical Review* 114:1–31.

2014. *Unruly Words: A Study of Vague Language*. Oxford: Oxford University Press.

Sanford, D. 1975. "Infinity and Vagueness," *Philosophical Review* 84:520–35.

Strawson, G. 2010. "Your Move: The Maze of Free Will," *The Stone, New York Times* online, July 22, 2010. Available at: www.scrfibd .com/doc/86763712/Week-2-Strawson-The-Maze-of-Free-Will.

Chapter 2

Bashour, B., and Muller, H. 2013. *Contemporary Philosophical Naturalism and Its Implications*. London: Routledge.

Bennett, J. 1976. *Linguistic Behavior*. Cambridge: Cambridge University Press.

Darwin, C. 1859. *On the Origin of Species*. London: John Murray.

Dennett, D. C. 1969. *Content and Consciousness*. London: Routledge.

1975. "Why the Law of Effect Will Not Go Away," *Journal for the Theory of Social Behaviour* 5:169–88.

1995. *Darwin's Dangerous Idea*. New York: Simon & Schuster.

2013. "The Evolution of Reasons," in Bashour and Muller (eds.), *Contemporary Philosophical Naturalism and Its Implications*. London: Routledge, pp. 13–47.

Dretske, F. 1989. *Explaining Behavior*. Cambridge, MA: MIT Press.

Fodor, J. 1990. *The Theory of Content*. Cambridge, MA: MIT Press.

Fraser, B., and Sterelny, K. 2013. "Evolution and Moral Realism." Available at: www.sas.upenn.edu/~weisberg/PBDB/PBDB7_files/Sterelny-Fraser .Evolution%20and%20%20Moral%20Realism.V7b.pdf.

Grice, P. 1957. "Meaning," *Philosophical Review* 66:377–88.

Kingsbury, J., Ryde, D., and Williford, K. 2012. *Millikan and Her Critics*. New York: Wiley Blackwell.

Leibniz, G. 1714. *Mondadology*, J. Bennett, trans. Available at: www .earlymoderntexts.com/pdf/leibmona.pdf.

Millikan, R. 1984. *Language, Thought and Other Biological Catagories*. Cambridge, MA: Bradford Books.

Neander, K. 2012. "Toward an Informational Teleosemantics," in Kingsbury, Ryde, and Williford (eds.), *Millikan and Her Critics*. New York: Wiley Blackwell, pp. 21–41.

Rosenberg, A. 2014. "How Physics Fakes Design," in Thompson and Walsh (eds.), *Evolutionary Biology: Conceptual, Ethical, and Religious Issues*. Cambridge University Press, pp. 217–38.

Ruskin, J. 2011. *The Modern Painters, 1856*. New York: National Library Association Facsimile.

Searle, J. 1980. "Minds, Brains and Programs," *Brain and Behavioral Science* 3:417–57.

Spinoza, B. 1677. *Ethics*, J. Bennett, trans. Available at: www
.earlymoderntexts.com/assets/pdfs/spinoza1665.pdf.
Taylor, C. 1964. *Explanation of Behavior*. London: Routledge.
Thompson, P., and Walsh, D. 2014. *Evolutionary Biology: Conceptual, Ethical, and Religious Issues*. Cambridge: Cambridge University Press.

Chapter 3

Allen, C. 2004. "Animal Pain," *Noûs* 38:617–43.
Baars, B. 1988. *A Cognitive Theory of Consciousness*. Cambridge: Cambridge University Press.
Baker, M., Wolanin, P., and Stock, J. 2006. "Signal Transduction in Bacterial Chemotaxis," *BioEssays* 28:9–22.
Bogdan, R. 1986. *Belief: Form, Content and Function*. Oxford: Oxford University Press.
Brook, A., and Akins, K. 2005. *Cognition and the Brain: The Philosophy and Neuroscience Movement*. Cambridge: Cambridge University Press.
Budd, G., and Jensen, S. 2015. "The Origin of the Animals and a 'Savannah' Hypothesis for Early Bilaterian Evolution," *Biological Reviews* doi: 10.1111/brv.12239.
Carruthers, P. 2015. *The Centered Mind: What the Science of Working Memory Shows Us About the Nature of Human Thought*. Oxford: Oxford University Press.
Chalmers, D. 1996. *The Conscious Mind: In Search of a Fundamental Theory*. Oxford: Oxford University Press.
Danbury, T., Weeks, C., Waterman-Pearson, A., et al. 2000. "Self-Selection of the Analgesic Drug Carprofen by Lame Broiler Chickens," *Veterinary Record* 146:307–11.
Darmaillacq, A.-S., Dickel, L., and Mather, J. 2014. *Cephalopod Cognition*. Cambridge: Cambridge University Press.
Dehaene, D. 2014. *Consciousness and the Brain: Deciphering How the Brain Codes Our Thoughts*. New York: Random House.
Denton, D., McKinley, M. J., Farrell, M., and Egan, G. F. 2009. "The Role of Primordial Emotions in the Evolutionary Origin of Consciousness," *Consciousness and Cognition* 18:500–14.
Dretske, F. 1986. "Misrepresentation," in Bogdan (ed.), *Belief: Form, Content and Function*. Oxford: Oxford University, pp. 17–36.
Eisemann, C. H., Jorgensen, K., Merritt, D. J., et al. 1984. "Do Insects Feel Pain? A Biological View," *Experientia* 40:164–7.
Elwood, R. 2012. "Evidence for Pain in Decapod Crustaceans," *Animal Welfare* 21:23–7.

Godfrey-Smith, P. Forthcoming. "Mind, Matter, and Metabolism," *Journal of Philosophy*.

Jékely, G. 2009. "Evolution of Phototaxis," *Philosophical Transactions of the Royal Society of London B* 364:2795–808.

Paps, J., and Nielsen, C. 2015. "The Phylogenetic Position of Ctenophores and the Origin(s) of Nervous Systems." *EvoDevo* 6. Available at: www.evodevojournal.com/content/6/1/1.

Keijzer, F., and Godfrey-Smith, P. 2015. "An Option Space for Early Neural Evolution," *Philosophical Transactions of the Royal Society of London B* 370. Available at: http://dx.doi.org/10.1098/rstb.2015.0181.

Jones, R. 2013. "Science, Sentience, and Animal Welfare," *Biology and Philosophy* 28:1–30.

Keijzer, F., van Duijn, M., and Lyon, P. 2013. "What Nervous Systems Do: Early Evolution, Input-Output, and the Skin-Brain Thesis," *Adaptive Behavior* 21:67–85.

Key, B. 2015. "Fish Do Not Feel Pain and Its Implications for Understanding Phenomenal Consciousness," *Biology and Philosophy* 30:149–65.

Lüttge, U., and Beyschlag, W. 2013. *Progress in Botany LXXVII*. New York: Springer.

Marshall, C. 2006. "Explaining the Cambrian 'Explosion' of Animals," *Annual Review of Earth and Planetary Sciences* 34:355–84.

McMenamin, M. 1998. *The Garden of Ediacara*. New York: Columbia University Press.

Milner, D., and Goodale, M. 2005. *Sight Unseen: An Exploration of Conscious and Unconscious Vision*. Oxford: Oxford University Press.

Moroz, L. 2015. "Convergent Evolution of Neural Systems in Ctenophores," *Journal of Experimental Biology* 218:598–611.

Nagel, T. 1974. "What Is It Like to Be a Bat?" *Philosophical Review* 83:435–50.

Nielsen, C. 2008. "Six Major Steps in Animal Evolution: Are We Derived Sponge Larvae?" *Evolution and Development* 10:241–57.

O'Malley, M. 2014. *Philosophy of Microbiology*. Cambridge: Cambridge University Press.

Pantin, C. 1956. "The Origin of the Nervous System," *Pubblicazioni della Stazione Zoologica di Napoli* 28:171–81.

Parker, A. 2003. *In the Blink of an Eye: How Vision Sparked the Big Bang of Evolution*. New York: Basic Books.

Pery, C., Barron, A., and Cheng, K. 2013. "Invertebrate Learning and Cognition: Relating Phenomena to Neural Substrate," *WIREs Cognitive Science* 4:561–82.

Peterson, K., Cotton, J., Gehling, J., and Pisani, D. 2008. "The Ediacaran Emergence of Bilaterians: Congruence Between the Genetic and the

Geological Fossil Records," *Philosophical Transactions of the Royal Society of London B* 363:1435–43.

Prinz, J. 2000. "A Neurofunctional Theory of Consciousness," in Brook and Akins (eds.), *Cognition and the Brain: The Philosophy and Neuroscience Movement*. Cambridge: Cambridge University Press, pp. 381–96.

Sneddon, L. 2011. "Pain Perception in Fish: Evidence and Implications for the Use of Fish," *Journal of Consciousness Studies* 18:209–29.

Spang, A., Saw, J., Jørgensen, S., et al. 2015. "Complex Archaea That Bridge the Gap Between Prokaryotes and Eukaryotes," *Nature* 521:173–79.

Trestman, M. 2013. "The Cambrian Explosion and the Origins of Embodied Cognition," *Biological Theory* 8:80–92.

Volkov, A., and Markin, V. 2014. "Active and Passive Electrical Signaling in Plants," in Lüttge and Beyschlag (eds.), *Progress in Botany LXXVII*. New York: Springer, pp. 143–76.

Chapter 4

Allman, J. 1999. *Evolving Brains*. New York: Scientific American Library.

Arstila, V., and Lloyd, D. 2014. *Subjective Time: The Philosophy, Psychology and Neuroscience of Temporality*. Cambridge, MA: MIT Press.

Baars, B. J., and Gage, N. M. 2007. *Cognition, Brain, and Consciousness*. San Diego: Academic Press.

Bickle, J. 2013. *The Oxford Handbook of Philosophy and Neuroscience*. Oxford: Oxford University Press.

Brozek, B. 2013. *Rule Following*. Krakow: Copernicus Center Press.

Chalmers, D. 1996. *The Conscious Mind: In Search of a Fundamental Theory*. Oxford: Oxford University Press.

Churchland, P. M. 1989. *A Neurocomputational Perspective: The Nature of Mind and the Structure of Science*. Cambridge, MA: MIT Press.

1996a. *The Engine of Reason, The Seat of the Soul*. Cambridge, MA: MIT Press.

1996b. "The Rediscovery of Light," *Journal of Philosophy* 93:211–28.

2007. *Philosophy at Work*. Cambridge: Cambridge University Press.

2013. *Plato's Camera*. Cambridge, MA: MIT Press.

Churchland, P. S. 1986. *Neurophilosophy: Towards a Unified Understanding of the Mind/Brain*. Cambridge, MA: MIT Press.

2002. *Brain-Wise: Studies in Neurophilosophy*. Cambridge, MA: MIT Press.

2013a. *Touching a Nerve*. New York: W.W. Norton.

2013b. "Exploring the Causal Underpinning of Determination, Resolve, and Will," *Neuron* 80:1337–8.

and Sejnowski, T. J. 1992. *The Computational Brain*. Cambridge, MA: MIT Press.

Craver, C. 2009. *Explaining the Brain*. Oxford: Oxford University Press.

Danks, D. 2014. *Unifying the Mind: Cognitive Representations as Graphical Models*. Cambridge, MA: MIT Press.

Dennett, D. C. 1987. *The Intentional Stance*. Cambridge, MA: MIT Press.

Eliasmith, C. 2013. *How to Build a Brain: A Neural Architecture for Biological Cognition*. Oxford: Oxford University Press.

Fodor, J. A. 1975. *The Language of Thought*. Cambridge, MA: Harvard University Press.

1980. "Methodological Solipsism Considered as a Research Strategy in Cognitive Psychology," *Behavioral and Brain Sciences* 3:63–109.

2000. *The Mind Doesn't Work That Way: The Scope and Limits of Computational Psychology*. Cambridge, MA: MIT Press.

1998. *In Critical Condition: Polemical Essays on Cognitive Science and Philosophy of Mind*. Cambridge, MA: MIT Press.

Frith, C. 2007. *Making Up the Mind: How the Brain Creates Our Mental World*. Oxford, UK: Blackwell.

Gazzaniga, M. S. 2015. *Tales from Both Sides of the Brain: A Life in Neuroscience*. New York: HarperCollins.

Gazzaniga, M., and LeDoux, J. 1978. *The Integrated Mind*. New York: Plenum Press.

Glimcher, P., and Fehr, E. 2013. *Neuroeconomics: Decision Making and the Brain*, 2nd edn. San Diego: Academic Press.

Glymour, C. 2001. *The Minds Arrows: Bayes Nets and Graphical Causal Models in Psychology*. Cambridge, MA: MIT Press.

Graziano, M. 2013. *Consciousness and the Social Brain*. Oxford: Oxford University Press.

Grens, K. 2014. "The Rainbow Connection," *The Scientist*. Available at: www.the-scientist.com/?articles.view/articleNo/41055/title/The-Rainbow-Connection/.

Grice, P. 1989. *Studies in the Way of Words*. Cambridge, MA: Harvard University Press.

Heller, M., Brozek, B., and Kurek, L. 2013. *Between Philosophy and Science*. Krakow: Copernicus Center Press.

Hinton, G. 2013. "Where Do Features Come From?" *Cognitive Science* 38:1078–111.

Lieberman, P. 2013. *The Unpredictable Species*. Princeton, NJ: Princeton University Press.

McGinn, C. 2014. "Storm over the Brain: Review of Patricia S. Churchland, Touching a Nerve," *New York Review of Books*, April 24.

2012. "All Machine and No Ghost," *New Statesman*, February, 141, p. 40.

Medawar, P. 1979. *Advice to a Young Scientist*. New York: Basic Books.

Mele, A. 2014. *Surrounding Free Will: Philosophy, Psychology, Neuroscience*. Oxford: Oxford University Press.

Moser, E. I., Roudi, Y., Witter, M. P., et al. 2014. "Grid Cells and Cortical Representation," *Nature Reviews Neuroscience* 15:466–81.

Nagel, T. 2012. *Mind and Cosmos: Why the Materialist Neo-Darwinian Conception of Nature Is Almost Certainly False*. Oxford: Oxford University Press.

Nanay, B. 2010. "A Modal Theory of Function," *Journal of Philosophy* 107: 412–31.

Pääbo, S. 2014. *Neanderthal Man: In Search of the Lost Genomes*. New York: Basic Books.

Pace-Schott, E. F., and Hobson, J. A. 2002. "The Neurobiology of Sleep: Genetics, Cellular Physiology and Subcortical Networks," *Nature Reviews Neuroscience* 3:591–600.

Parvizi, J., Rangarajan, V., Shirer, W. R., Desai, N., and Greicius, M. D. 2013. "The Will to Persevere Induced by Electrical Stimulation of the Human Cingulate Gyrus," *Neuron* 80:1359–67.

Petersen, S. E., and Posner, M. I. 2012. "The Attention System of the Human Brain: 20 Years After," *Annual Review of Neuroscience* 35:73–89.

Quine, W. V. O. 1960. *Word and Object*, 2nd edn. Cambridge, MA: MIT Press.

Ryvlin, R., Cross, J. H., and Rheims, S. 2014. "Epilepsy Surgery in Children and Adults," *Lancet Neurology* 13:1114–26.

Schooler, J., Nadelhoffer, T., Nahmias, E., and Vohs, K. 2014. "Measuring and Manipulating Beliefs About Free Will and Related Concepts: The Good, the Bad, and the Ugly," in Mele (ed.), *Surrounding Free Will: Philosophy, Psychology, Neuroscience*. Oxford: Oxford University Press, pp. 72–94.

Scruton, R. 2014. *The Soul of the World*. Princeton, NJ: Princeton University Press.

Shannon, C., and Weaver, W. 1998. *The Mathematical Theory of Communication*. Champaign, IL: University of Illinois Press.

Silva, A. J., Landreth, A., and Bickle, J. 2014. *Engineering the Next Revolution in Neuroscience: The New Science of Experiment*. Oxford: Oxford University Press.

Smith, D. L. 2011. *Less Than Human: Why We Demean, Enslave, and Exterminate Others*. New York: St. Martin's Press.

Solomon, S. G., and Lennie, P. 2007. "The Machinery of Colour Vision," *Nature Reviews Neuroscience* 8:276–86.

Squire, L. R., Stark, C. E., and Clark, R. E. 2004. "The Medial Temporal Lobe," *Annual Review of Neuroscience* 27:279–306.

Berg, D., Bloom, F. E., et al. 2012. *Fundamental Neuroscience*, 4th edn. San Diego: Academic Press.

Striedter, G. F., Belgard, T. G., Chen, C. C., et al. 2014. "NSF Workshop Report: Discovering General Principles of Nervous System Organization by Comparing Brain Maps Across Species," *Brain, Behavior and Evolution* 83:1–8.

Thagard, P. 2014. "Explanatory Identities and Conceptual Change," *Science and Education* 23:1531–48.

Weinberg, S. 2015. *To Explain the World: The Discovery of Modern Science*. New York: HarperCollins.

Yu, S., Gao, B., Fang, Z., et al. 2013. "Stochastic Learning in Oxide Binary Synaptic Devices for Neuromorphic Computing," *Frontiers in Neuroscience* 7:1–9.

Chapter 5

Abrams, M. 2005. "Teleosemantics Without Natural Selection," *Biology and Philosophy* 20:97–116.

Bigelow, J., and Pargetter, R. 1987. "Functions," *Journal of Philosophy* 84: 181–96.

Bogdan, R. 1986. *Belief: Form, Content and Function*. New York: Oxford University Press.

Boorse, C. 1976. "Wright on Functions," *Philosophical Review* 85:70–86.

Burge, T. 2010. *Origins of Objectivity*. New York: Oxford University Press.

Cummins, R. 1975. "Functional Analysis," *Journal of Philosophy* 72:741–64.

Davidson, D. 1987. "Knowing One's Own Mind," *Proceedings and Addresses of the American Philosophical Association* 60:441–58.

Dretske, F. 1986. "Misrepresentation" in Bogdan (ed.), *Belief: Form, Content and Function*. New York: Oxford University Press, pp. 17–36.

　1988. *Explaining Behavior*. Cambridge, MA: Bradford Books.

Fodor, J. 1984. "Semantics, Wisconsin Style," *Synthese* 59:231–50.

　1987. *Psychosemantics: The Problem of Meaning in the Philosophy of Mind*. Cambridge, MA: Bradford Books.

　1990. *A Theory of Content*. Cambridge, MA: Bradford Books.

Jablonka, E., and Lamb, M. 1999. *Epigenetic Inheritance and Evolution*. Oxford: Oxford University Press.

Kriegel, U. 2013. *Phenomenal Intentionality*. Oxford: Oxford University Press.

Lewis, D. 1969. *Convention*. London: Wiley.

Mameli, M. 2004. "Nongenetic Selection and Nongenetic Inheritance," *British Journal for the Philosophy of Science* 55:35–71.

Millikan, R. G. 1984. *Language, Thought and Other Biological Categories*. Cambridge, MA: Bradford Books.

1989. "In Defense of Proper Functions," *Philosophy of Science* 56:288–302.

1991. "Speaking Up for Darwin," in Rey and Loewer (eds.), *Meaning and Mind: Fodor and His Critics*. Oxford, UK: Blackwell, pp. 151–64.

1996. "On Swampkinds," *Mind and Language* 11:103–17.

Milner, A., and Goodale, M. 1995. *The Visual Brain in Action*. Oxford: Oxford University Press.

Nanay, B. 2014. "Teleosemantics Without Etiology," *Philosophy of Science* 81:798–810.

Neander, K 1991. "The Teleological Notion of 'Function,'" *Australasian Journal of Philosophy* 69:454–68.

1995. "Misrepresenting and Malfunctioning," *Philosophical Studies* 79: 109–41.

1996. "Swampman Meets Swampcow," *Mind and Language* 11:118–29.

Papineau, D. 1984. "Representation and Explanation," *Philosophy of Science* 51:550–72.

1987. *Reality and Representation*. Oxford, UK: Basil Blackwell.

1993. *Philosophical Naturalism*. Oxford, UK: Blackwell.

1996. "Doubtful Intuitions," *Mind and Language* 11:130–2.

2001. "The Status of Teleosemantics, or How to Stop Worrying About Swampman," *Australasian Journal of Philosophy* 79:79–89.

2003. "Is Representation Rife?" *Ratio* 16:107–23.

2014. "Sensory Experience and Representational Properties," *Proceedings of the Aristotelian Society* 114:1–33.

2016. "Against Representationalism (about Conscious Sensory Experience)," *International Journal of Philosophical Studies* 24:324–47.

Pietrowski, P. 1992. "Intentionality and Teleological Error," *Pacific Philosophical Quarterly* 73:267–82.

Plantinga, A. 1993. *Warrant and Proper Function*. Oxford: Oxford University Press.

Ramsey, F. 1927. "Facts and Propositions," *Aristotelian Society Supplementary Volume* 7:153–206.

Rey, G., and Loewer, B. 1991. *Meaning and Mind: Fodor and His Critics* Oxford, UK: Blackwell.

Seyfarth, R., Cheney, D., and Marler, P. 1980. "Monkey Responses to Three Different Alarm Calls: Evidence of Predator Classification and Semantic Communication," *Science* 210:801–3.

Skyrms, B. 1996. *Evolution of the Social Contract*. Cambridge: Cambridge University Press.

2010. *Signals: Evolution, Learning and Information*. Oxford: Oxford University Press.

Wright, L. 1973. "Functions," *Philosophical Review* 82:139–68.

Chapter 6

Adriaans, P., and van Benthem, J. 2008. *Philosophy of Information.* Amsterdam: Elsevier.

Ariew, A., Cummins, R. and Perlman, M. 2002. *Functions: New Essays in the Philosophy of Psychology and Biology.* Oxford: Oxford University Press.

Bogdan, R. 1986. *Belief: Form, Content and Function.* Oxford: Oxford University Press.

Boorse, C. 1977. "Health as a Theoretical Concept," *Philosophy of Science* 44: 542–73.

　　2002. "A Rebuttal on Functions," in Ariew, A., Cummins, R. and Perlman, M. (eds.), *Functions: New Essays in the Philosophy of Psychology and Biology.* Oxford: Oxford University Press, pp. 63–112.

　　1997. "A Rebuttal on Health," in Humber and Almeder (eds.), *What Is Disease?* Totowa, NJ: Humana Press, pp. 3–143.

Brandon, R. 2013. "A General Case for Function Pluralism," in Huneman (ed.), *Functions: Selection and Mechanisms,* New York: Springer, pp. 97–104.

Caramazza, A. 1986. "On Drawing Inferences About the Structure of Normal Cognitive Systems from the Analysis of Patterns of Impaired Performance: The Case for Single-Patient Studies," *Brain and Cognition* 5:41–66.

　　1992. "Is Cognitive Neuroscience Possible?" *Journal of Cognitive Neuroscience* 4:80–95.

　　and Coltheart, M. 2006. "Cognitive Neuropsychology Twenty Years On," *Cognitive Neuropsychology* 23:3–12.

Coltheart, M. 2004. "Brain Imaging, Connectionism and Cognitive Neuropsychology," *Cognitive Neuropsychology* 21:21–5.

Craver, C. 2001. "Role Functions, Mechanisms, and Hierarchy," *Philosophy of Science* 68:53–74.

Couch, M., and Pfeifer, J. 2016. *Kitcher and His Critics.* Oxford: Oxford University Press.

Cummins, R. 1975. "Functional Analysis," *Journal of Philosophy* 72: 741–65.

Davies, P. S. 2001. *Norms of Nature: Naturalism and the Nature of Functions.* Cambridge, MA: MIT Press.

Dayal, S., Rodionov, R. N., Arning, E., et al. 2008. "Tissue-Specific Downregulation of Dimethylarginine Dimethylaminohydrolase in Hyperhomocysteinemia," *American Journal of Physiology – Heart and Circulatory Physiology* 295:H816–25.

Dretske, F. 1986. "Misrepresentation," in Bogdan (ed.), *Belief: Form, Content and Function,* Oxford: Oxford University Press, pp. 17–36.

2008. "*Epistemology and Information,*" in Adriaans and van Benthem (eds.), *Philosophy of Information*, Amsterdam: Elsevier, pp. 29–48.

Figdor, C. 2010. "Neuroscience and the Multiple Realization of Cognitive Functions," *Philosophy of Science* 77:419–56.

Garson, J. 2013. "The Functional Sense of Mechanism," *Philosophy of Science* 80:317–33.

Godfrey-Smith, P. 1993. "Functions: Consensus Without Unity," *Pacific Philosophical Quarterly* 74:196–208.

Humber, J. M., and Almeder, R. F. 1997. *What Is Disease?* Totowa, NJ: Humana Press.

Huneman, P. 2013. *Functions: Selection and Mechanisms.* New York: Springer.
"Introduction," in Huneman (ed.), Functions: Selection and Mechanisms. New York: Springer, pp. 1–16.

Jacob, P. 1997. *What Minds Can Do: Intentionality in a Non-Intentional World.* Cambridge: Cambridge University Press.

Kitcher, P. 1993. "Function and Design," *Midwest Studies in Philosophy* 18: 379–97.

McCloskey, M. 2009. *Visual Reflections: A Perceptual Deficit and Its Implications.* Oxford: Oxford Psychology.

McGeer, V. 2007. "Why Neuroscience Matters to Cognitive Neuropsychology," *Synthese* 159:347–71.

Millikan, R. G. 1989 "An Ambiguity in the Notion 'Function,'" *Biology and Philosophy* 4:172–6.

2002. "Biofunctions: Two Paradigms," in Ariew, Cummins, and Perlman (eds.), *Functions: New Essays in the Philosophy of Psychology and Biology.* Oxford: Oxford University Press, pp. 113–43.

Neander, K. 1991a. "Functions as Selected Effects: The Conceptual Analyst's Defense," *Philosophy of Science* 58:168–84.

1991b. "The Teleological Notion of 'Function'," *Australasian Journal of Philosophy* 69:454–68.

1995. "Misrepresenting and Malfunctioning," *Philosophical Studies* 79: 109–41.

2012. "Biological Function," in *Routledge Encylopedia of Philosophy.* Available at: www.rep.routledge.com/articles/biological-function.

2013. "Toward an Informational Teleosemantics," in D. Ryder, J. Kingsbury, and K. Williford (eds.), *Millikan and Her Critics.* Oxford, UK: Wiley Blackwell, pp. 21–40.

2015. "Functional Analysis and the Species Design," *Synthese* doi: 10.1007/s 11229-015-0940-9.

2016. "Kitcher's Two Design Stances," in Couch and Pfeifer (eds.), *Kitcher and His Critics.* Oxford: Oxford University Press, pp. 45–73.

and Rosenberg, A. 2012. "Solving the Circularity Problem for Functions: A Response to Nanay," *Journal of Philosophy* 109:613–22.

Phillips, C. G., Zeki, S., and Barlow, H. B. 2012. "Localization of Function in the Cerebral Cortex: Past, Present and Future," *Brain* 107:328–61.

Ryder, D., Kingsbury, J., and Williford, K. 2013. *Millikan and Her Critics.* Oxford, UK: Wiley-Blackwell.

Scarantino, A. 2013 "Animal Communication as Information-Mediated Influence," in Stegman (ed.), *Animal Communication Theory: Information and Influence.* Cambridge: Cambridge University Press, pp. 63–88.

Shea, N. 2007. "Consumers Need Information: Supplementing Teleosemantics with an Input Condition," *Philosophy and Phenomenological Research* 75:404–35.

Shulte, P. 2012. "How Frogs See the World: Putting Millikan's Teleosemantics to the Test," *Philosophia* 40:483–96.

Squire, L. R., and Kandel, E. R. 2003. *Memory: From Mind to Molecules.* New York: Macmillan.

Stampe, D. 1977. "Toward a Causal Theory of Linguistic Representation," *Midwest Studies in Philosophy* 2:42–63.

Stegman, U. 2013. *Animal Communication Theory: Information and Influence.* Cambridge: Cambridge University Press.

Wright, L. 1973. "Functions," *Philosophical Review* 82:139–46.

Chapter 7

Allen, C., Bekoff, M., and Lauder, G. 1998. *Nature's Purposes: Analyses of Function and Design in Biology.* Cambridge, MA: MIT Press.

Aristotle. 1984a. *The Complete Works: The Revised Oxford Translation,* J. Barnes (ed.). Princeton, NJ: Princeton University Press.

Aristotle. 1984b. "Nicomachean Ethics," in J. Barnes (ed.), *The Complete Works of Aristotle.* Princeton: Princeton University Press, 1729–1867.

Atkinson, A. B. 2015. *Inequality: What Can Be Done?* Cambridge, MA: Harvard University Press.

Baker, R. R., and Bellis, M. A. 1995. *Human Sperm Competition: Copulation, Masturbation and Infidelity.* London: Chapman Hall.

Barash, D. P., and Lipton, J. E. 2001. *The Myth of Monogamy: Fidelity and Infidelity in Animals and People.* New York: Holt.

Benatar, D. 2006. *Better Never to Have Been.* Oxford: Oxford University Press.

Boyd, R., and Richerson, P. J. 1992. "Punishment Allows the Evolution of Cooperation (or Anything Else) in Sizable Groups," *Ethology and Sociobiology* 13:166–88.

Boyd, R., and Richerson, P. J. 2005. *The Origin and Evolution of Cultures.* New York: Oxford University Press.

Broadie, S. 2007. "Nature and Craft in Aristotelian Teleology," in S. Broadie (ed.), *Aristotle and Beyond: Essays in Metaphysics and Ethics.* Cambridge: Cambridge University Press.

Brun, G., Doğuoğlu, U., and Kuenzle, D. 2008. *Epistemology and Emotions.* Aldershot: Ashgate.

Burton, F. D. 1971. "Sexual Climax in the Female Macaca Mulatta," *Proceedings of the 3rd International Congress of Primatology* 3: 181–91.

Buss, D. M. 1994. *The Evolution of Desire: Strategies of Human Mating.* New York: Basic Books.

Churchland, P. S. 2012. *Braintrust: What Neuroscience Tells Us About Morality.* Princeton, NJ: Princeton University Press.

Clarke, E. 2012. "Plant Individuality: A Solution to the Demographer's Dilemma." *Biology and Philosophy* 27:321–61.

Conway Morris, S. 2003. *Life's Solution: Inevitable Humans in a Lonely Universe.* Cambridge: Cambridge University Press.

Crisp, R. 1998. *How Should One Live? Essays on the Virtues.* Oxford: Oxford University Press.

de Sousa, R. 2005. "Biological Individuality," *Croatian Journal of Philosophy* 54:195–218.

  2008. "Epistemic Feelings," in Brun, Doğuoğlu, and Kuenzle (eds.), *Epistemology and Emotions.* Aldershot: Ashgate, pp. 185–204.

Deonna, J. A., and Teroni, F. 2011. *In Defense of Shame.* Oxford: Oxford University Press.

Easton, D., and Hardy, J. W. 2009. *The Ethical Slut: A Practical Guide to Polyamory, Open Relationships and Other Adventures,* 2nd edn. Berkeley, CA: Celestial Arts.

Fine, C. 2011. *Delusions of Gender: How Our Minds, Society, and Neurosexism Create Difference.* New York: W.W. Norton.

Fisher, H. 1998. "Lust, Attraction and Attachment in Mammalian Reproduction," *Human Nature* 9:23–52.

  2004. *Why We Love: The Nature and Chemistry of Romantic Love.* New York: Holt.

Forber, P., and Smead, R. 2014. "The Evolution of Fairness Through Spite," *Proceedings of Biological Science* doi: 10.1098/rspb.2013.2439.

Forster, E. M. 1951. "What I Believe," in *Two Cheers for Democracy.* New York: Harcourt Brace, pp. 65–76.

Gide, A. 1942. *Les Nourritures Terrestres.* Paris: Gallimard.

Goodman, N. 1983. *Fact, Fiction, and Forecast,* 4th edn. Cambridge, MA: Harvard University Press.

Gould, S. J. 1981. *The Mismeasure of Man.* New York: W.W. Norton.

Haidt, J., and Bjorklund, F. 2008. "Social Intuitionists Answer Six Questions About Moral Psychology," in Sinnot-Armstrong (ed.), *Moral Psychology*, Vol. II. Cambridge, MA: MIT Press, pp. 181–217.

Harris, C. R. 2004. "The Evolution of Jealousy," *American Scientist* 92:62–71.

Harris, S. 2011. *The Moral Landscape: How Science Can Determine Human Values.* New York: Free Press.

Hume, D. 1975. *Enquiry Concerning Human Understanding; A Letter from a Gentleman to His Friend in Edinburgh.* Indianapolis, IN: Hackett.

Hursthouse, R. 1998. "Normative Virtue Ethics," in Crisp (ed.), *How Should One Live? Essays on the Virtues.* Oxford: Oxford University Press, pp. 19–38.

Huxley, T. H., and Huxley, J. 1947. *Evolution and Ethics, 1893–1943.* London: Pilot Press.

Kauppinen, A. 2014. "Moral Sentimentalism," in E. N. Zalta (ed.), *Stanford Encyclopedia of Philosophy* (Spring Edition). Available at: http://plato.stanford.edu/archives/spr2014/entries/moral-sentimentalism/.

Kreisberg, J. C. 1995. "A Globe, Clothing Itself with a Brain." *Wired*, June.

Langton, C. G., 1992. "Life on the Edge of Chaos," in Langton, Taylor, Farmer, and Rasmussen (eds.), *Artificial Life II.* Redwood City, CA: Addison-Wesley, pp. 41–92.

Taylor, C., Farmer, J. D., and Rasmussen, S. 1992. *Artificial Life II.* Redwood City, CA: Addison-Wesley.

Lloyd, E. 2005. *The Case of the Female Orgasm: Bias in the Science of Evolution.* Cambridge, MA: Harvard University Press.

Maynard Smith, J. 1984. "Game Theory and the Evolution of Behavior," *Behavioral and Brain Sciences* 7:95–126.

and Szathmáry, E. 1999. *The Origins of Life: From the Birth of Life to the Origins of Language.* Oxford: Oxford University Press.

Mill, J. S. 1874. *Nature, the Utility of Religion, Theism, Being Three Essays on Religion.* London: Longman, Green, Reader, and Dyer.

1991. *Utilitarianism: Collected Works of John Stuart Mill*, Vol. X. Toronto: University of Toronto Press.

Miller, J. 2005. "March of the Conservatives: Penguin Film as Political Fodder," *New York Times.* Available at: www.nytimes.com/2005/09/13/science/13peng.html?pagewanted=print.

Millikan, R. G. 1984. *Language, Thought, and Other Biological Categories.* Cambridge, MA: MIT Press.

1993. *White Queen Psychology and Other Essays for Alice.* Cambridge, MA: MIT Press.

Nagel, T. 2012. *Mind and Cosmos: Why the Materialist Neo-Darwinian Conception of Nature Is Almost Certainly False.* Oxford: Oxford University Press.

Nietzsche, F. 1967. *On the Genealogy of Morality,* M. Clark and A. J. Swensen, trans. and notes. Indianapolis, IN: Hackett.

Nowak, M. A., Tarnita, C. E., and Wilson, E. O. 2010. "Inclusive Fitness Theory and Eusociality," *Nature* 466:1057–62.

Nussbaum, M. C. 2000. *Women and Human Development: The Capabilities Approach.* Cambridge: Cambridge University Press.

Penrose, R. 1994. *Shadows of the Mind: A Search for the Missing Science of Consciousness.* Oxford: Oxford University Press.

Rawls, J. 1977. *A Theory of Justice.* Cambridge, MA: Harvard University Press.

Ridley, M. 2000. *Mendel's Demon: Gene Justice and the Complexity of Life.* London: Weidenfeld and Nicolson.

Ryan, C., and Jethá, C. 2010. *Sex at Dawn: The Prehistoric Origins of Modern Sexuality.* New York: Harper.

Sade, D. A., Marquis de. 1810. *La Philosophie dans le Boudoir* (facsimile). Whitefish, MT: Kessinger.

Shaw, G. B. 1986 [1908]. "Getting Married: A Disquisitory Play with Preface," in D. H. Laurence (ed.), *Getting Married and Press Cuttings by Bernard Shaw: Definitive Text.* Harmondsworth, UK: Penguin.

Sinnot-Armstrong, W. 2008. *Moral Psychology,* Vol. II. Cambridge, MA: MIT Press.

Sober, E., and Wilson, D. S. 1998. *Unto Others: The Evolution and Psychology of Unselfish Behavior.* Cambridge, MA: Harvard University Press.

Tavris, C. 1992. *The Mismeasure of Woman.* New York: Touchstone.

Teilhard de Chardin, P. 1961. *The Phenomenon of Man.* New York: Harper & Row.

Tennov, D. 1979. *Love and Limerence: The Experience of Being in Love.* New York: Stein and Day.

Thompson, P. 1995. *Issues in Evolutionary Ethics.* Albany, NY: SUNY Press.

2002. "The Evolutionary Biology of Evil," *Monist* 85:239–59.

Wilkinson, R., and Pickett, K. 2010. *The Spirit Level: Why Equality Is Better for Everyone.* London: Penguin.

Wilson, D. S. 2015. *Does Altruism Exist? Culture, Genes, and the Welfare of Others.* New Haven, CT: Yale University Press.

Yang, E. N., and Mathieu, C. 2007. *Leaving Mother Lake: A Girlhood at the Edge of the World.* Boston: Little, Brown.

Chapter 8

Andrews, K. 2014. "Animal Cognition," in E. N. Zalta (ed.), *The Stanford Encyclopedia of Philosophy* (Fall Edition). Available at: http://plato.stanford.edu/archives/fall2014/entries/cognition-animal/.

Ariely, D. 2008. *Predictably Irrational: the Hidden Forces That Shape Our Decisions.* London: HarperCollins.

Binmore, K. 2005. *Natural Justice.* Oxford: Oxford University Press.

Bisin, A., and Jackson, M. 2011. *Handbook of Social Economics.* Amsterdam: North-Holland.

Charlseworth, B. 1994. *Evolution in Age: Structured Populations.* Cambridge: Cambridge University Press.

Chater, N. 2012. "Building Blocks of Human Decision Making," in Hammerstein and Stevens (eds.), *Evolution and the Mechanisms of Decision Making.* Cambridge MA: MIT Press, pp. 53–68.

Clayton, N., Emery, N., and Dickinson, A. 2006. "The Rationality of Animal Memory," in Nudds and Hurley (eds.), *Animal Minds.* Oxford: Oxford University Press, pp. 197–216.

Cosmides, L., and Tooby, J. 1994. "Better Than Rational: Evolutionary Psychology and the Invisible Hand," *American Economic Review* 84: 327–32.

Curry, P. 2001. "Decision Making Under Uncertainty and the Evolution of Interdependent Preferences," *Journal of Economic Theory* 98:57–69.

Danielson, P. *Modelling Rationality, Morality and Evolution.* Oxford: Oxford University Press.

   2004. "Rationality and Evolution," in Rawling and Mele (eds.), *The Oxford Handbook of Rationality.* Oxford: Oxford University Press, pp. 417–37.

Davidson, D. 1984. *Inquiries into Truth and Interpretation.* Oxford: Oxford University Press.

Dennett, D. C. 1987. *The Intentional Stance.* Cambridge, MA: MIT Press.

Gardner, A., and Grafen, A. 2009. "Capturing the Superorganism: A Formal Theory of Group Adaptation," *Journal of Theoretical Biology* 22:659–71.

Gigerenzer, G. 2010. *Rationality for Mortals: How People Cope with Uncertainty.* Oxford: Oxford University Press.

   and Selten, R. 2001. *Bounded Rationality: the Adaptive Toolbox.* Cambridge, MA: MIT Press.

Gintis, H. 2009. *The Bounds of Reason.* Princeton: Princeton University Press.

Godfrey-Smith, P. 1996. *Complexity and the Function of Mind in Nature.* Cambridge: Cambridge University Press.

Grafen, A. 1999. "Formal Darwinism, the Individual-as-Maximizing-Agent Analogy, and Bet-Hedging," *Proceedings of the Royal Society B* 266:799–803.

2006a. "Optimization of Inclusive Fitness," *Journal of Theoretical Biology* 238:541–63.

2006b. "A Theory of Fisher's Reproductive Value," *Journal of Mathematical Biology* 53:15–60.

2007. "The Formal Darwinism Project: A Mid-Term Report," *Journal of Evolutionary Biology* 20:1243–54.

Haig, D. 2012. "The Strategic Gene," *Biology and Philosophy* 27:461–79.

Hamilton, W. D. 1964. "The Genetical Evolution of Social Behavior I and II," *Journal of Theoretical Biology* 7:1–52.

Hammerstein, P., and Stevens, J. 2014a. "Six Reasons for Invoking Evolution in Decision Theory," in Hammerstein and Stevens (eds.), *Evolution and the Mechanisms of Decision Making*. Cambridge MA: MIT Press, pp. 1–20.

and Stevens, J. 2014b. *Evolution and the Mechanisms of Decision Making*. Cambridge, MA: MIT Press.

Houston, A., and McNamara. J. 1999. *Models of Adaptive Behavior*. Cambridge: Cambridge University Press.

McNamara, J., and Steer, M. 2007. "Do We Expect Natural Selection to Produce Rational Behavior?" *Philosophical Transactions of the Royal Society B* 362:1531–43.

Kacelnik, A. 2006. "Meanings of Rationality," in Nudds and Hurley (eds.), *Animal Minds*. Oxford: Oxford University Press, pp. 87–106.

Kahneman, D. 2011. *Thinking Fast and Slow*. London: Penguin.

and Tversky, A. 2000. *Choices, Values and Frames*. Cambridge: Cambridge University Press

Kahneman, D., Slovic, P., and Tversky, A. 1982. *Judgment Under Uncertainty: Heuristics and Biases*. Cambridge: Cambridge University Press.

Kennedy, J. 1992. *The New Anthropomorphism*. Cambridge: Cambridge University Press.

Lewis, D. 1981. "Causal Decision Theory," *Australasian Journal of Philosophy* 59:5–30.

Martens, J. Forthcoming. "Hamilton Meets Causal Decision Theory," *British Journal for the Philosophy of Science*.

Maynard Smith, J. 1974. "The Theory of Games and the Evolution of Animal Conflicts," *Journal of Theoretical Biology* 47:209–21.

1982. *Evolution and the Theory of Games*. Cambridge: Cambridge University Press.

McDowell, J. 1994. *Mind and World*. Cambridge, MA: Harvard University Press.

Mylius, S., and Diekmann, O. 1995. "On Evolutionarily Stable Life Histories, Optimization and the Need to Be Specific About Density Dependence," *Oikos* 74:218–24.

Nudds, M., and Hurley, S. 2006. *Animal Minds*. Oxford: Oxford University Press.

Okasha, S. 2006. *Evolution and the Levels of Selection*. Oxford: Oxford University Press.

2011. "Optimal Choice in the Face of Risk: Decision Theory Meets Evolution," *Philosophy of Science* 78:83–104.

and Binmore, K. 2014. *Evolution and Rationality*. Cambridge: Cambridge University Press.

Orr, A. 2007. "Absolute Fitness, Relative Fitness, and Utility," *Evolution* 61:2997–3000.

Quine, W. V. O. 1969. "Epistemology Naturalized," in *Ontological Relativity and Other Essays*. New York: Columbia University Press.

Ramsey, F. 1931. "Truth and Probability," in R. B. Braithwaite (ed.), *Foundations of Mathematics and Other Logical Essays*. New York: Harcourt, pp. 156–198.

Rawling, P., and Mele, A. 2004. *The Oxford Handbook of Rationality*. Oxford: Oxford University Press.

Rayo, L., and Robson, A. 2013. "Biology and the Arguments of Utility," *Cowles Foundation Discussion Papers*. Available at: http://ssrn.com /abstract=2254895.

Robson, A. 1996. "A Biological Basis for Expected and Non-expected Utility," *Journal of Economic Theory* 68:397–424.

and Samuelson, L. 2011. "The Evolutionary Foundations of Preferences," in Bisin and Jackson (eds.), *Handbook of Social Economics*. Amsterdam: North-Holland, pp. 221–310.

Samuelson, L., and Swinkels, J. 2006. "Information, Evolution and Utility," *Theoretical Economics* 1:119–42.

Savage, L. 1954. *The Foundations of Statistics*. New York: Wiley.

Seeley, T. 1996. *The Wisdom of the Hive*. Cambridge, MA: Harvard University Press.

2010. *Honey-Bee Democracy*. Princeton, NJ: Princeton University Press.

Skyrms, B. 1996. *Evolution of the Social Contract*. Cambridge: Cambridge University Press.

Sober, E. 1998. "Three Differences Between Evolution and Deliberation," in Danielson (ed.), *Modelling Rationality, Morality and Evolution*. Oxford: Oxford University Press, pp. 408–22.

Stearns, S. 2000. "Daniel Bernoulli (1738): Evolution and Economics Under Risk," *Journal of Biosciences* 25:221–8.

Stephens, C. 2001. "When Is It Selectively Advantageous to Have True Beliefs?" *Philosophical Studies* 105:161–89.

Sterelny, K. 2003. *Thought In a Hostile World*. Oxford, UK: Blackwell.

2012. "From Fitness to Utility," in Okasha and Binmore (eds.), *Evolution and Rationality*. Cambridge: Cambridge University Press, pp. 246–73.

Todd, P., Gigerenzer, G., and the ABC Research Group. 2012. *Ecological Rationality: Intelligence in the World*. Oxford: Oxford University Press.

Weibull, J. 1995. *Evolutionary Game Theory*. Cambridge, MA: MIT Press.

Wilson, D. S. 2002. *Darwin's Cathedral*. Chicago: University of Chicago Press.

Chapter 9

Boehm, C. 1999. *Hierarchy in the Forest*. Cambridge, MA: Harvard University Press.

2012. *Moral Origins*. New York: Basic Books.

2014. "The Moral Consequences of Social Selection," *Behavior* 151:167–83.

Boyd, R., and Richerson, P. J. 1992. "Punishment Allows the Evolution of Cooperation (and Anything Else) in Sizable Groups," *Ethology and Sociobiology* 13:171–95.

Churchland, P. S. 2011. *Braintrust*. Princeton, NJ: Princeton University Press.

2014. "The Neurobiological Platform for Moral Values," *Behavior* 151: 283–96.

Clarke-Doane, J. 2012. "Morality and Mathematics: The Evolutionary Challenge," *Ethics* 122:313–40.

Darwin, C. 1859. *On the Origin of Species by Means of Natural Selection, or the Preservation of Favoured Races in the Struggle for Life*. London: John Murray.

1871. *The Descent of Man, and Selection in Relation to Sex*. London: John Murray.

de Waal, F. 1989. *Peacemaking Among Primates*. Cambridge, MA: Harvard University Press.

1992. *Chimpanzee Politics*. Baltimore, MD: Johns Hopkins University Press.

1996. *Good Natured*. Cambridge, MA: Harvard University Press.

2006. *Primates and Philosophers*. Princeton, NJ: Princeton University Press.

2009. *The Age of Empathy: Nature's Lessons for a Kinder Society*. New York: Three Rivers Press.

2013. *The Bonobo and the Atheist*. New York: Norton.

2014. "Natural Normativity," *Behavior* 151:185–204.

Dworkin, R. 2013. *Religion Without God*. Cambridge, MA: Harvard University Press.

Gould, S. J., and Lewontin, R. 1979. "The Spandrels of San Marco and the Panglossian Paradigm: A Critique of the Adaptationist Programme," *Proceedings of the Royal Society B*, 205:581–98.

James, W. 1978. Preface to *The Meaning of Truth*, published together with *Pragmatism*. Cambridge, MA: Harvard University Press.

Kitcher, P. S. 2010. "Varieties of Altruism," *Economics and Philosophy* 26: 121–48.

2011. *The Ethical Project*. Cambridge, MA: Harvard University Press.

2014. "Is a Naturalized Ethics Possible?" *Behavior* 151:245–60.

McBrearty, S., and Brooks, A. S. 2000. "The Revolution That Wasn't: A New Interpretation of the Origin of Modern Human Behavior," *Journal of Human Evolution* 39:453–563.

Nagel, T. 2012. *Mind and Cosmos*. New York: Oxford University Press.

Parfit, D. 2011. *On What Matters*, Vol. II. Oxford: Oxford University Press.

Peirce, C. S. 1934. "How to Make Our Ideas Clear," in C. Hartshorne and P. Weiss (eds.), *Collected Papers of C. S. Peirce*, Vol. V. Cambridge, MA: Harvard University Press, pp. 248–71.

Renfrew, C., and Shennan, S. 1982. *Ranking, Resource, and Exchange*. Cambridge: Cambridge University Press.

Shafer-Landau, R. 2012. "Evolutionary Debunking, Moral Realism, and Moral Knowledge," *Journal of Ethics and Social Philosophy* 7:1–38.

Sterelny, K. 2012a. *The Evolved Apprentice*. Cambridge, MA: MIT Press.

2012b. "Morality's Dark Past," *Analyse und Kritik* 34:95–115.

Street, S. 2005. "A Darwinian Dilemma for Realist Theories of Value," *Philosophical Studies* 127:109–66.

Tomasello, M. 2009. *Why We Cooperate*. Cambridge, MA: MIT Press.

Westermarck, E. 1926. *The Origin and Development of the Moral Ideas*, 2nd edn. London: Macmillan.

Wilson, E. O. 1975. *Sociobiology: The New Synthesis*. Cambridge, MA: Harvard University Press.

Chapter 10

Antony, L. M. 1998. "Human Nature and Its Role in Feminist Theory," in Kourany (ed.), *Philosophy in a Feminist Voice: Critiques and Reconstructions*. Princeton, NJ: Princeton University Press, pp. 63–91.

2000. "Natures and Norms," *Ethics* 111:8–36.

Aviezer, O., Sagi, A., and Van Ijzendoorn, M. 2002. "Balancing the Family and the Collective in Raising Children: Why Communal Sleeping in Kibbutzim Was Predestined to End," *Family Process* 41:435–54.

Bechtel, W. 1986. *Integrating Scientific Disciplines*. Dordrecht, Netherlands: Springer.

Beit-Hallahmi, B., and Rabin, A. I. 1977. "The Kibbutz as a Social Experiment and as a Child-Rearing Laboratory," *American Psychologist* 32:532–54.

Bloom, P. 2013. *Just Babies: The Origins of Good and Evil.* New York: Random House.

Buss, D. M., Larsen, R. J., Westen, D., and Semmelroth, J. 1992. "Sex Differences in Jealousy: Evolution, Physiology, and Psychology," *Psychological Science* 3:251–5.

Carey, S. 2009. *The Origin of Concepts.* Oxford: Oxford University Press.

Chomsky, N., and Foucault, M. 2006. *The Chomsky-Foucault Debate: On Human Nature.* New York: New Press.

Curtiss, S. 1977. *Genie: A Psycholinguistic Study of a Modern-Day "Wild Child."* New York: Academic Press.

Darwin, C. 2002. *The Expression of the Emotions in Man and Animals.* Oxford: Oxford University Press.

De Waal, F. 2009. *The Age of Empathy: Nature's Lessons for a Kinder Society.* New York: Three Rivers Press.

Downes, S., and Machery, E. 2014. Arguing About Human Nature. New York: Routledge.

Ekman, P. 1993. "Facial Expression and Emotion," *American Psychologist* 48:384.

and Friesen, W. V. 1971. "Constants Across Cultures in the Face and Emotion," *Journal of Personality and Social Psychology* 17:124.

and Friesen, W. V. 1979. "Nonverbal Leakage and Clues to Deception," *Psychiatry* 32:88–105.

and Friesen, W. V. 1969. "The Repertoire of Nonverbal Behavior: Categories, Usage, and Coding," *Semiotica* 1:49–98.

Evans, N., and Levinson, S. C. 2009. "The Myth of Language Universals: Language Diversity and Its Importance for Cognitive Science," *Behavioral and Brain Sciences* 32:429–48.

Fitch, W. T. 2011. "Unity and Diversity in Human Language," *Philosophical Transactions of the Royal Society B* 366:376–88.

Foot, P. 2001. *Natural Goodness.* Oxford: Oxford University Press.

Garfinkel, A. 1981. *Forms of Explanation: Rethinking the Questions in Social Theory.* New Haven, CT: Yale University Press.

Gendron, M., Roberson, D., van der Vyver, J. M., and Barrett, L. F. 2014. "Perceptions of Emotion from Facial Expressions Are Not Culturally Universal: Evidence from a Remote Culture," *Emotion* 14:251.

Ghiselin, M. T. 1997. *Metaphysics and the Origins of Species.* Albany, NY: SUNY Press.

Gintis, H. 2008. "Punishment and Cooperation," *Science* 319:1345–6.

Gissis, S., and Jablonka, E. 2011. *Transformations of Lamarckism: From Subtle Fluids to Molecular Biology*. Cambridge, MA: MIT Press.

Golan, S. 1958. "Behavior Research in Collective Settlements in Israel: 2. Collective Education in the Kibbutz," *American Journal of Orthopsychiatry* 28:549–56.

Griffiths, P. E. 2009. "Reconstructing Human Nature," *Journal of the Sydney University Arts Association* 31:30–57.

2011. "Our Plastic Nature," in Gissis and Jablonka (eds.), *Transformations of Lamarckism: from Subtle Fluids to Molecular Biology*. Cambridge, MA: MIT Press, pp. 319–30.

and Machery, E. 2008. "Innateness, Canalization, and 'Biologicizing the Mind,'" *Philosophical Psychology* 21:397–414.

Machery, E., and Linquist, S. 2009. "The Vernacular Concept of Innateness," *Mind and Language* 24:605–30.

Hassin, R. R., Aviezer, H., and Bentin, S. 2013. "Inherently Ambiguous: Facial Expressions of Emotions, in Context," *Emotion Review* 5:60–5.

Hempel, C. 1965. *Aspects of Scientific Explanation and Other Essays in the Philosophy of Science*. New York: Free Press.

Henney, J. E., Taylor, C. L., and Boon, C. S. 2010. *Strategies to Reduce Sodium Intake in the United States*. Washington, DC: National Academies Press.

Herrmann, B., Thöni, C., and Gächter, S. 2008. "Antisocial Punishment Across Societies." *Science* 319:1362–7.

Hull, D. L. 1986. "On Human Nature," *PSA: Proceedings of the Biennial Meeting of the Philosophy of Science Association* 2:3–13.

Jaggar, A. M. 1983. *Feminist Politics and Human Nature*. Oxford, UK: Rowman & Littlefield.

Kant, I. 2011. *Observations on the Feeling of the Beautiful and Sublime*, P. Frierson and P. Guyer, trans. Cambridge University Press.

Kitcher, P. 1999. "Essence and Perfection," *Ethics* 110:59–83.

Kourany, J. A. 1998. *Philosophy in a Feminist Voice: Critiques and Reconstructions*. Princeton, NJ: Princeton University Press.

Kronfeldner, M., Roughley, N., and Toepfer, G. 2014. "Recent Work on Human Nature: Beyond Traditional Essences," *Philosophy Compass* 9: 642–52.

Lennox, J. G. 2001. *Aristotle's Philosophy of Biology: Studies in the Origins of Life Science*. Cambridge: Cambridge University Press.

Lewens, T. 2012. "Human Nature: The Very Idea," *Philosophy and Technology* 25:459–74.

Linquist, S., Machery, E., Griffiths, P. E., and Stotz, K. 2011. "Exploring the Folkbiological Conception of Human Nature," *Philosophical Transactions of the Royal Society of London B* 366:444–53.

Machery, E. 2008. "A Plea for Human Nature," *Philosophical Psychology* 21:321–30.

2012. "Reconceptualizing Human Nature: Response to Lewens," *Philosophy and Technology* 25:475–8.

and Barrett, C. 2006. "Debunking Adapting Minds," *Philosophy of Science* 73:232–46.

Nelson, N. L., and Russell, J. A. 2013. "Universality Revisited," *Emotion Review* 5:8–15.

Ramsey, G. 2013. "Human Nature in a Post-Essentialist World," *Philosophy of Science* 80:983–93.

Rapaport, D. 1958. "Behavior Research in Collective Settlements in Israel: VII. The Study of Kibbutz Education and Its Bearing on the Theory of Development," *American Journal of Orthopsychiatry* 28:587–97.

Richerson, P. J., and Boyd, R. 2005. *Not by Genes Alone: How Culture Transformed Human Evolution*. Chicago: University of Chicago Press.

Rousseau, J.-J. 1979. *Emile or: On Education*, A. Bloom, trans. New York: Basic Books.

Safdar, S., Friedlmeier, W., Matsumoto, D., et al. 2009. "Variations of Emotional Display Rules Within and Across Cultures: A Comparison Between Canada, USA, and Japan," *Canadian Journal of Behavioral Science/Revue Canadienne des Sciences du Comportement* 41:1.

Samuels, R. 2012. "Science and Human Nature," *Royal Institute of Philosophy Supplement* 70:1–28.

Setiya, K. 2012. *Knowing Right from Wrong*. Oxford: Oxford University Press.

Smith, E. A. 2011. "Endless Forms: Human Behavioral Diversity and Evolved Universals," *Philosophical Transactions of the Royal Society B* 366:325–32.

Sterelny, K. 2003. *Thought in a Hostile World: The Evolution of Human Cognition*. Oxford, UK: Blackwell.

2012. *The Evolved Apprentice*. Cambridge, MA: MIT Press.

Stotz, K. 2010. "Human Nature and Cognitive-Developmental Niche Construction," *Phenomenology and the Cognitive Sciences* 9:483–501.

Thompson, M. 2008. *Life and Action*. Cambridge, MA: Harvard University Press.

Tooby, J., and Cosmides, L. 1990. "On the Universality of Human Nature and the Uniqueness of the Individual: The Role of Genetics and Adaptation," *Journal of Personality* 58:17–67.

Walsh, D. 2006. "Evolutionary Essentialism," *British Journal for the Philosophy of Science*, 57:425–48.

Wilde, S., Timpson, A., Kirsanow, K., et al. 2014. "Direct Evidence for Positive Selection of Skin, Hair, and Eye Pigmentation in Europeans

During the Last 5000 y," *Proceedings of the National Academy of Sciences* 111:4832–7.

Wilson, E. O. 1978. *On Human Nature.* Cambridge, MA: Harvard University Press.

Wimsatt, W. C. 1986. "Developmental Constraints, Generative Entrenchment, and the Innate-Acquired Distinction," in Bechtel (ed.), *Integrating Scientific Disciplines.* Dordrecht, Netherlands: Springer, pp. 185–208.

Winsor, M. P. 2003. "Non-Essentialist Methods in Pre-Darwinian Taxonomy," *Biology and Philosophy* 18:387–400.

2006. "The Creation of the Essentialism Story: An Exercise in Metahistory," *History and Philosophy of the Life Sciences* 28:149–74.

Chapter 11

Barnes, B., and Dupré, J. 2008. *Genomes and What to Make of Them.* Chicago: University of Chicago Press.

Bird, A., and Tobin, E. 2012. "Natural Kinds," in E. N. Zalta (ed.), *The Stanford Encyclopedia of Philosophy* (Winter Edition). Available at: http://plato.stanford.edu/archives/win2012/entries/natural-kinds/.

Brennan, J., and Capel, B. 2004. "One Tissue, Two Fates: Molecular Genetic Events That Underlie Testis Versus Ovary Development," *Nature Reviews Genetics* 5:509–21.

Buss, D. M. 1999. *Evolutionary Psychology: The New Science of the Mind.* Needham Heights, MA: Allyn and Bacon.

Butler, J. 1990. *Gender Trouble: Feminism and the Subversion of Identity.* New York: Routledge.

Cahill, L. 2006. "Why Sex Matters for Neuroscience." *Nature Reviews Neuroscience* 7:1–8.

Champagne, F. A., and Meaney, M. J. 2006. "Stress During Gestation Alters Postpartum Maternal Care and the Development of the Offspring in a Rodent Model," *Biological Psychiatry* 59:1227–35.

Champagne, F. A., Weaver, I. C., Diorio, J., et al. 2006. "Maternal Care Associated with Methylation of the Estrogen Receptor-Alpha 1b Promoter and Estrogen Receptor-Alpha Expression in the Medial Preoptic Area of Female Offspring," *Endocrinology* 147:2909–15.

Dawkins, R. 1976. *The Selfish Gene.* Oxford University Press.

Doidge, N. 2007. *The Brain That Changes Itself: Stories of Personal Triumph from the Frontiers of Brain Science.* London: Penguin.

Dupré, J. 2001. *Human Nature and the Limits of Science.* Oxford: Oxford University Press.

2003. *Human Nature and the Limits of Science.* Oxford: Oxford University Press.

2012. *Processes of Life: Essays in the Philosophy of Biology*. Oxford: Oxford University Press.

Eicher, E. M., and Washburn, L. L. 1986. "Genetic Control of Primary Sex Determination in Mice," *Annual Review of Genetics* 20:327–60.

Ellis, B. 2001. *Scientific Essentialism*. Cambridge: Cambridge University Press.

Fausto-Sterling, A. 1985. *Myths of Gender*. New York: Basic Books.

2000. *Sexing the Body: Gender Politics and the Construction of Sexuality*. New York: Basic Books.

2012. *Sex/Gender: Biology in a Social World*. New York: Routledge.

Fine, C. 2000. *Delusions of Gender: How Our Minds, Society, and Neurosexism Create Difference*. London: Icon Books.

Fisher, R. A. 1930. *The Genetical Theory of Natural Selection*. Oxford: Oxford University Press.

Foucault, Michel (1979) [1976]. *The History of Sexuality Volume I: An Introduction*, R. Hurley, trans. London: Allen Lane.

Friedan, B. 1963. *The Feminine Mystique*. New York: W. W. Norton.

Gilbert, S. F., and Epel, D. 2009. *Ecological Developmental Biology: Integrating Epigenetics, Medicine, and Evolution*. Sunderland, MA: Sinauer Associates.

Griffiths, P., and Stotz, K. 2013. *Genetics and Philosophy: An Introduction*. Cambridge: Cambridge University Press.

Hooks, B. 1984. *Feminist Theory: From Margin to Center*. Boston: South End Press.

Kinsey, A. C., Pomeroy, W. B., and Martin, C. E. 1948. *Sexual Behavior in the Human Male*. Philadelphia: W. B. Saunders.

Kinsey, A. C., Pomeroy, W. B., Martin, C. E., and Gebhard, P. H. 1953. *Sexual Behavior in the Human Female*. Philadelphia: W. B. Saunders.

Kohler, R. E. 1994. *Lords of the Fly: Drosophila Genetics and the Experimental Life*. Chicago: University of Chicago Press.

Liebers, D., De Knijff, P., and Helbig, A. J. 2004. "The Herring Gull Complex Is Not a Ring Species," *Proceedings of the Royal Society B* 271:893.

Meloni, M., and Testa, G. 2014. "Scrutinizing the Epigenetics Revolution," *BioSocieties* 9: 431–56.

Nanney, D. L. 1980. *Experimental Ciliatology*. New York: Wiley.

Reiter, R. 1975. *Toward an Anthropology of Women*. New York: Monthly Review Press.

Rubin, G. 1975. "The Traffic in Women: Notes on the 'Political Economy' of Sex," in Reiter (ed.), *Toward an Anthropology of Women*. New York: Monthly Review Press, pp. 157–210.

Odling-Smee, F. J., Laland, K. N., and Feldman, M. W. 2003. *Niche Construction: The Neglected Process in Evolution*. Princeton, NJ: Princeton University Press.

Singh, D. 1993. "Adaptive Significance of Waist-to-Hip Ratio and Female Physical Attractiveness," *Journal of Personality and Social Psychology* 65:293–307.

Stoller, R. J., 1968. *Sex and Gender: On the Development of Masculinity and Femininity*. New York: Science House.

Unger, Rhoda, K. 1979. "Toward a Redefinition of Sex and Gender," *American Psychologist* 34:1085–94.

Zerjal, T., Xue, Y., Bertorelle, G., et al. 2003. "The Genetic Legacy of the Mongols," *American Journal of Human Genetics* 72:717–21.

Chapter 12

Allport, G. 1954. *The Nature of Prejudice*. Reading, MA: Addison-Wesley.

Andreasen, R. O. 1998. "A New Perspective on the Race Debate," *British Journal of Philosophy of Science* 49:199–225.

   2000. "Race: Biological Reality or Social Construct?" *Philosophy of Science* 67:S653–66.

   2004. "The Cladistic Race Concept: A Defense," *Biology and Philosophy* 19:425–42.

   2005. "The Meaning of Race," *Journal of Philosophy* 102:94–106.

Appiah, K. 1992. *In My Father's House: Africa in the Philosophy of Culture*. Oxford: Oxford University Press.

   1996. "Race, Culture, Identity: Misunderstood Connections," in Appiah and Gumann (eds.), *Color Conscious: The Political Morality of Race*. Princeton, NJ: Princeton University Press, pp. 53–136.

   2000. "Racial Identity and Racial Identification," in Back and Solomos (eds.), *Theories of Race and Racism: A Reader*. London: Routledge, pp. 607–15.

   2006. "How to Decide If Races Exist?" *Proceedings of the Aristotelian Society* 106:363–80.

   and Gumann, A. 1996. *Color Conscious: The Political Morality of Race*. Princeton, NJ: Princeton University Press.

Astuti, R. 1995. "'The Vezo Are Not a Kind of People': Identity, Difference and 'Ethnicity' Among a Fishing People of Western Madagascar," *American Ethnologist* 22:462–82.

   Solomon, G. E., and Carey, S. 2003. "Constraints on Conceptual Development: A Case Study of the Acquisition of Folkbiological and Folksociological Knowledge in Madagascar," in *Monographs of the Society for Research in Child Development*. Hoboken, NJ: Wiley.

Atran, S. 1998. "Folk Biology and the Anthropology of Science: Cognitive Universals and Cultural Particulars," *Behavioral and Brain Sciences* 21: 547–69.

Back, L., and Solomos, J. 2000. *Theories of Race and Racism: A Reader.* London: Routledge.

Bamshad, M., Wooding, S., Salisbury, B., and Stephens, J. C. 2004. "Deconstructing the Relationship Between Genetics and Race," *Nature Reviews Genetics* 5:598–608.

Barrett, C. L. 2001. "On the Functional Origins of Essentialism," *Mind and Society* 3:1–30.

Bastian, B., and Haslam, N. 2006. "Psychological Essentialism and Stereotype Endorsement," *Journal of Experimental Social Psychology* 42:228–35.

and Haslam, N. 2007. "Psychological Essentialism and Attention Allocation: Preferences for Stereotype-Consistent Versus Stereotype Inconsistent Information," *Journal of Social Psychology* 147:531–41.

Bigler, R., and Liben, L. 2007. "Developmental Intergroup Theory: Explaining and Reducing Children's Social Stereotyping and Prejudice," *Current Directions in Psychological Science* 16:162–6.

Bolnick, D. 2008. "Individual Ancestry Inference and the Reification of Race as a Biological Phenomenon," in Koenig, Lee, and Richardson (eds.), *Revisiting Race in a Genomic Age.* New Brunswick, NJ: Rutgers University Press, pp. 70–85.

Bouchard, F., and Rosenberg, A. 2004. "Fitness, Probability and the Principles of Natural Selection," *British Journal for the Philosophy of Science* 55:693–712.

Boyd, R., and Richerson, P. J. 1985. *Culture and the Evolutionary Process.* Chicago: University of Chicago Press.

Brown, R. A., and Amelagos, G. J. 2001. "Apportionment of Racial Diversity: A Review," *Evolutionary Anthropology* 10:34–40.

Buss, D. 2005. *Handbook of Evolutionary Psychology.* New York: Wiley.

Churchland, P. 1986. *Neurophilosophy: Toward a Unified Science of the Mind-Brain.* Cambridge, MA: MIT Press.

Cohen, H., and Lefèbvre, C. 2005. *Handbook of Categorization in Cognitive Science.* New York: Elsevier.

Condit, C., Parrott, R., Harris, T., Lynch, J., and Dubriwny, T. 2004. "The Role of 'Genetics' in Popular Understandings of Race in the United States," *Public Understanding of Science* 13:249–72.

Coop, G., Eisen, N. B., Nielsen, R., Przeworski, M., and Rosenberg, N. 2014. "Letters: 'A Troublesome Inheritance,'" *New York Times*, August 8.

Cottrell, C. A., and Neuberg, S. L. 2005. "Different Emotional Reactions to Different Groups: A Sociofunctional Threat-Based Approach to Prejudice," *Journal of Personality and Social Psychology* 88:770–89.

Richards, D. A., and Nichols, A. L. 2010. "Predicting Policy Attitudes from General Prejudice Versus Specific Intergroup Emotions," *Journal of Experimental Social Psychology* 46:247–54.

Crandall, C. S., and Schaller, M. 2004. *The Social Psychology of Prejudice: Historical and Contemporary Issues.* Lawrence, MA: Lewinian Press.

Dasgupta, N., DeSteno, D., Williams, L. A., and Hunsinger, M. 2009. "Fanning the Flames of Prejudice: The Influence of Specific Incidental Emotions on Implicit Prejudice," *Emotion* 9:585–91.

Dar-Nimrod, I., and Heine, S. 2011. "Genetic Essentialism: On the Deceptive Determinism of DNA," *Psychological Bulletin* 137:800–18.

Diamond, J. 1994. "Race Without Color," *Discover* November:82–9.

Dobzhansky, T., Hecht, M. K., and Steere, W. C. 1972. *Evolutionary Biology VI.* New York: Appelton-Centry-Crofts.

Dubreuil, B. 2010. *Human Evolution and the Origins of Hierarchies: The State of Nature.* Cambridge: Cambridge University Press.

Ekman, P. and Davidson, R. J. 1994. *The Nature of Emotions: Fundamental Questions.* New York: Oxford University Press.

Faucher, L., and Machery, E. 2009. "Racism: Against Jorge Garcia's Moral and Psychological Monism," *Philosophy of Social Sciences* 39:41–62.

Faulkner, J., Schaller, M., Park, J. H., and Duncan, L. A. 2004. "Evolved Disease-Avoidance Mechanisms and Contemporary Xenophobic Attitudes," *Group Processes and Intergroup Relations* 7:333–53.

Fausto-Sterling, A. 2008. "The Bare Bones of Race," *Social Studies of Science* 38:657–94.

Feldman, M. W., Lewontin, R. C., and King, M. C. 2003. "Race: A Genetic Melting-Pot," *Nature* 424:374.

and Lewontin, R. C. 2008. "Race, Ancestry and Medicine," in Koenig, Lee, and Richardson (eds.), *Revisiting Race in a Genomic Age.* New Brunswick, NJ: Rutgers University Press, pp. 89–101.

Fishbein, H. D. 1996. *Peer Prejudice and Discrimination: Evolutionary, Cultural, and Developmental Dynamics.* Boulder, CO: WestView.

Fiske, S. T., Cuddy, A. J. C., and Glick, P. 2006. "Universal Dimensions of Social Cognition: Warmth and Competence," *Trends in Cognitive Sciences* 11:77–83.

Gannett, L. 2005. "Group Categories and Pharmacogenetics Research," *Philosophy of Science* 72:1232–45.

Gelman, S. A. 2010. "Modules, Theories, or Islands of Expertise? Domain-Specificity in Socialization," *Child Development* 81:715–19.

and Heyman, G. D. 1999. "Carrott-Eaters and Creature-Believers: The Effects of Lexicalization on Children's Inferences About Social Categories," *Psychological Science* 10:490–3.

Gil-White, F. 1999. "How Thick Is blood? The Plot Thickens: If Ethnic Actors Are Primordialists, What Remains of the Circumstantialists/Primordialists Controversy?" *Ethnic and Racial Studies* 22:789–820.

2001a. "Are Ethnic Groups Biological 'Species' to the Human Brain," *Current Anthropology* 42:515–54.

2001b. "Sorting Is Not Categorization: A Critique of the Claim That Brazilians Have Fuzzy Racial Categories," *Cognition and Culture* 1: 219–50.

2005. "How Conformism Creates Ethnicity Creates Conformism (And Why This Matters to Lots of Things." *The Monist* 88(2):189–237.

Glasgow, J. M. 2003. "On the New Biology of Race," *Journal of Philosophy* 9:456–74.

2009. *A Theory of Race.* New York: Routledge.

2011. "Another Look at the Reality of Race, By Which I Mean Race-f," in Hazlett (ed.), *New Waves of Metaphysics.* London: Palgrave Macmillan, pp. 54–71.

Shulman, J. L., and Covarrubias, E. G. 2009. "The Ordinary Conception of Race in the United States and Its Relation to Racial Attitudes: A New Approach," *Journal of Cognition and Culture* 9:15–38.

Green, J. D. 2014. "Beyond Point-and-Shoot Morality: Why Cognitive (Neuro)Science Matters for Ethics," *Ethics* 124:695–726.

Hale, T. 2015. "A Non-Essentialist Theory of Race: The Case of an Afro-Indigenous Village in Northern Peru," *Social Anthropology* 23:135–51.

Hardimon, M. O. 2003. "The Ordinary Concept of Race," *Journal of Philosophy* 100:437–55.

2012. "The Idea of a Scientific Concept of Race," *Journal of Philosophical Research* 37:249–82.

Haslam, N., Rothschild, L., and Ernst, D. 2000. "Essentialist Beliefs About Social Categories," *British Journal of Social Psychology* 39:113–27.

Bastian, B., Bain, P., and Kashima, Y. 2006. "Psychological Essentialism, Implicit Theories, and Intergroup Relations," *Group Processes and Intergroup Relations* 9:63–76.

Hazlett, A. 2011. *New Waves of Metaphysics.* London: Palgrave Macmillan.

Henrich, J., and Boyd, R. 1998. "The Evolution of Conformist Transmission and the Emergence of Between-Group Differences," *Evolution and Human Behavior* 19:215–41.

Hochman, A. 2013. "Racial Discrimination: How Not to Do It," *Studies in History and Philosophy of Science Part C* 3:278–86.

Huang, J., Sedlovaskaya, A. Acherman, J., and Bargh, J. 2011. "Immunizing Against Prejudice: Effects of Disease Protection on Attitudes Toward Out-Groups," *Psychological Science* 22:1550–6.

Hudson, N. 1996. "From 'Nation' to 'Race': The Origin of Racial Classification in Eighteenth-Century Thought." *Eighteenth-Century Studies* 29:247–64.

Jarayatne, T., Ybarra, O., Shledon, J., et al. 2006. "White Americans' Genetic Lay Theories of Race Differences and Sexual Orientation: Their Relationship with Prejudice Toward Blacks, Gay Men and Lesbians," *Group Processes and Intergroup Relations* 9:77–94.

Kanovski, M. 2007. "Essentialism and Folksociology: Ethnicity Again," *Journal of Cognition and Culture* 7:241–81.

Kaplan, J. M. 2010. "When Socially Determined Categories Make Biological Realities," *The Monist* 93:283–99.

2011. "Race: What Biology Can Tell Us About a Social Construct," in *Encyclopedia of Life Sciences*. Hoboken, NJ: Wiley.

2014. "Ignorance, Lies, and Ways of Being Racist," *Critical Race Theory* 2:160–82.

Keller, J. 2005. "In Genes We Trust: The Biological Component of Psychological Essentialism and Its Relationship to Mechanisms of Motivated Social Cognition," *Journal of Personality and Social Psychology* 88:686–702.

Kelly, D., Faucher, L., and Machery, E. 2010. "Getting Rid of Racism: Assessing Three Proposals in Light of Psychological Evidence," *Journal of Social Philosophy* 41:293–322.

Keltner, D., and Haidt, J. 1999. "Social Functions of Emotions at Four Levels of Analysis," *Cognition and Emotion* 13:505–21.

Kincaid, H. 2012. *The Oxford Handbook of Philosophy of Social Science*. Oxford: Oxford University Press.

Kinzler, K. D., and Spelke, E. 2011. "Do Infants Show Social Preferences for People Differing in Race?" *Cognition* 119:1–9.

Kitcher, P. 2003. "Race, Ethnicity, Biology, Culture," in Philip Kitcher (ed.), *In Mendel's Mirror: Philosophical Reflections on Biology*. New York: Oxford University Press, pp. 230–56.

2007. "Does 'Race' Have a Future?" *Philosophy and Public Affairs* 35:293–317.

Klein, C. 2010. "Images Are Not the Evidence in Neuroimaging," *British Journal for the Philosophy of Science* 61:265–78.

Keltner, D., and Haidt, J. 2001. "Social Functions of Emotions," in Mayne and Bonanno (eds.), *Emotions: Current Issues and Future Directions*. New York: Guilford Press, pp. 192–213.

Koenig, B., Lee, S., and Richardson, S. 2008. *Revisiting Race in a Genomic Age*. New Brunswick, NJ: Rutgers University Press.

Knobe, J. 2007. "Experimental Philosophy," *Philosophy Compass* 2:81–92.

Kripke, S. 1980. *Naming and Necessity*. Cambridge, MA: Harvard University Press.

Kurzban, R., and Neuberg, S. L. 2005. "Managing Ingroup and Outgroup Relationships," in Buss (ed.), *Handbook of Evolutionary Psychology*. New York: Wiley, pp. 653–75.

LaFollette, H. 2003. *The Oxford Handbook of Practical Ethics*. New York: Oxford University Press.

Leslie, S.-J. 2014. "Carving Up the Social World with Generics," *Oxford Studies in Experimental Philosophy* 1:208–32.

Forthcoming. "The Original Sin of Cognition: Fear, Prejudice, and Generalization," *Journal of Philosophy*.

Levenson, R. W. 1994. "Human Emotion: A Functional View," in Ekman and Davidson (eds.), *The Nature of Emotions: Fundamental Questions*. New York: Oxford University Press, pp. 123–6.

Lewis, J., Haviland-Jones, M., and Barrett, L. F. 2008. *Handbook of Emotions*, 3rd edn. New York: Guilford.

Lewontin, R. C. 1972. "The Apportionment of Human Diversity," in Dobzhansky, Hecht, and Steere (eds.), *Evolutionary Biology VI*. New York: Appelton-Centry-Crofts, pp. 381–98.

Lieberman, D., Tooby, J., and Cosmides, L. 2007. "The Architecture of Human Kin Detection," *Nature* 445:727–31.

Long, J., and Kittle, R. 2009. "Human Genetic Diversity and the Nonexistence of Biological Races," *Human Biology* 81:777–98.

Lorusso, L., and Bacchini, F. 2015. "A Reconsideration of the Role of Self-Identified Races in Epidemiology and Biomedical Research," *Studies of History and Philosophy of Biological and Biomedical Sciences* 52:56–64.

Machery, E. 2008. "A Plea for Human Nature," *Philosophical Psychology* 21:321–29.

2014. "In Defense of Reverse Inference," *British Journal for the Philosophy of Science* 65:251–67.

and Faucher, L. 2005a. "Social Construction and the Concept of Race," *Philosophy of Science* 72:1208–19.

and Faucher, L. 2005b. "Why Do We Think Racially? Culture, Evolution and Cognition," in Cohen and Lefèbvre (eds.), *Handbook of Categorization in Cognitive Science*. New York: Elsevier, pp. 1009–33.

Mackie, D. M., and Smith, E. R. 2002. *From Prejudice to Intergroup Emotions: Differentiated Reactions to Social Groups*. New York: Psychology Press.

Maglo, K. N. 2010. "Genomics and the Conundrum of Race: Some Epistemological and Ethical Considerations," *Perspectives in Biology and Medicine* 53:357–72.

Mallon, R. 2006. "Race: Normative, Not Metaphysical or Semantic," *Ethics* 116:525–51.

2013. "Was Race Thinking Invented in the Modern West?" *Studies in History and Philosophy of Science Part A* 44:77–88.

and Kelly, D. 2012. "Making Race Out of Nothing: Psychological Constrained Social Roles," in Kincaid (ed.), *The Oxford Handbook of Philosophy of Social Science*. Oxford University Press, pp. 507–29.

Martinovic, B., and Verkuyten Ercomer, M. 2012. "Host National and Religious Identification Among Turkish Muslims in Western Europe: The Role of Ingroup Norms, Perceived Discrimination and Value Incompatibility," *European Journal of Social Psychology* 42:893–903.

Mayne, T., and Bonanno, G. A. 2001. *Emotions: Current Issues and Future Directions*. New York: Guilford Press.

McDonald, M., Asher, B., Kerr, N., and Navarrette, C. D. 2011. "Fertility and Intergroup Bias in Racial and Minimal-Group Contexts: Evidence for Shared Architecture," *Psychological Science* 22:860–5.

McElreath, R., Boyd, R., and Richerson, P. J. 2003. "Shared Norms Can Lead to the Evolution of Ethnic Markers," *Current Anthropology* 44:122–30.

Montagu, A. 1962. "The Concept of Race," *American Anthropologist* 64: 919–28.

Moya, C., and Boyd, R. 2015. "A Functionalist Framework with Illustrations from the Peruvian Altiplano," *Human Nature* 26:1–27.

Neander, K. 1991. "Functions as Selected Effects: The Conceptual Analyst's Defense," *Philosophy of Science* 58:168–84.

Nesse, R. 2001. *The Evolution of Subjective Commitment*. New York: Russell Sage Foundation.

Neuberg, S. L., and Cottrell, C. A. 2002. "Intergroup Emotions: A Biocultural Approach," in Mackie and Smith (eds.), *From Prejudice to Intergroup Emotions: Differentiated Reactions to Social Groups*. New York: Psychology Press, pp. 265–84.

2006. "Evolutionary Bases of Prejudices," in Schaller, Simpson, and Kenrick (eds.), *Evolution and Social Psychology*. New York: Psychology Press, pp. 163–87.

2008. "Managing the Threats and Opportunities Afforded by Human Sociality," *Group Dynamics* 21:63–72.

Kenrick, D. T. and Schaller, M. 2011. "Human Threat Management Systems: Self-Protection and Disease Avoidance," *Neuroscience and Biobehavioral Reviews* 35:1042–51.

Novembre, J., Johnson, T., Bryc, K., et al. 2008. "Genes Mirror Geography in Europe," *Nature* 456:98–101.

Omi, M., and Winant, H. 1994. *Racial Formation in the United-States: From the 1960s to the 1990s*, 2nd edn. New York: Routledge.

Pääbo, S. 2003. "The Mosaic That Is Our Genome," *Nature* 421:409–12.

Parker Tapias, M., Glaser, J., and Keltner, D. 2007. "Emotion and Prejudice: Specific Emotions Toward Outgroups," *Group Processes and Intergroup Relations* 10:27–39.

Pauker, K., Ambady, N., and Apfelbaum, E. P. 2010. "Race Salience and Essentialist Thinking in Racial Stereotype Development," *Child Development* 81:1799–1813.

Peralta, C. A., Risch, N., Lin, F., et al. 2009. "The Association of African Ancestry and Elevated Creatinine in the Coronary Artery Risk Development in Young Adults (CARDIA) Study," *American Journal of Nephrology* 31:202–8.

Phelan, J. C., Link, B. G., and Feldman, N. M. 2013. "The Genomic Revolution and Beliefs About Essential Racial Differences: A Backdoor to Eugenics," *American Sociological Review*, 78:167–91.

Plante, C., Roberts, S., Snider, J., et al. 2015. "'More Than Skin-Deep': Biological Essentialism in Response to a Distinctiveness Threat in a Stigmatized Fan Community," *British Journal of Social Psychology* 54: 359–70.

Piglucci, M. 2013. "What Are We to Make of the Concept of Race? Thoughts of a Philosopher-Scientist," *Studies in History and Philosophy of Biological and Biomedical Sciences* 44:272–7.

and Kaplan, J. M. 2003. "On the Concept of Biological Race and Its Applicability to Humans," *Philosophy of Science* 70:1161–72.

Prentice, D., and Miller, D. 2007. "Psychological Essentialism of Human Categories," *Current Directions in Psychological Science* 16:202–6.

Regnier, D. 2015. "Clean People, Unclean People: The Essentialisation of 'Slaves' Among the Southern Betsileo of Madagascar," *Social Anthropology* 23:152–68.

Rhodes, M., Leslie, S.-J., and Tworek, C. 2012. "Cultural Transmission of Social Essentialism," *Proceedings of the National Academy of Sciences* 109:13526–31.

Richerson, P. J., and Boyd, R. 2001. "The Evolution of Subjective Commitment to Groups: The Tribal Instincts Hypothesis," in Nesse (ed.), *The Evolution of Subjective Commitment*. New York: Russell Sage Foundation, pp. 186–220.

Risch, N., Burchard, E., Ziv, E., and Tang, H. 2002. "Categorization of Humans in Biomedical Research: Genes, Race and Disease," *Genome Biology* 3:1–12.

Rosenberg, N. 2011. "A Population-Genetic Perspective on Similarities and Differences Among Worldwide Human Populations," *Human Biology* 83:59–64.

Pritchard, J. K., Weber, J. L., et al. 2002. "Genetic Structure of Human Populations," *Science* 298:2381–5.

Ramachandran, M. S., Zhao, C., Pritchard, J. K., and Feldman, M. W. 2005. "Clines, Clusters, and the Effect of Study Design on the Inference of Human Population Structure," *PLoS Genetic* 1:660–71.

Rosenberg, A., and Bouchard, F. 2003. "Fitness," in *Stanford Encyclopedia of Philosophy* (Winter 2007 Edition). Available at: http://plato.stanford.edu/archives/win2007/entries/fitness/.

Roskies, A. 2010. "How Does Neuroscience Affect Our Conception of Volition," *Annual Review of Neuroscience* 33:109–30.

Schaller, M., and Conway, L. G. 2004. "The Substance of Prejudice: Biological- and Social-Evolutionary Perspectives on Cognition, Culture and the Contents of Stereotypical Beliefs," in Crandall and Schaller (eds.), *The Social Psychology of Prejudice: Historical and Contemporary Issues*. Lawrence, MA: Lewinian Press, pp. 149–64.

and Neuberg, S. L. 2012. "Danger, Disease, and the Nature of Prejudice(s)," *Advances in Experimental Social Psychology* 46:1–54.

Simpson, J. A., and Kenrick, D. T. 2006. *Evolution and Social Psychology*. New York: Psychology Press.

Serre, D., and Pääbo, S. 2004. "Evidence for Gradients of Human Genetic Diversity Within and Among Continents," *Genome Reasearch* 14:1679–85.

Sesardic, N. 2010. "Race: A Social Destruction of a Biological Concept," *Biology and Philosophy* 25:143–62.

Shiao, J., Bode, T., Beyer, A., and Selvig, D. 2012. "The Genomic Challenge to the Social Construction of Race," *Sociological Theory* 30:67–88.

Shulman, J. L., and Glasgow, J. 2010. "Is Race-Thinking Biological or Social, and Does It Matter for Racism? An Exploratory Study," *Journal of Social Philosophy* 41:244–59.

Sousa, P., Atran, S., and Medin, D. 2002. "Essentialism and Folkbiology: Evidence from Brazil," *Journal of Cognition and Culture* 2:195–223.

Spencer, Q. 2012. "What 'Biological Racial Realism' Should Mean," *Philosophical Studies* 159:181–204.

2013. "Biological Theory and the Metaphysics of Race: A Reply to Kaplan and Winther," *Biological Theory* 8:114–20.

2014a. "A Radical Solution to the Race Problem," *Philosophy of Science*, 81:1025–38.

2014b. "The Unnatural Racial Naturalism," *Studies in History and Philosophy of Biological and Biomedical Sciences* 46:79–87.

2015. "Philosophy of Race Meets Population Genetics," *Studies in History and Philosophy of Science Part C* 52:46–55.

Taylor, P. 2011. "Rehabilitating a Biological Notion of Race? A Response to Sesardic," *Biology and Philosophy* 26:469–73.

Templeton, A. 1998. "Human Races: A Genetic and Evolutionary Perspective," *American Anthropologist* 100:632–50.

Tishkoff, S. A., and Kidd, K. K. 2004. "Implications of Biogeography in Human Populations for 'Race' and 'Medicine,'" *Nature Genetics* 36 (Suppl):S21–7.

Tooby, J., and Cosmides, L. 2008. "The Evolutionary Psychology of Emotions and Their Relationship to Internal Regulatory Variables," in Lewis, Haviland-Jones, and Barrett (eds.), *Handbook of Emotions*, 3rd edn. New York: Guilford, pp. 114–37.

Weiss, K. M., and Fullerton, S. M. 2005. "Racing Around, Getting Nowhere," *Evolutionary Anthropology* 14:165–9.

Williams, M. J., and Eberhardt, J. L. 2008. "Biological Conceptions of Race and the Motivation to cross Racial Boundaries," *Journal of Personality and Social Psychology* 94:1033–47.

Wilson, J. F., Weale, M. E., Smith, A. C., et al. 2001. "Population Genetic Structure of Variable Drug Response," *Nature Genetics* 29:265–9.

Zack, N. 1998. *Thinking About Race*. Belmont, CA: Wadsworth.

2002. *Philosophy of Science and Race*. New York: Routledge.

2003. "Race and Racial Discrimination," in LaFollette (ed.): 245–71.

Zakharia, F., Basu, A., Absher, D., et al. 2009. "Characterizing the Admixed African Ancestry of African Americans," *Genome Biology* 10:R141.

Chapter 13

Alcock, J. 2001. *The Triumph of Sociobiology*. Oxford: Oxford University Press.

Alexander, R. 1979. *Darwinism and Human Affairs*. Seattle: University of Washington Press.

Baghramian, M. 2012. *Reading Putnam*. New York: Routledge.

Barash, D. 1979. *The Whisperings Within*. New York: Harper & Row.

Beebee, H., and Sabbarton-Leary, N. 2010. *The Semantics and Metaphysics of Natural Kinds*. New York: Routledge.

Block, N. 1980. *Readings in Philosophy of Psychology, Vol. I*. Cambridge, MA: Harvard University Press.

Bowles, S., and Gintis, H. 2011. *A Cooperative Species: Human Reciprocity and its Evolution*. Princeton, NJ: Princeton University Press.

Boyd, R. N. 1980. "Materialism Without Reductionism: What Physicalism Does Not Entail," in Block (ed.), *Readings in Philosophy of Psychology, Vol. I*. Cambridge, MA: Harvard University Press, pp. 1–67.

1999. "Homeostasis, Species, and Higher Taxa," in Wilson (ed.), *Species: New Interdisciplinary Essays*. Cambridge, MA: MIT Press, pp. 141–85.

2001. "Reference, (In)Commensurability, and Meanings: Some (Perhaps) Unanticipated Complexities," in Hoyningen-Huene and Sankey (eds.),

Incommensurability and Related Matters. Dordrecht, Netherlands: Kluwer, pp. 1–64.

2010a. "Realism, Natural Kinds, and Philosophical Methods," Beebee and Sabbarton-Leary (eds.), *The Semantics and Metaphysics of Natural Kinds*. New York: Routledge, pp. 212–34.

2010b. "Homeostasis, Higher Taxa and Monophyly," *Philosophy of Science* 77:686–701.

2012. "What of Pragmatism with the World Here?" in Baghramian (ed.), *Reading Putnam*. New York: Routledge, pp. 39–94.

Brown, S., and Lewis, B. P. 2004. "Relational Dominance and Mate-Selection Criteria: Evidence That Males Attend to Female Dominance," *Evolution and Human Behavior* 25:406–15.

Buller, D. 2005. *Adapting Minds: Evolutionary Psychology and the Persistent Quest for Human Nature*. Cambridge, MA: MIT Press.

and Hardcastle, V. G. 2000. "Evolutionary Psychology, Meet Developmental Neurobiology: Against Promiscuous Modularity," *Brain and Mind* 1:307–25.

Buss, D. M. 1989. "Sex Differences in Human Mate Preferences: Evolutionary Hypotheses Tested in 37 Cultures," *Behavior and Brain Sciences* 12:1–14.

Carruthers, P. 2006. *The Architecture of the Mind: Massive Modularity and the Flexibility of Thought*. Oxford: Oxford University Press.

Cashdan, E. 2013. "What Is a Human Universal? Human Behavioral Ecology and Human Nature," in Downes and Machery (eds.), *Arguing About Human Nature*. New York: Routledge.

Cimino, A., and Andrew, W. D. 2010. "On the Perception of Newcomers: Toward an Evolved Psychology of Intergenerational Coalitions." *Human Nature* 21:186–202.

Cosmides, L., and Tooby, J. 1997. *Evolutionary Psychology: A Primer*. Center for Evolutionary Psychology, University of California, Santa Barbara.

Daly, M., and Wilson, M. 1985. "Child Abuse and Other Risks of Not Living with Both Parents," *Ethology and Sociobiology* 6:197–210.

Davidson, D., and Harman, G. 1972. *The Semantics of Natural Language*. Dordrecht, Netherlands: D. Reidel.

Downes, S. M., and Machery, E. 2013. *Arguing About Human Nature*. New York: Routledge.

Earley, J. 2008. "How Philosophy of Mind Needs Philosophy of Chemistry." *Hyle: The International Journal for Philosophy of Chemistry* 14:1–26.

Fedyk, M. 2014. "How (Not) to Bring Psychology and Biology Together," *Philosophical Studies* 172:949–67.

Goodman, N. 1954. *Fact Fiction and Forecast*. Cambridge, MA: Harvard University Press.

Gould, S. J., and Veba, E. J. 1982. "Exaptation – A Missing Term in the Science of Form," *Paleobiology* 8:4–15.

Haidt, J., and Bjorklund, F. 2008. "Social Intuitionists Answer Six Questions About Moral Psychology," in Sinnott-Armstron (ed.), *Moral Psychology*, Vol. II. Cambridge, MA: MIT Press, pp. 181–217.

Hauser, M. 2006. *Moral Minds: How Nature Designed Our Universal Sense of Right and Wrong*. New York: HarperCollins.

Hazan, C., and Diamond, L. 2000. "The Place of Mating in Human Attachment," *Review of General Psychology* 4:186–204.

Heil, J., and Mele, A. 1993. *Mental Causation*. Oxford: Oxford University Press.

Hoyningen-Huene, P., and Sankey, H. 2001. *Incommensurability and Related Matters*. Dordrecht, Netherlands: Kluwer.

Hull, D. 1978. "A Matter of Individuality," *Philosophy of Science* 45:335–60.

Jablonka, E., and Lamb, M. J. 2006. *Evolution in Four Dimensions: Genetic, Epigenetic, Behavioral, and Symbolic Variation in the History of Life*. Cambridge, MA: MIT Press.

Joyce, R. 2005. *The Evolution of Morality*. Cambridge, MA: MIT Press.

Kim, J. 1993. "The Non-Reductivist's Troubles with Mental Causation," in Heil and Mele (eds.), *Mental Causation*. Oxford: Oxford University Press, pp. 189–210.

Kripke, S. A. 1972. "Naming and Necessity," in Davidson and Harman (eds.), *The Semantics of Natural Language*. Dordrecht Netherlands: D. Reidel, pp. 253–5.

Lehrman, D. S. 1953. "Critique of Konrad Lorenz's Theory of Instinctive Behavior," *Quarterly Review of Biology* 28:337–63.

Magnus, P. D. 2011. "Drakes, Seadevils, and Similarity Fetishism," *Biology and Philosophy* 26:857–70.

Mayr, E. 1969. *Principles of Systematic Zoology*. New York: McGraw-Hill.
1963. *Populations, Species, and Evolution*. Cambridge, MA: Harvard University Press.

Millikan, R. G. 1984. *Language Thought and Other Biological Categories*. Cambridge, MA: MIT Press.

Neander, K. 1991. "The Teleological Notion of 'Function'," *Australasian Journal of Philosophy* 69:454–68.

Pedersen, W. C., Miller, L. C., Putcha-Bhagavatula, A. D., and Yang, Y. 2002. "Evolved Sex Differences in the Number of Partners Desired? The Long and the Short of It." *Psychological Science* 13:157–61.

Richardson, R. C. 2007. *Evolutionary Psychology as Maladapted Psychology*. Cambridge, MA: MIT Press.

Rieppel, O. 2005a. "Modules, Kinds and Homology," *Journal of Experimental Zoology B* 304:18–27.

2005b. "Monophyly, Paraphyly and Natural Kinds," *Biology and Philosophy* 20:465–87.

Singh D. 1993. "Adaptive Significance of Female Physical Attractiveness: Role of Waist-to-Hip Ratio," *Journal of Personality and Social Psychology* 65:293–307.

Sinnott-Armstrong, W. 2008. *Moral Psychology*, Vol. II: *The Cognitive Science of Morality: Intuition and Diversity*. Cambridge, MA: MIT Press.

Street, S. 2006. "A Darwinian Dilemma for Realist Theories of Value," *Philosophical Studies* 127:109–66.

Sturgeon, N. 1992. "Nonmoral Explanations," in Tomberlin (ed.), *Philosophical Perspectives* 6. Atascadero, CA: Ridgeview, pp. 98–9.

Tinbergen, N. 1951. *The Study of Instinct*. Oxford: Oxford University Press.

Tomberlin, J. E. 1992. *Philosophical Perspectives* 6. Atascadero, CA: Ridgeview.

Wagner G. P. 2001a. "Characters, Units and Natural Kinds: An Introduction," in Wagner (ed.), *The Character Concept in Evolutionary Biology*. San Diego: Academic Press, pp. 1–10

2001b. *The Character Concept in Evolutionary Biology*. San Diego: Academic Press.

West-Eberhard, M. J. 2003. *Developmental Plasticity and Evolution*. Oxford: Oxford University Press.

Wilson, E. O. 1975. *Sociobiology: The New Synthesis*. Cambridge, MA: Harvard University Press.

1978. *On Human Nature*. Cambridge, MA: Harvard University Press.

Wilson, J. 2006. "On Characterizing the Physical," *Philosophical Studies* 131:61–99.

Wilson, R. A. 1999. *Species: New Interdisciplinary Essays*. Cambridge, MA: MIT Press.

Further Reading

Ariew, A., and Cummins, R. 2002. *Functions: New Essays in the Philosophy of Psychology and Biology*. Oxford: Oxford University Press.

Churchland, P. S. 1986. *Neurophilosophy: Towards a Unified Understanding of the Mind/Brain*. Cambridge, MA: MIT Press.

Churchland, P. S. 2002. *Brainwise: Studies in Neurophilosophy*. Cambridge, MA: MIT Press.

Dennett, D. C. 1995. *Darwin's Dangerous Idea*. New York: Simon & Schuster.

2004. *Freedom Evolves*. London: Penguin.

Downes, S., and Machery, E. 2014. *Arguing About Human Nature*. New York: Routledge.

Dretske, F. 1989. *Explaining Behavior*. Cambridge, MA: MIT Press.

Joyce, R. 2005. *The Evolution of Morality*. Cambridge, MA: MIT Press.

Kahane, G. 2011. "Evolutionary Debunking Arguments," *Noûs* 45:103–25.

Kingsbury, J., Ryde, D., and Williford, K. 2012. *Millikan and Her Critics*. New York: Wiley Blackwell.

Kitcher, P. S. 2011. *The Ethical Project*. Cambridge, MA: Harvard University Press.

Kornblith, H. 1993. *Inductive Inference and Its Natural Ground*. Cambridge, MA: MIT Press.

Macdonald, G., and Papineau, D. 2006. *Teleosemantics: New Philosophical Essays*. Oxford, UK: Clarendon.

Machery, E. 2008. "A Plea for Human Nature," *Philosophical Psychology* 21:321–30.

and Faucher, L. 2005. "Why Do We Think Racially? Culture, Evolution and Cognition," in H. Cohen and C. Lefèbvre (eds.), *Handbook of Categorization in Cognitive Science*. New York: Elsevier.

Millikan, R. 1984. *Language, Thought and Other Biological Catagories*. Cambrdige, MA: Bradford Books.

1993. *White Queen Psychology and Other Essays for Alice*. Cambridge, MA: MIT Press.

Munz, P. 1993. *Philosophical Darwinism: On the Origin of Knowledge by Means of Natural Selection*. London: Routledge.

Neander, K. 1991. "The Teleological Notion of 'Function,'" *Australasian Journal of Philosophy* 69:454–68.

Okasha, S., and Binmore, K. 2014. *Evolution and Rationality*. Cambridge: Cambridge University Press.

Papineau, D. 2001. "The Status of Teleosemantics, or How to Stop Worrying About Swampman," *Australasian Journal of Philosophy* 79:79–89.

Rescher, N. 1990. *A Useful Inheritance: Evolutionary Aspects of the Theory of Knowledge*. Lanham, MD: Rowman.

Ruse, M. 1986. *Taking Darwin Seriously: A Naturalistic Approach to Philosophy*. Oxford, UK: Blackwell.

Skyrms, B. 1996. *Evolution of the Social Contract*. Cambridge: Cambridge University Press.

2010. *Signals: Evolution, Learning, and Information*. Oxford: Oxford University Press.

Smith, D. L. 2013. "Self-Deception: A Teleofunctional Approach," *Philosophia* 42:181–99.

Sterelny, K. 2003. *Thought in a Hostile World*. Oxford, UK: Blackwell.

2012. *The Evolved Apprentice*. Cambridge, MA: MIT Press.

Street, S. 2005. "A Darwinian Dilemma for Realist Theories of Value," *Philosophical Studies* 127:109–66.

Index

aboutness. *See* intentionality
accommodationism, 277, 278, 280–2, 283, 290
adaptation, 16, 26, 27, 31, 32, 34, 88, 127, 128, 162, 169, 177, 208, 209, 219, 220, 224, 247, 268, 272, 288, 292, 293
adaptiveness, 33, 34, 43, 128, 134, 136, 154, 163, 166, 167, 168, 170, 171, 172, 177, 262, 263, 268, 272, 273, 295, 297, 299
African Americans, 258, 272, 273, 274
Africans, 253, 255
alarm calls, 97, 102
alleles, 254, 257
altruism, 21, 150, 153, 155, 162, 215, 217, 293, 294
Andrews, K., 164
androgen, 234
anemones, 59
anesthesia, 77, 85
anthropology, 30, 72, 159, 184, 251
antibody selection, 128
anti-reductionism, 7, 276, 277, 283, 286, 296
Antony, L., 211, 213
apes, 19, 191, 300, *See* chimpanzees, bonobos
Aquinas, T., 146, 147, 157
archaea, 53, 54
Aristotle, 9, 10, 23, 26, 28, 29, 141, 146, 147, 148, 153, 209, 211, 218, 221, 234
arthropods, 61, 62, 63, 64, 70
Artificial Intelligence, 18, *See* strong AI
Asians, 255, 273
Astuti, R., 264
attention, 25, 67, 76, 78, 85, 87, 88, 262, 272

Australopithecus, 14, 15
autonomy of psychology, 78

bacteria, 15, 19, 49, 53, 54, 58, 80, 87, 166
Barash, D., 158, 293, 294
Bayesianism, 161, 163
behavioral ecology, 163
behavioral economic, 30
behavioral economics, 72
behaviorism, 38
Benatar, D., 150
Bennett, J., 38
Berkeley, G., 198
biases, 5, 171, 261, 264, 270
bilaterians, 65
Binmore, K., 162
biological individuals, 279
biological racial realism, 252–9
biological rationality, 166, 168, 169, 170, 172, 176, 183
biophilosophy, 1–8, 229, 247, 248, 249, 259, 275
biostatistical theory, 128
bipedalism, 208, 214, 218, 221, 222
Birds, 41, 43, 47, 68, 69, 103, 164, 167, 173, 174, 212, 217, 223, *See* eagles, gulls, penguins, ravens, scrub jays
bivalent logic, 11
Black, J., 28
Bloom, P., 205, 208, 211
bluehead wrasse, 232
Boehm, C., 196
bonobos, 77
Boorse, C., 128
Boyd, R., 206, 210, 264
Boyle, R., 23
brain, 19, 31, 32, 37, 38, 39, 41, 43, 44, 69, 72–94, 98, 99, 105, 121, 122, 124, 127, 129, 130, 131, 132, 137,

342

Printed in the United States
By Bookmasters